Encyclopaedia of Mathematical Sciences

Volume 3

Editor-in-Chief: R.V. Gamkrelidze

V.I. Arnold (Ed.)

Dynamical Systems III

With 81 Figures

Springer-Verlag
Berlin Heidelberg NewYork
London Paris Tokyo

Title of the Russian original edition:
Itogi nauki i tekhniki, Sovremennye problemy matematiki,
Fundamental'nye napravleniya, Vol. 3, Dinamicheskie sistemy 3
Publisher VINITI, Moscow 1985

Mathematics Subject Classification (1980):
Primary 34-02, 58-02, 70-02, Secondary 34C35, 34C40, 34D10, 53C57, 58Fxx,
70D05, 70D10, 70Exx, 70Fxx, 70Hxx, 70Jxx

ISBN 3-540-17002-2 Springer-Verlag Berlin Heidelberg New York
ISBN 0-387-17002-2 Springer-Verlag New York Berlin Heidelberg

Library of Congress Cataloging-in-Publication Data.
Dinamicheskie sistemy. English. Dynamical systems.
(Encyclopaedia of mathematical sciences; v.)
Translation of: Dinamicheskie sistemy, issued as part of the series:
Itogi nauki i tekhniki. Seriia Sovremennye problemy matematiki.
Bibliography: p. Includes index.
1. Mechanics, Analytic. 2. Mechanics, Celestial.
I. Arnol'd, V.I. (Vladimir Igorevich), 1937–.
II. Iacob, A. III. Title, IV. Series.
QA805.D5613 1988 531 87-20655
ISBN 0-387-17002-2 (v. 3)

Printed in Germany
Typesetting, printing and bookbinding:
Universitätsdruckerei H. Stürtz AG, D-8700 Würzburg
2141/3140-543210

List of Editors, Contributors and Translators

Editor-in-Chief

Revaz V. Gamkrelidze, Academy of Sciences of the USSR, Institute for Scientific Information (VINITI), Baltiiskaya ul. 14, 125219 Moscow, USSR

Consulting Editor

Vladimir I. Arnold, Steklov Mathematical Institute, ul. Vavilova 42, 117966 Moscow, USSR

Contributors

Vladimir I. Arnold, Steklov Mathematical Institute, ul. Vavilova 42, 117966 Moscow, USSR

V.V. Kozlov, Department of Mathematics and Mechanics, University of Moscow, 119899 Moscow, USSR

A.I. Neishtadt, Space Research Institute, Moscow, USSR

Translator

Andrei Iacob, Department of Mathematics, Brandeis University, Waltham, MA 02254, USA

Mathematical Aspects of Classical and Celestial Mechanics

V.I. Arnold V.V. Kozlov A.I. Neishtadt

Translated from the Russian
by A. Iacob

Contents

Preface

This work describes the fundamental principles, problems, and methods of classical mechanics focussing on its mathematical aspects. The authors have striven to give an exposition stressing the working apparatus of classical mechanics, rather than its physical foundations or applications. This apparatus is basically contained in Chapters 1, 3, 4 and 5.

Chapter 1 is devoted to the fundamental mathematical models which are usually employed to describe the motion of real mechanical systems. Special consideration is given to the study of motion under constraints, and also to problems concerned with the realization of constraints in dynamics.

Chapter 3 is concerned with the symmetry groups of mechanical systems and the corresponding conservation laws. Also discussed are various aspects of the theory of the reduction of order for systems with symmetry, often used in applications.

Chapter 4 contains a brief survey of various approaches to the problem of the integrability of the equations of motion, and discusses some of the most general and effective methods of integrating these equations. Various classical examples of integrated problems are outlined. The material presented in this chapter is used in Chapter 5, which is devoted to one of the most fruitful branches of mechanics – perturbation theory. The main task of perturbation theory is the investigation of problems of mechanics which are "close" to exactly integrable problems. Elements of this theory, in particular, the widely used "averaging principle", have emerged in celestial mechanics from attempts to take into account the mutual gravitational perturbations of planets in the solar system.

Chapter 6 is related to Chapters 4 and 5, and studies the theoretical possibility of integrating (in a precisely defined sense) the equations of motion. Approximate integration methods are discussed in Chapter 5: their significance is increased by the fact that integrable systems occur so rarely in reality. Also in this chapter there is a study of the n-body problem with special consideration given to the problem of the stability of the solar system. Some of the classical problems of celestial mechanics are treated in Chapter 2, including the integrable 2-body problem, and the classification of final motions in the 3-body problem. This chapter also contains an analysis of collisions, various aspects of regularization in the general problem of n points interacting gravitationally, and various limiting variants of this problem. Elements of the theory of oscillations are given in Chapter 7.

This text is not a complete exposition of these topics and we do not give detailed proofs. Our main purpose is to acquaint the reader with classical mechanics as a whole, in both its classical and its contemporary aspects. The interested reader will find the necessary proofs, and more detailed information, in the works listed at the end of this volume.

Chapter 1
Basic Principles of Classical Mechanics

To describe the motion of mechanical systems one uses a variety of mathematical models which are based on different "principles" – laws of motion. In this chapter we list the basic objects and principles of classical mechanics. The simplest and most important model of motion of real bodies is Newtonian mechanics, which describes the motion of a free system of interacting point masses in three-dimensional Euclidean space. In §6 we discuss the extent to which Newtonian mechanics is useful in describing complicated models of motion.

§1. Newtonian Mechanics

1.1. Space, Time, Motion. *Space*, in which motion takes place, is three-dimensional and *Euclidean*, with a fixed orientation. We shall denote it by E^3. Fix a point $o \in E^3$ – an "origin" or "reference point". Then the position of each point s in E^3 is uniquely specified by its *position (radius) vector* $\overrightarrow{os} = r$ (with its tail and tip at o and s, respectively). The set of all position vectors is the three-dimensional linear space R^3. This space is equipped with the scalar product $\langle\ ,\ \rangle$.

Time is one-dimensional; we denote it uniformly by t. The set $R = \{t\}$ is called the *time axis*.

A *motion* (or *path*) of the point s is a smooth mapping $\Delta \to E^3$, where Δ is a time interval. We say that the motion is defined on the interval Δ. To each motion there corresponds a unique smooth vector-function $r: \Delta \to R^3$.

The *velocity* v of the point s at time $t \in \Delta$ is the derivative $dr/dt = \dot{r}(t) \in R^3$. Velocity is clearly independent of the choice of the reference point.

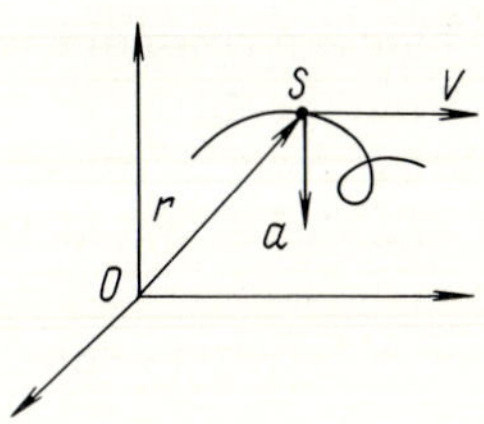

Fig. 1

The *acceleration* of the point s is the vector $a=\dot{v}=\ddot{r}\in R^3$. It is customary to represent the velocity and acceleration as vectors with tail at the point s.

The set E^3 is also known as the *position* (or *configuration*) *space* of the point s. The pair (s, v) is called a *state* of s, and the space $E^3 \times R^3\{v\}$ is the *state space* (or the *velocity phase space*).

Now consider the more general case in which n points $s_1, \ldots, s_n$ are moving in E^3. The set $E^{3n}=E^3\{s_1\}\times\ldots\times E^3\{s_n\}$ is called the position (configuration) space of this "free" system. In case it is necessary to exclude collisions of points, we must reduce E^{3n} by removing the diagonal

$$\Delta=\bigcup_{i<j}\{s_i=s_j\}.$$

Let $(r_1, \ldots, r_n)=r\in R^{3n}$ be the position vectors of the points $s_1, \ldots, s_n$. The motion of the free system is described by smooth vector-functions $r(t) =(r_1(t), \ldots, r_n(t))$. As above, we define the velocity

$$v=\dot{r}=(\dot{r}_1, \ldots, \dot{r}_n)=(v_1, \ldots, v_n)\in R^{3n}$$

and the acceleration

$$a=\ddot{r}=(\ddot{r}_1, \ldots, \ddot{r}_n)=(a_1, \ldots, a_n)\in R^{3n}.$$

The set $E^{3n}\times R^{3n}\{v\}$ is called the state space (or velocity phase space) of the system, and each pair (s, v) is referred to as a state of the given system.

1.2. The Newton-Laplace Principle of Determinacy. This principle asserts that the state of a mechanical system at any fixed moment of time uniquely determines all of its (future and past) motion.

Suppose we know the state (r_0, v_0) of the system at the moment of time t_0. Then by the principle of determinacy we also know the motion $r(t)$, with $r(t_0)=r_0$ and $\dot{r}(t_0)=\dot{r}_0=v_0$, for all $t\in\Delta\subset R$. In particular, we can calculate the acceleration $\ddot{r}$ at $t=t_0$.[1] The result is $\ddot{r}(t_0)=f(t_0, r_0, \dot{r}_0)$, where f is a function whose existence follows from the *Newton-Laplace principle.* Since we may choose an arbitrary value for t_0, we conclude that the equation

$$\ddot{r}=f(t, r, \dot{r})$$

holds for all t. This differential equation is known as the *equation of motion* or as *Newton's equation.* The existence of Newton's equation (with a smooth vector-function $f\colon R\{t\}\times R^{3n}\{r\}\times R^{3n}\{r\}\to R^{3n}$) and the determinacy principle are actually equivalent. This is a consequence of the theorem of existence and uniqueness of solutions from the theory of ordinary differential equations. The function f in Newton's equation is usually determined experimentally. Its specification is part of the definition of the mechanical system under consideration.

We will now give examples of Newton's equation.

---

[1] We will assume that all functions we encounter in dynamics are smooth.

a) The equation describing the fall of a point (small body) in vacuum near the Earth's surface (obtained experimentally by Galileo) has the form: $\ddot{r} = -g e_z$, where $g \approx 9.8\,\mathrm{m/s^2}$ (the free fall or gravitational acceleration), and e_z is the vertical unit vector. △

b) R. Hooke showed that the equation governing the small oscillations of a body attached to the extremity of an elastic spring has the form: $\ddot{x} = -\alpha x$, $\alpha > 0$. The constant coefficient α depends on the body and on the spring in question. This mechanical system is known as the *harmonic oscillator*. △

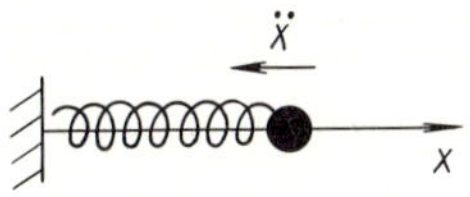

Fig. 2. Harmonic oscillator

As it turns out, in experiments, rather than determining the acceleration f appearing in the right-hand side of Newton's equation, it is more convenient to determine the product $mf = F$, where m is a positive number called the *mass* of the point (unraveling the physical meaning of the mass concept is not among the tasks of dynamics). Thus, in Hooke's experiments the constant $m\alpha = c$ depends on the properties of the elastic spring, but not on the choice of the attached body; c is called the elasticity constant (or coefficient).

The pair (s, m) (or (r, m), where r is the position vector of the point s) is called a *material point* (or *point mass*, or *particle*) of mass m. Hereafter we shall often use the letter m to denote both the point s and its mass m. If a system consists of n material points with masses $m_1, \ldots, m_n$, then Newton's equations

$$\ddot{r}_i = f_i(t, r_1, \ldots, r_n, \dot{r}_1, \ldots, \dot{r}_n), \qquad 1 \leq i \leq n,$$

may be rewritten as

$$m_i \ddot{r}_i = F_i(t, r, \dot{r}), \qquad 1 \leq i \leq n.$$

The vector $F_i = m_i f_i$ is called the *force* acting on the point m_i. "The word force does not appear in the basic laws of dynamics that we just indicated. As a matter of fact, we may also manage without it."[2] The last equations will be also referred to as Newton's equations.

c) Newton established that if one considers n point masses $(r_1, m_1), \ldots, (r_n, m_n)$ in space, then the force acting on the i-th point is $F_i = \sum_{i \neq j} F_{ij}$, where

[2] P.E. Appell [5], Vol. I, p. 94. In Newton's time the term "force" ("vis" in Latin) was used for various quantities, among them the acceleration of a point. Leibnitz used the term "vis viva" (alive force) for the product of the mass of the point and the square of its velocity. The modern term "force" corresponds to what Newton called "vis motrix" (accelerating force).

$$F_{ij} = \frac{\gamma m_i m_j}{|r_{ij}|^3} r_{ij}, \qquad r_{ij} = r_i - r_j, \ \gamma = \text{const} > 0.$$

This is the *law of universal gravitation (attraction)*. △

d) The resistance force acting on a body moving rapidly in air is proportional to the square of its velocity (Stoke's law). Accordingly, the equation describing the motion of a body falling in air is: $m\ddot{z} = mg - c\dot{z}^2$, with $c > 0$. One can show that the limit $\lim_{t\to\infty} v(t)$ always exists and equals $\sqrt{mg/c}$, regardless of the initial state of the body. △

The determinacy principle holds also in relativistic mechanics. Newton's classical mechanics is distinguished from relativistic mechanics by *Galileo's principle of relativity*.

1.3. The Principle of Relativity. The direct product $E^3 \times R\{t\}$ (space-time) carries a natural structure of affine space. The *Galilean group* is the group of all affine transformations of $E^3 \times R$ which preserve time intervals and which for every fixed $t \in R$ are *isometries* of E^3. Thus, if $g: (s, t) \mapsto (s', t')$ is a Galilean transformation, then

1) $t_\alpha - t_\beta = t'_\alpha - t'_\beta$,

2) if $t_\alpha = t_\beta$, then $|s_\alpha - s_\beta| = |s'_\alpha - s'_\beta|$.

Obviously, the Galilean group acts on $R^3\{r\} \times R\{t\}$. We mention three examples of Galilean transformations of this space. First, uniform motion with velocity v:

$$g_1(r, t) = (r + tv, t).$$

Next, translation of the reference point (origin) is space-time:

$$g_2(r, t) = (r + x, t + \alpha).$$

Finally, rotation of the coordinate axes:

$$g_3(r, t) = (Gr, t),$$

where $G: R^3 \to R^3$ is an orthogonal transformation.

Proposition 1. *Every Galilean transformation $g: R^3 \times R \to R^3 \times R$ can be uniquely represented as a composition (product) $g_1 g_2 g_3$ of transformations of the type indicated above.*

Let us introduce in E^3 a "fixed" *coordinate system (reference frame)*: we fix a point $o \in E^3$ and pick three mutually orthogonal axes through o. Every Galilean transformation takes this coordinate system into a new coordinate system which is in uniform rectilinear motion with respect to the original system. Such coordinate systems are called *inertial*.

The action of the Galilean group on $E^3 \times R$ extends to an action on $E^3 \times \ldots \times E^3 \times R$ by the rule: if $g: (s, t) \to (s', t')$, then $g: (s_1, \ldots, s_n, t) \to (s'_1, \ldots, s'_n, t')$.

The principle of relativity states that Newton's equations, written in inertial systems, are invariant with respect to the Galilean transformation group.

This principle imposes a series of conditions on the right-hand side of Newton's equation, written in an inertial coordinate system. Thus, since among the Galilean transformations there are the time translations, the forces do not depend on the time t:

$$m_i \ddot{r}_i = F_i(r, \dot{r}), \qquad 1 \leqq i \leqq n.$$

Forces that do depend on t may arise in Newtonian mechanics only in simplified models of motion.

Translations in three-dimensional space E^3 are also Galilean transformations. From the homogeneity of E^3 it follows that in inertial coordinate systems forces can depend only on the relative coordinates $r_k - r_l$. Also, from the invariance of Newton's equations with respect to the subgroup of uniform motions g_1 it follows that forces can depend only on the relative velocities of the points:

$$m_i \ddot{r}_i = F_i(r_k - r_l, \dot{r}_k - \dot{r}_l), \qquad i, k, l = 1, \ldots, n.$$

Finally, from the isotropy of E^3 (that is, the invariance under the subgroup of rotations g_3) it follows that

$$F(Gr, G\dot{r}) = GF(r, \dot{r}).$$

If a mechanical system consists of only one point, then its motion in any inertial coordinate system is uniformly and rectilinear.[3] In fact, in this case the force F does not depend on t, r, $\dot{r}$, and is invariant under rotations. Consequently, $F \equiv 0$.

If the given system consists of two points, then the forces F_1 and F_2 acting on these points are directed along the straight line connecting them. Moreover, according to the principle asserting the *equality of action and reaction*, $F_1 = -F_2$. This principle, which is independent of the principle of relativity, leads to the general notions of *forces of interaction* and *closed mechanical system*. Thus, a system of n material points (r_i, m_i), $i = 1, \ldots, n$, on which the forces F_i act, is said to be closed if

$$F_i = \sum_{\substack{i \neq j \\ 1 \leqq j \leqq n}} F_{ij}, \qquad F_{kl} = -F_{lk}.$$

The vector F_{ij} is called the force with which the j-th point acts on the i-th. An important example of interaction is universal gravitation.

We note that if a system consists of three material points, then from the principle of relativity if follows that the forces acting on the points lie in the plane which contains them.

Among the examples of laws of motion given in §1.2, only universal gravitation is Galilean-invariant. If, however, in a system of material points interact-

[3] This is the Galileo-Newton *law of inertia*.

ing gravitationally, one of the masses is infinitesimally small (say, a speck of dust in the Solar system), then its influence on the motion of the other points can be neglected. This leads to a "restricted" problem (with many important applications in astronomy) for which Galileo's principle of relativity is no longer valid. All laws of motions, encountered in Newtonian mechanics, which are not Galilean-invariant, are obtained from invariant laws by making similar simplifying assumptions.

1.4 Basic Dynamical Quantities. Conservation Laws. The following characteristics of motion are important in dynamics:

$p = mv$ – the *momentum of* the point mass

$k = r \times p = m(r \times v)$ – the *angular momentum*

$M = r \times F$ – the *moment of the force F (torque)*

$T = \dfrac{mv^2}{2}$ – the *kinetic energy*

$I = mr^2$ – the *moment of inertia*.

If the given system consists of several point masses, the corresponding dynamical quantities are additive functions.

Proposition 2. *Let* $P = \sum p_i$ *and* $F = \sum F_i$. *Then* $\dot{P} = F$.

The point

$$\xi = \frac{\sum m_i r_i}{\sum m_i}$$

is called the *center of mass* (or *barycenter*) of the system. It is readily seen that the position of the center of mass does not depend on the choice of the origin of coordinates.

Corollary. *The center of mass of a closed system moves uniformly and rectilinearly*: $\ddot{\xi} = 0$.[4]

Proposition 3. *Let* $K = \sum k_i = \sum m_i(r_i \times v_i)$ *and* $M = \sum r_i \times F_i$. *Then* $\dot{K} = M$.

Corollary 1. *If the system is closed, then* $K = \text{const.}$[5]

The force acting on a material point is said to be *central* if the straight line along which it acts always passes through the point $o \in E^3$.

Corollary 2. *The motion under the action of a central force occurs in a fixed plane passing through the point* o.

Proposition 4. *Let* $T = \sum m_i v_i^2/2$. *Then* $\dot{T} = \sum \langle F_i, v_i \rangle$.

The forces $F_i(r_1, \ldots, r_n)$ are called *conservative* if the differential 1-form

$$\sum_{i=1}^{n} \langle F_i(r), dr_i \rangle,$$

[4] This fact was observed by Newton.

[5] This fact was established independently by Euler, D. Bernoulli, and d'Arcy.

known as the *(total) work* of the forces F_i in the *infinitesimal displacements* dr_i, is exact, i.e., it is the differential of some function $V(r_1, \ldots, r_n)$ defined everywhere in

$$E^{3n} \setminus \Delta = E^{3n} \setminus \bigcup_{i<j} \{r_i = r_j\}.$$

The functions V and $U = -V$ are called respectively the *force function*[6] and the *potential energy* (or simply the *potential*) of the system of point masses.

Corollary. *If the forces are conservative, then for every motion the total energy $T+U$ is constant.*[7]

The existence of the laws of conservation of momentum, angular momentum, and total energy for a closed system of point masses is related to the invariance of Newton's equations with respect to the Galilean transformation group.

Proposition 5. *If the forces of interaction in a system depend only on the relative distances of the points, i.e.,*

$$F_{ij} = f_{ij}(|r_{ij}|)\, e_{ij}, \qquad r_{ij} = r_i - r_j,\ e_{ij} = \frac{-r_{ij}}{|r_{ij}|},$$

then they are conservative.[8]

In this case the potential energy is $\sum_{i<j} U_{ij}$, where

$$U_{ij} = \int f_{ij}(|r_{ij}|)\, d|r_{ij}|$$

designates the potential energy of the interaction of the point masses m_i and m_j.

For example, in the case of universal gravitation the force function equals

$$V = \sum_{i<j} \frac{\gamma m_i m_j}{|r_{ij}|}.$$

Proposition 6. *Let $I = m_i r_i^2$ be the moment of inertia of the system with respect to the point $o \in E^3$. Then*

$$\ddot{I} = 4T + 2\sum_i \langle F_i, r_i \rangle.$$

If the forces are conservative and the force function is homogeneous of degree k, then

$$\ddot{I} = 4T + 2\sum_i \left\langle \frac{\partial V}{\partial r_i}, r_i \right\rangle = 4T + 2kV = 4h + 2(k+2)V$$

---

[6] The force function was introduced by J.L. Lagrange.

[7] In special cases this fact was already known to Huygens, Newton, and J. and D. Bernoulli.

[8] This was established by Lagrange.

depends only on the position of the points and the total energy. In the case of gravitational attraction, $k=-1$, and hence $\ddot{I}=4h-2U$; this formula was obtained by Lagrange.

As an application of the dynamical quantities and conservation laws introduced above we consider now the *regions of possible motions* of a closed system in which the forces of interaction depend only on distances.

First we attach to the center of mass a new inertial system and take it to be already fixed: $\sum m_i r_i = 0$. It is clear that the forces of interaction are not affected by such a change of origin of coordinates, because they depend only on the differences $r_i - r_j$.

The total angular momentum $K=\sum m_i(r_i \times v_i)$ with respect to the center of mass is constant. The plane passing through the center of mass and perpendicular to the constant vector K is usually referred to as Laplace's invariable plane.

Proposition 7. *The following inequality holds:*

$$K^2 \leqq 2IT. \tag{1}$$

◁ In fact,

$$K^2 = |\sum m_i(r_i \times v_i)|^2 \leqq (\sum m_i(r_i \times v_i))^2 \leqq (\sum m_i r_i^2)(\sum m_i v_i^2) = 2IT. \ ▷$$

Since $T=h-U$, we further obtain from (1) the inequality $K^2 \leqq 2I(h-U)$ or, equivalently, $U+\frac{K^2}{2I} \leqq 2h$. Consider the hypersurface

$$\Gamma = \{r=(r_1, \ldots, r_n) \in R^{3n} : \sum m_i r_i = 0\}$$

and write $B_{K,h}$ to denote the region of possible motions, i.e., the set of points of Γ at which the system with the given values, h and K, of energy and total angular momentum may be found. We have the obvious inclusion:

$$B_{K,h} \subset \left\{ r \in \Gamma : U + \frac{K^2}{2I} \leqq h \right\}.$$

However, the opposite inclusion is, generally speaking, not true.

Proposition 8. *In the planar problem, where the motion takes place in the Laplace plane,* $B_{K,h} = \left\{ r \in \Gamma : U + \frac{K^2}{2I} \leqq h \right\}$.

◁ Let $m_i r_i = 0$ and $v_i = \frac{1}{I}(K \times r_i) + \frac{1}{I}\alpha r_i$. Then $\sum m_i(r_i \times v_i) = K$ and $T = \frac{K^2}{2I} + \frac{\alpha^2}{2I}$. We can always choose α to ensure that $r \in B_{K,h}$. ▷

§2. Lagrangian Mechanics

2.1. Preliminary Remarks. We start with a simple example. Suppose a point of mass m moves on the smooth[9] regular surface Σ, given by the equation

$$f(x, y, z)=0, \tag{2}$$

under the action of a known force F. It is natural to identify the influence of surface Σ upon the motion of the point as the action of a force N perpendicular to Σ. If we now regard m as a free point, its motion is described by Newton's equation

$$m\ddot{r}=F+N. \tag{3}$$

From this equation, taking account of the *constraint equation* (2), we can determine the force N uniquely as a function of state and time.

Equation (3) may be rewritten as

$$\langle m\ddot{r}-F, \xi\rangle=0, \tag{4}$$

where ξ is an arbitrary vector tangent to Σ, and then interpreted as Newton's law of motion in the tangent plane to the surface Σ.

In mechanics the force N is usually called the *pressure* or, more generally, the *reaction of constraint* (2) (or the *constraint force*), and the tangent vectors ξ are known as the *virtual displacements, variations*, or *velocities* of the constrained point m.

In the general case of the nonfree motion of n points $(m_1, r_1), \ldots, (m_n, r_n)$, the *constraints* are defined by a smooth manifold M embedded in the configuration space of the free system, $R^{3n}=R^3\{r_1\}\times \ldots \times R^3\{r_n\}$. These constraints allow only those motions for which $(r_1(t), \ldots, r_n(t))\in M$ for all t. If known forces $F_1, \ldots, F_n$ act on the given points, equation (4) is naturally generalized to

$$\sum_{i=1}^{n} \langle m_r\ddot{r}_i-F_i, \xi_i\rangle=0, \tag{5}$$

where $(\xi_1, \ldots, \xi_n)$ is an arbitrary vector tangent to M. Equation (5) is called the "general equation of dynamics", or the *d'Alembert-Lagrange principle.* In the case of a free system of points, the vectors ξ_i are completely arbitrary, and hence equation (5) is equivalent to the system of Newton's equations.

Let $q=(q_1, \ldots, q_k)$ be local coordinates on M. Then r_i are smooth functions of q and

$$\dot{r}_i=\sum_{j=1}^{k} \frac{\partial r_i}{\partial q_j}\dot{q}_j.$$

The kinetic energy

$$T(\dot{q}, q)=\tfrac{1}{2}\sum m_i\dot{r}_i^2$$

[9] Here "smooth" has a twofold meaning: infinite differentiability and absence of friction.

is a positive definite quadratic form on M. We also introduce the *"generalized forces"* as the covectors $Q(q)$ defined by means of the equality

$$\sum_{i=1}^{n} \langle F_i, dr_i \rangle = \sum_{j=1}^{k} Q_j \, dq_j.$$

Theorem 1 (Lagrange). *The functions $q(t)$ which describe the motion of the constrained system satisfy the equation*

$$(T'_{\dot q})^{\cdot} - T'_q = Q.$$

If the forces $F_1, \ldots, F_n$ are conservative (in the sense of the definition given in §1), then the form $\sum Q_j(q)\, dq_j$ is the total differential of a smooth function $V(q)$. In this situation it is natural to consider the function $L = T + V$ and reexpress the equation of motion in the form of *Lagrange's equation*

$$(L'_{\dot q})^{\cdot} = L'_q.$$

This shows immediately that the motions of the mechanical system coincide with the extremals of the variational problem

$$\delta \int_{t_1}^{t_2} L\, dt = 0.$$

"It is astonishing that in Lagrange's work this statement may be read only between the lines. This explains the strange situation that this relation – mainly through Jacobi's influence – is generally known in Germany, and therefore also in France, as *Hamilton's principle.* In England no one understands this expression; there this equation is known under the correct if intuitive name of *principle of stationary action.*" (Felix Klein, Vorlesungen über die Entwicklung der Mathematik im 19. Jahrhundert, I. Mechanik, Ch. 5, Springer-Verlag, Berlin, 1926.)

2.2. Variations and Extremals. A *Lagrangian system* on the smooth manifold M is specified by a single function $L: TM \times \Delta \to R$, where Δ is an interval of the time axis $R = \{t\}$. We call the points $q \in M$ *configurations* (or *positions*) of the system, and we call each tangent vector $v \in T_q M$ a velocity at q. A pair q, v will also be referred to as a *state* of the system. In Lagrangian mechanics it is customary to call M the *configuration (position) space*, the tangent bundle TM the *state space* (or *velocity phase space*), L the *Lagrange function* or the *Lagrangian*, and $\dim M$ the *number of degrees of freedom.*

Example 1. A *natural mechanical system* is a triplet (M, T, V) consisting of a smooth configuration manifold M, a Riemannian metric T on M (the kinetic energy of the system), and a smooth function V on M (the force function of the force field). Recall that a Riemannian metric is a smooth function on the tangent bundle whose restriction to each tangent space is a positive definite quadratic form. The Lagrange function is $L = T + V$ (function $V: M \to R$ is lifted in the obvious manner to a function $TM \to R$). △

Let a_1 and a_2 be two points of M (not necessarily distinct). We call *path* from a_1 to a_2, starting at time t_1 and ending at time t_2 ($t_1, t_2 \in \Delta$), any C^∞-map

$$\omega: [t_1, t_2] \to M$$

such that $\omega(t_1) = a_1$ and $\omega(t_2) = a_2$. We let $\Omega(M; a_1, a_2; t_1, t_2)$, or simply Ω if there is no danger of confusion, denote the set of all such paths.

One may think of Ω as some kind of "infinite dimensional manifold". We define the tangent space to Ω at the point ω as the vector space consisting of the smooth vector fields W on M along the path ω such that $W(a_1)=0$, $W(a_2)=0$. A *vector field* W_t, given at the points $\omega(t) \in M$, will be said to be smooth if it is the restriction to $\omega(t)$ of a smooth vector field defined in a neighborhood of the path ω in M. We denote the tangent space to Ω at the point ω by $T_\omega \Omega$.

A *variation* (with fixed extremities) of the path ω is a map $\hat{\alpha}: (-\varepsilon, \varepsilon) \to \Omega$ such that

1) $\hat{\alpha}(0) = \omega$;

2) the map $\alpha: (-\varepsilon, \varepsilon) \times [t_1, t_2] \to M$, defined by the formula $\alpha(u, t) = \hat{\alpha}(u)(t)$, is smooth.

Since $\hat{\alpha}(u) \in \Omega(M; a_1, a_2; t_1, t_2)$, it follows that

3) $\alpha(u, t_1) = a_1$, $\alpha(u, t_2) = a_2$ for all $u \in (-\varepsilon, \varepsilon)$.

By variation of the path ω we shall often mean also the map α.

One may think of variation $\hat{\alpha}$ as a "smooth" path on Ω. Then it is natural to define its velocity vector

$$\frac{d\hat{\alpha}}{du}(0) \in T_\omega \Omega$$

as the following vector field $W_t \in T_{\omega(t)} M$ along ω:

$$W_t = \frac{\partial \alpha}{\partial u}(0, t).$$

Since

$$\frac{\partial \alpha}{\partial u}(0, t) \in T_{\omega(t)} M$$

for every $t_1 \leqq t \leqq t_2$, and

$$\frac{\partial \alpha}{\partial u}(0, t_1) = 0, \qquad \frac{\partial \alpha}{\partial u}(0, t_2) = 0,$$

W_t indeed belongs to $T_\omega \Omega$.

Lemma 1. *For every $W \in T_\omega \Omega$ there is a variation $\hat{\alpha}(u)$ such that*

$$\frac{d\hat{\alpha}}{du}(0) = W.$$

W_t is called a *variation vector field.* Clearly, it does not uniquely determine a variation of the given path.

Example 2. Let $q_1, \ldots, q_n$ be local coordinates on M, and let $W_1, \ldots, W_n$ be the coordinates of the tangent vector field W in the basis $\partial/\partial q_1, \ldots, \partial/\partial q_n$. If the path $\omega(t)$ is represented by smooth functions $q_1^*(t), \ldots, q_n^*(t)$, then the variation $\alpha(u, t)$ can be defined through the formulas

$$q_1(u, t)=q_1^*(t)+uW_1(t), \ldots, q_n(u, t)=q_n^*(t)+uW_n(t). \quad \triangle$$

Let $F: \Omega \to R$ be a function on Ω whose values are real numbers (in the classical terminology F is called a *functional*). We next define the differential $\delta F: T_\omega \Omega \to R$; it will be referred to as the *variation of the functional F*.

Pick $W \in T_\omega \Omega$. According to Lemma 1, there is a variation $\hat{\alpha}: (-\varepsilon, \varepsilon) \to \Omega$ with $\hat{\alpha}=\hat{\alpha}(u)$, such that

$$\hat{\alpha}(0)=\omega, \qquad \frac{d\hat{\alpha}}{du}(0)=W.$$

By definition,

$$\delta F(W)=\frac{dF(\hat{\alpha}(u))}{du}(0).$$

We should check that this definition is correct, i.e., that $\delta F(W)$ is linear in W and does not depend on the choice of a variation $\hat{\alpha}(u)$ of the path ω. However, we shall not do this here because in the situations discussed below it will be obvious that these requirements are met.

The point $\omega \in \Omega$ is critical (or stationary) for F if at this point $\delta F \equiv 0$. Suppose, for example, that F takes a minimal value on the path ω_0 and the derivative $dF(\hat{\alpha}(u))/du$ exists. Then clearly ω_0 is a critical path.

2.3. Lagrange's Equations. Let $q: [t_1, t_2] \to M$ be a smooth path belonging to Ω. The velocity v of q at time t equals the derivative $\dot{q}(t)$. At each t, $t_1 \leqq t \leqq t_2$, one can consider the sets of numbers $L'_{\dot{q}_i}$, $1 \leqq i \leqq n$, and $(L'_{\dot{q}_i})^{\cdot} - L_{q_i}$, $1 \leqq i \leqq n$, called the *momentum* of the system and the *Lagrangian derivative* of function L, respectively, and denoted here by p and $[L]$.

We now pass to new coordinates $\bar{q}$ by the rule $q=q(\bar{q})$. Let $J=\partial q/\partial \bar{q}$ designate the nonsingular Jacobi matrix of this change of variables. In the new coordinates, the Lagrange function is given by the formula $\bar{L}(\dot{\bar{q}}, \bar{q}, t)=L(\dot{q}, q, t)$.

Lemma 2. $p=J^{*-1}\bar{p}$, $[L]=J^{*-1}[\bar{L}]$.
These formulas are established by straightforward computation.

Since p and $[L]$ transform according to a covariant law under coordinate changes, we may regard them as covectors at the point $q(t) \in M$. Consequently, the two expressions $p \cdot w$ and $[L] \cdot w$ are well defined for tangent vectors $w \in T_{q(t)}M$.

Definition. The functional

$$F(\omega)=\int_{\omega} L\,dt=\int_{t_1}^{t_2} L(\dot{\omega}(t), \omega(t), t)\,dt$$

is called *action.*

Theorem 2 (first variation formula).

$$\delta F(W)=-\int_{t_1}^{t_2}([L]\cdot W)\,dt. \tag{6}$$

◁ Let $\hat{\alpha}(u)$ be a variation of the path ω. Then

$$\frac{dF(\hat{\alpha}(u))}{du}=\int_{t_1}^{t_2}(L'_{\dot{q}}q''_{ut}+L'_q q'_u)\,dt.$$

Integration by parts of the first term of the integral yields

$$\frac{dF}{du}(0)=p\cdot W\Big|_{t_1}^{t_2}-\int_{t_1}^{t_2}([L]\cdot W)\,dt.$$

It remains to use the fact that $W(t_1)=0$ and $W(t_2)=0$. ▷

Definition (Hamilton's principle). The path $\omega\in\Omega$ is called a *motion* of the Lagrangian system (M, L) if it is a critical point of the action functional.

It follows from formula (6) that the critical paths of the functional F are in fact the solutions of Lagrange's equation $[L]_{\omega(t)}=0$. In particular, the restriction of a motion $\omega(t)$ to any subinterval of $[t_1, t_2]$ is again a motion.

In local coordinates q on M, the Lagrange equation can be written in the explicit form: $A(\dot{q}, q, t)\ddot{q}+\Phi(\dot{q}, q, t)=0$, where $A=L''_{\dot{q}\dot{q}}$.

Lemma 3. *Let* $\bar{A}=\bar{L}''_{\dot{\bar{q}}\dot{\bar{q}}}$. *Then* $A=J^{*-1}\bar{A}J^{-1}$.

If $\det A\neq 0$, then Lagrange's equation can be resolved with respect to accelerations. This implies, in particular, that the state of the system at time $t_0\in\Delta$ uniquely determines its motion. Notice that according to Lemma 3 the condition that matrix A be nonsingular is independent of the choice of local coordinates on M.

2.4. Poincaré's Equations. Let $v_1, \ldots, v_n$ be linearly independent vector fields on the n-dimensional manifold M. At each point $q\in M$ the commutators $[v_i, v_j]$ can be written in the basis given by the vectors $v_1, \ldots, v_n$: $[v_i, v_j]=\sum c_{ij}^k(q)\,v_k$. If $q\colon \Delta\to M$ is a smooth path and f a smooth function on M, then

$$\dot{f}=f'_q\cdot\dot{q}=\sum v_i(f)\,\omega_i, \tag{7}$$

where $v_i(f)$ is the derivative of f in the direction of the field v_i. The variables ω_i are linear functions of the velocities, and are called *"quasi-velocities"*. Now write the Lagrangian as a function of q and ω: $\hat{L}(\omega, q)=L(\dot{q}, q)$. Suppose

$q(u, t)$ is a variation of the path $q(t)$. Set

$$\frac{\partial f(q(u,t))}{\partial u} = \sum_j v_j(f)\, w_j.$$

Since the differentiations with respect to q and t commute, we get

$$\frac{\partial \omega_k}{\partial u} = \frac{\partial w_k}{\partial t} + \sum_{i,j} c_{ij}^k \omega_i w_j.$$

Let us calculate the variation of the action in quasi-velocities:

$$\delta \int_{t_1}^{t_2} \hat{L}(\omega, q)\, dt = \frac{\partial \hat{L}}{\partial \omega_k} \cdot w_k \Big|_{t_1}^{t_2} + \int_{t_1}^{t_2} \sum_k \left[-\frac{d}{dt}\frac{\partial \hat{L}}{\partial \omega_k} + \sum c_{ik}^j \frac{\partial \hat{L}}{\partial \omega_j} \omega_i + v_k(\hat{L}) \right] \cdot w_k\, dt.$$

Since the variations w_k are independent inside the interval (t_1, t_2) and vanish at its endpoints, Hamilton's principle gives the equations of motions in coordinates q, ω:

$$(\hat{L}'_{\omega_k})^{\cdot} = \sum c_{ik}^j \hat{L}'_{\omega_j} \omega_i + v_k(\hat{L}).$$

These equations were first obtained by H. Poincaré in 1901. If we take for v_k the independent vector fields $\partial/\partial q_k$, then the Poincaré equations written above become the ordinary Lagrange equations.

Now let M be a *Lie group* G and let $v_1, \ldots, v_n$ be independent *left-invariant* vector fields on G. In this case $c_{ij}^k = \text{const}$. Suppose that the Lagrangian L is invariant under left translations on G. Then $v_k(\hat{L}) \equiv 0$, and hence $\hat{L}$ depends only on the quasi-velocities ω_k; the latter should be regarded as coordinates in the *Lie algebra* $\mathcal{G}$ of the group G. In these circumstances Poincaré's equations form a closed system of differential equations on $\mathcal{G}$.

Example 3. By *rigid body* fixed at a point we mean a system of point masses subjected to the following constraints (in the sense of §2.1):

a) the distances between points are constant;

b) one of the points of the body coincides with a fixed point o of E^3.

It is clear that the set of all positions of a rigid body can be obtained from a fixed one by rotations of E^3 around the point o; moreover, the correspondence positions↔rotations is one-to-one. It follows that the configuration space of this system can be identified with the group $SO(3)$. Every rotation of the body is given by a function $B(t)$ whose values are orthogonal matrices, i.e., elements of $SO(3)$. The velocity of rotation $\dot{B}(t)$ is a vector tangent to the group at the point $B(t)$. It is natural to translate this vector to the tangent space of $SO(3)$ at the identity, i.e., to the Lie algebra $so(3)$. This can be achieved by either left or right translations. Accordingly, the result is the skew-symmetric matrix $B^{-1}\dot{B} \in so(3)$ or $\dot{B}B^{-1} \in so(3)$.

Let $R(t)$ be the position vector of a point of the body in the fixed reference frame. Then $R(t) = B(t)\, R(0)$, whence

$$V(t) = \dot{R}(t) = B(t)\, R(0) = \dot{B}(t)\, B^{-1}(t)\, R(t).$$

In three-dimensional oriented Euclidean space every skew-symmetric operator is the operator of cross (vector) multiplication by a vector Ω: $\Omega \times (\cdot)$. This gives *Euler's formula* $V = \Omega \times R$. The vector Ω is called the *angular velocity* in *space coordinates.*

If r and v are the position vector and the velocity vector of the same point, but in a moving reference frame attached to the body, then again $v = \omega \times r$, where $\omega = B^{-1}\Omega$ is the angular velocity vector in *body coordinates.* The rule f: $B^{-1}\dot{B} \mapsto \omega$ defines an isomorphism of the Lie algebra $so(3)$ (which may be regarded as the Lie algebra of left-invariant vector fields on $SO(3)$) onto the Lie algebra of vectors of three-dimensional oriented Euclidean space with the commutator given by ordinary cross multiplication.

Let m denote the mass distribution in the rigid body. The angular momentum of the body (in body coordinates) is by definition the vector

$$k = \int (r \times v)\, dm = \int (r \times (\omega \times r))\, dm.$$

The linear symmetric operator A: $\omega \to k$ is called the *inertia operator*, and its mutually perpendicular eigendirections l_i are called the *principal* or *inertia axes.* The eigenvalues of A can be calculated by the formula

$$A_i = \int r_i^2\, dm,$$

where r_i are the distances from the points of the body to the axes l_i. The numbers A_i are called the *moments of inertia* of the body relative to the axes l_i, or the *principal moments of inertia.*

The kinetic energy of the body,

$$T = \tfrac{1}{2} \int V^2\, dm = \tfrac{1}{2} \int v^2\, dm$$

is, in view of the formula $T = \langle k, \omega \rangle / 2 = \langle A\omega, \omega \rangle / 2$, a quadratic form in the angular velocity.

Let e_i be unit vectors of the principal axes, indexed so that $e_1 \times e_2 = e_3$, $e_2 \times e_3 = e_1$, and $e_3 \times e_1 = e_2$. Let v_1, v_2, v_3 be the left invariant vector fields defined as the preimages of the vectors e_1, e_2, e_3 under the isomorphism f: $so(3) \to R^3$. Obviously,

$$[v_1, v_2] = v_3, \qquad [v_2, v_3] = v_1, \qquad [v_3, v_1] = v_2. \tag{8}$$

Let $\omega = \sum \omega_i e_i$. Then

$$T = \tfrac{1}{2}(A_1 \omega_1^2 + A_2 \omega_2^2 + A_3 \omega_3^2). \tag{9}$$

Using formulas (8) and (9), we write Poincaré's equations in the absence of external forces in explicit form, as follows:

$$A_1 \dot{\omega}_1 = (A_2 - A_3)\omega_2 \omega_3, \qquad A_2 \dot{\omega}_2 = (A_3 - A_1)\omega_3 \omega_1,$$
$$A_3 \dot{\omega}_3 = (A_1 - A_2)\omega_1 \omega_2.$$

These equations, obtained first by Euler in 1758, can be written as a single vector equation: $A\dot{\omega} + \omega \times A\omega = 0$.

Consider now the more general case in which the body is placed in an axially-symmetric force field with force function V. In view of Poincaré's equations, we must add the terms $v_i(V)$ to the right-hand side of Euler's equations. Let $\gamma=\sum\gamma_i e_i$ be the unit vector of the field's symmetry axis. Then clearly $V=V(\gamma_1, \gamma_2, \gamma_3)$. The condition that the vector γ be constant in the fixed reference frame is expressed by the equation

$$\dot{\gamma}=\gamma\times\omega, \tag{10}$$

known as Poisson's equation. From the formula

$$\dot{V}=\left\langle\frac{\partial V}{\partial\gamma}, \dot{\gamma}\right\rangle=\left\langle\frac{\partial V}{\partial\gamma}, \gamma\times\omega\right\rangle=\left\langle\omega, \frac{\partial V}{\partial\gamma}\times\gamma\right\rangle$$

and relation (7) we obtain the equations of motion of the rigid body:

$$A\dot{\omega}+\omega\times A\omega=\gamma\times\frac{\partial V}{\partial\gamma}. \tag{11}$$

We call the closed system of equations (10)–(11) the *Euler-Poisson equations.*

Example 4. We also consider the problem of the motion of a rigid body in a boundless ideal fluid. In this case the state space of the rigid body is the group $L(3)$ of motions of three-dimensional Euclidean space. The Lie algebra $l(3)$ of $L(3)$ is the semidirect product of the algebra of rotations, $so(3)$, and the three-dimensional commutative algebra of translations. The rotation of the body is described by *Kirchhoff's equations* (1870)

$$\dot{k}=k\times\omega+e\times u, \qquad \dot{e}=e\times\omega,$$

where $\omega=H'_k$, $u=H'_e$, and $H(k, e)=\frac{1}{2}\langle Ak, k\rangle+\langle Bk, e\rangle+\frac{1}{2}\langle Ce, e\rangle$ is the kinetic energy of the "body-fluid" system, with A, B, and C symmetric operators. The vectors ω and k are the angular velocity and the angular momentum, respectively; e and u are the "impulsive force" and the "impulsive momentum" of the body in fluid.

One can show that Kirchhoff's equations are actually Poincaré equations on the Lie algebra $l(3)$. A detailed analysis of Kirchhoff's problem is given in [57, 121].

2.5. Constrained Motion. We shall say that *constraints* are imposed upon the Lagrangian system (M, L) if at each moment of time $t\in\Delta$ in the state space TM there is singled out a submanifold S given locally by equations

$$f_1(\dot{q}, q, t)=\ldots=f_m(\dot{q}, q, t)=0$$

with linearly independent covectors $f'_{1\dot{q}}, \ldots, f'_{m\dot{q}}$. The constraints allow only those paths $\omega\colon \Delta\to M$ for which $(\omega(t), \dot{\omega}(t))\in S$ for all $t\in\Delta$. Usually one considers linear constraints, for which the functions f_k are linear in the velocities.

A *constrained Lagrangian system* is a triplet (M, L, S), described as above.

The tangent vectors $\xi \in T_q M$ satisfying the conditions

$$f'_{1\dot{q}} \cdot \xi = \ldots = f'_{m\dot{q}} \cdot \xi = 0,$$

are called the *virtual velocities* of the system (M, L, S) at time t and at state $(q, \dot{q}) \in S$. The correctness of this definition follows from Lemma 2.

Definition (the d'Alembert-Lagrange principle). The admissible (i.e., allowed by the constraints) smooth path $q: \Delta \to M$ is called a *motion* of the constrained Lagrangian system (M, L, S) if for every $t \in \Delta$, $[L]_{q(t)} \cdot \xi = 0$ for all virtual velocities ξ at the state $(q(t), \dot{q}(t))$.

Using this principle we can write a closed system of equations of motion:

$$[L] = \sum_{j=1}^{n} \mu_j f'_{j\dot{q}}, \qquad f_1 = \ldots = f_m = 0. \tag{12}$$

These are called *Lagrange's equations with multipliers*. If the matrix

$$(f'_{i\dot{q}} \cdot (L''_{\dot{q}\dot{q}})^{-1} f'_{j\dot{q}})_{ij} \tag{13}$$

is nonsingular, then the multipliers μ_j can be expressed as functions of the state of the system and time. In this case (12) are (possibly nonautonomous) differential equations on S. Consequently, constrained Lagrangian systems obey the principle of determinacy.

The d'Alembert-Lagrange principle has several equivalent formulations. We give two of them which belong to Gauss and Hölder.

Following Gauss, we introduce the set of *conceivable* motions: these are smooth paths $q_\mu: \Delta \to M$, allowed by the constraints, which all pass at some fixed time $t_0 \in \Delta$ through one and the same state $(x, v) \in S$. A path $q_0: \Delta \to M$ with the same state at $t = t_0$ will be called a *released motion* if $[L]_{q_0(t)} = 0$ for all $t \in \Delta$. Finally, an *actual motion* $q_a: \Delta \to M$ is a path satisfying the d'Alembert-Lagrange principle and the initial condition: $q_a(t_0) = x$, $\dot{q}_a(t_0) = v$.

Again, let $A = L''_{\dot{q}\dot{q}}$. Suppose $q_\alpha(t)$ and $q_\beta(t)$ are arbitrary smooth paths with the same initial state (x, v) at $t = t_0$. Following Gauss, we call the quadratic form

$$Z = \tfrac{1}{2}(A(\ddot{q}_\alpha - \ddot{q}_\beta), (\ddot{q}_\alpha - \ddot{q}_\beta))|_{t_0}$$

the *constraint*. It is readily verified that under coordinate changes on M the differences of accelerations $\ddot{q}_\alpha - \ddot{q}_\beta$ at $t = t_0$ behave like tangent vectors. Therefore, according to Lemma 3 the constraint is invariantly defined. The value of the constraint may be taken as a measure of the "deviation" of motions.

Theorem 3. *The deviation of conceivable motions from a released motion takes a stationary value on an actual motion.*

The proof rests on the well-known rule of Lagrange multipliers.

Usually the matrix A is positive definite (as for natural systems). In this case Theorem 3 admits the following:

Corollary (Gauss' principle of least constraint [38]). *Among conceivable motions the one that deviates least from the released motion is the actual motion.*

Example 5. A *mathematical pendulum* of length l is a heavy point mass which moves without friction along a vertical circle of radius l. Let φ denote the angle of deviation of the pendulum from the vertical (Fig. 3). Pick a state $(\varphi, \dot{\varphi})$; the accelerations a_μ of conceivable motions of the pendulum lie on a line which is parallel to its velocity and whose distance to the point m is $l\dot{\varphi}^2$. The acceleration of the released motion is clearly identical with the free-fall acceleration g. The constraint $Z = m\langle a_\mu - g, a_\mu - g\rangle/2$ coincides, modulo a constant term, with the function $m(l\ddot{\varphi} + g\sin\varphi)^2/2$. The condition that Z be minimal leads to the equation governing the oscillations of the mathematical pendulum: $l\ddot{\varphi} + g\sin\varphi = 0$. △

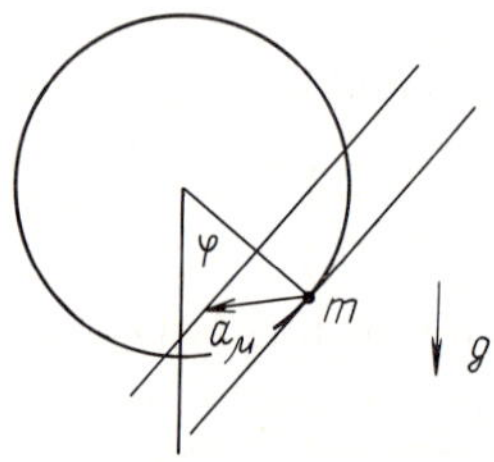

Fig. 3

To formulate *Hölder's principle* we need a number of new definitions. Let $\omega \in \Omega$ be a smooth admissible path. In the linear space $T_\omega\Omega$ of variation vector fields we single out the subspace Γ of fields $\tilde{W}$ with the property that for all t the vectors $\tilde{W}_t$ are virtual velocities. We call the path ω a critical point (in the sense of Hölder) of the action functional F if the restriction of δF to the subspace Γ vanishes.

Theorem 4 (Hölder's principle [38]). *An admissible path is a motion of the given constrained Lagrangian system if and only if it is a critical point (in the sense of Hölder) of the action functional.*

This is a direct consequence of Theorem 2 and the d'Alembert-Lagrange principle.

Herz has classified Lagrangian systems with linear constraints into *holonomic* and *nonholonomic* according to whether the imposed constraints are *(completely) integrable* or not. Integrability is most easily defined in the case of homogeneous constraints which do not depend explicitly on time:

$$a_1(q)\cdot\dot{q} = \ldots = a_m(q)\cdot\dot{q} = 0. \tag{14}$$

In conformity with the general definition of constraints, the covectors $a_1, \ldots, a_m$ are assumed to be linearly independent at all points of M^n.

The constraint (14) is said to be *completely integrable* if on M^n there is a smooth $(n-m)$-dimensional foliation whose leaves (which are $(n-m)$-dimensional submanifolds of M^n) are tangent at all their points to the planes defined by equations (14). We give one of the variants of the Frobenius criterion for the complete integrability of a distribution of tangent planes (14): Let $\varphi_s = a_s dq$, $s=1, \ldots, m$, be 1-forms on M^n; the constraint (14) is completely integrable if and only if the 2-forms $d\varphi_s$, $s=1, \ldots, m$, vanish on the space of admissible velocities.

Let $q: \Delta \to M^n$ be an admissible path of a system with holonomic constraints. Then for every $t \in \Delta$ the point $q(t)$ belongs to some leaf N^{n-m}. Consider the restriction $\hat{L}: TN^{n-m} \to R$ of the Lagrangian L to $TN^{n-m} \subset TM^n$.

Proposition 9. *The admissible path* $q: \Delta \to M^n$ *is a motion of the holonomic system* (M^n, L, S) *if and only if the path* $\hat{q}: \Delta \to N^{n-m}$, $\hat{q}(t) \equiv q(t)$, *is a motion of the Lagrangian system* $(N^{n-m}, \hat{L})$.

Thus, holonomic systems are practically indistinguishable from ordinary unconstrained Lagrangian systems.

Corollary. *The motion of a holonomic system* (M^n, L, S) *is completely determined by the restriction of the Lagrangian* L *to the manifold* $S \subset TM$.

This is not true, of course, in the nonholonomic case.

The best known examples of motion with nonintegrable constraints are skating on ice and a ball rolling on a rough surface. In the first case the velocity of the contact point in the direction perpendicular to the plane of the skate is equal to zero, whereas in the second the velocity of the contact point vanishes. We conclude this section with two examples which illustrate the "paradoxical" behavior of nonholonomic systems.

Example 6. Consider a skate on an inclined plane C with Cartesian coordinates x, y; we shall assume that the y-axis is horizontal, while the x-axis is directed downward. Let (x, y) be the coordinates of the point of contact of the "balanced" skate with the plane, and let φ be its rotation angle, measured from the x-axis. The equation of the integrable constraint is $\dot{x}\sin\varphi - \dot{y}\cos\varphi = 0$. For an appropriate choice of mass, length, and time units, we can write the Lagrangian as $L = (\dot{x}^2 + \dot{y}^2 + \dot{\varphi}^2)/2 + x$. The corresponding Lagrange equations with a multiplier are readily integrated. If, for example, at the initial time $\varphi = 0$ and we let the skate spin with angular velocity $\dot{\varphi}(0) = \omega$, then

$$x = \frac{\sin^2 \omega t}{2\omega^2}, \qquad y = \frac{1}{2\omega^2}(\omega t - \tfrac{1}{2}\sin 2\omega t), \qquad \varphi = \omega t.$$

An interesting feature of this solution is that on the average the skate does not slide down the inclined plane: $0 \leqq x(t) \leqq 1/2\omega^2$. $\triangle$

Let us also consider the problem of a homogeneous ball oscillating inside a vertical tube. It seems natural to expect the ball to descend on a spiral trajectory with increasing steepness. However, in reality (if the velocity of the center of mass at the initial time is not vertical) the ball will perform harmonic oscillations between two fixed horizontal planes.

§3. Hamiltonian Mechanics

3.1. Symplectic Structures and Hamilton's Equations. Let M^{2n} be a smooth even-dimensional manifold. There are several equivalent ways of endowing M with a *symplectic structure.* We list the best known.

a) A symplectic structure on M is defined as a closed nondegenerate 2-form ω^2. According to *Darboux's theorem,* in a small neighborhood of each point of M the symplectic structure ω^2 can be reduced, in suitable local coordinates $p_1, \ldots, p_n, q_1, \ldots, q_n$, to the "canonical" form

$$\sum_{i=1}^{n} dp_i \wedge dq_i.$$

Such local coordinates p, q are usually termed *symplectic* or *canonical.*

The form ω^2 permits us to construct a natural isomorphism from the tangent space $T_x M$ to the cotangent space $T_x^* M$, which sends each vector $\xi \in T_x M$ into the 1-form $\omega_\xi^1 \in T_x^* M$ by the rule: $\omega_\xi^1(\eta) = \omega^2(\eta, \xi)$, $\eta \in T_x M$. Since the 2-form ω^2 is bilinear and nondegenerate, the mapping $\xi \mapsto \omega_\xi^1$ is indeed a linear isomorphism. We denote the inverse mapping by $\mathfrak{I}: T_x^* M \to T_x M$. Let H be a smooth function on M (which may depend on time). Since the differential dH is a covector field, $\mathfrak{I}\, dH$ is a smooth vector field on M. Such fields are termed *Hamiltonian* vector fields. The corresponding differential equation

$$\dot{x} = \mathfrak{I}\, dH(x) \tag{15}$$

is called *Hamilton's equation.*

If F and G are smooth functions on M, then $\omega^2(\mathfrak{I}\, dG, \mathfrak{I}\, dF)$ is a correctly defined smooth function, called the *Poisson bracket* of the functions F and G, and denoted here $\{F, G\}$. The Poisson bracket has the following properties:

1) bilinearity,

2) skew-symmetry,

3) $\{F_1 F_2, G\} = F_1 \{F_2, G\} + F_2 \{F_1, G\}$ (the Leibnitz rule),

4) $\{\{H, F\}, G\} + \{\{F, G\}, H\} + \{\{G, H\}, F\} = 0$ (the Jacobi identity),

5) nondegeneracy (if $x \in M$ is not a critical point of F, then there exists a smooth function G such that $\{F, G\}(x) \neq 0$).

In symplectic local coordinates p, q,

$$\{F, G\} = \sum_{i=1}^{n} (G'_{p_i} F'_{q_i} - G'_{q_i} F'_{p_i}).$$

The Poisson bracket $\{F, G\}$ may also be calculated by the formula $dF(\Im dG)$, i.e., as the value of the covector dF on the vector $\Im dG$. Therefore, the derivative of function F in the direction of the Hamiltonian vector field $\Im dH$ is in fact $\{F, H\}$. Thus, Hamilton's equation (15) can be written in the equivalent form $\dot{F}=\{F, H\}$. Since the coordinate functions $p_1, \ldots, p_n$, $q_1, \ldots, q_n$ form a "complete" set of independent functions, the equations

$$\dot{p}_i=\{p_i, H\}, \quad \dot{q}_i=\{q_i, H\}, \qquad 1 \leqq i \leqq n$$

or, equivalently,

$$\dot{p}_i=-H'_{q_i}, \qquad \dot{q}_i=H'_{p_i}, \qquad 1 \leqq i \leqq n,$$

form a closed system. They are called Hamilton's *canonical* equations.

b) According to P. Dirac, the manifold M is endowed with a symplectic structure if there is given a mapping $\{\,,\,\}: C^\infty(M) \times C^\infty(M) \to C^\infty(M)$ which possesses properties 1)–5) given in a.

Suppose F is a smooth function on M. It follows from properties 1) and 3) that $v_F=\{F, \cdot\}$ is a derivation, i.e., a vector field on M. Every tangent vector is representable in this form. Let G and $v_G=\{G, \cdot\}$, be another smooth function and the corresponding tangent vector. Define the 2-form ω^2 by the formula

$$\omega^2(v_G, v_F)=\{F, G\}.$$

This form is obviously bilinear, skew-symmetric, and nondegenerate; the latter follows from the nondegeneracy of the Poisson bracket. From Jacobi's identity one can deduce that ω^2 is also closed. Therefore, Dirac's definition of a symplectic structure is equivalent to the definition given in a).

c) Finally, in the "classical" approach a symplectic structure on M is introduced by means of a *symplectic atlas*, that is, a collection of pairwise compatible charts such that the transition maps from one chart to another are *canonical transformations.*

Let P, Q and p, q be two sets of local coordinates on M. The transformation $p, q \mapsto P, Q$ is called *canonical* if

$$P\,dQ - p\,dq = dS(p, q),$$

where S is a smooth function, which will be referred to as a *primitive function* of the given canonical transformation.

It follows from the definition of canonical transformations that the form $\omega^2_{p,q}=dp \wedge dq$ is correctly defined on the entire manifold M. In fact,

$$\omega^2_{p,q}=d(p\,dq)=d(P\,dQ-dS)=dP \wedge dQ - d\,dS=\omega^2_{P,Q}.$$

Let us indicate criteria for transformation $p, q \mapsto P, Q$ to be canonical:

α) Let

$$\Gamma=\begin{bmatrix} Q'_q & Q'_p \\ P'_q & P'_p \end{bmatrix}$$

be the Jacobi matrix of the given transformation. Then the transformation $p, q \mapsto P, Q$ is canonical if and only if $\Gamma^* \mathfrak{J} \Gamma = \mathfrak{J}$, where

$$\mathfrak{J} = \begin{bmatrix} 0 & -E \\ E & 0 \end{bmatrix}$$

is the unit symplectic matrix.

β) $\oint_\gamma P\,dQ = \oint_\gamma p\,dq$ for every closed contour which is contractible to a point.

γ) $\{F, G\}_{P,Q} = \{F, G\}_{p,q}$ for all smooth functions F and G.

This shows, in particular, that canonical transformations preserve the canonical form of Hamilton's equations. In fact, $\dot{F}_{P,Q} = \dot{F}_{p,q} = \{F, H\}_{p,q} = \{F, H\}_{P,Q}$.

3.2. Generating Functions. Let $g: (p, q) \mapsto (x, y)$ be a canonical change of variables such that

$$\det\left(\frac{\partial x(p, q)}{\partial p}\right) \neq 0. \tag{16}$$

In this case we can solve equation $x = x(p, q)$ (at least locally) for p and regard x and q as independent variables. Then

$$p = p(x, q), \qquad y = y(p, q) = y(p(x, q), q).$$

The condition that transformation g be canonical,

$$p\,dq - x\,dy = dF(p, q),$$

can be put in the form

$$p\,dq + y\,dx = d(F + xy) = dS(x, q),$$

whence $p = S'_q$, $y = S'_x$. Inequality (16) implies that

$$\det\left(\frac{\partial^2 S}{\partial x\,\partial q}\right) \neq 0. \tag{17}$$

Function $S(x, q)$ is called a *generating function* of the transformation g. If, for example, g is the identical transformation, then $S = xq$.

Proposition 19. *For every function* $S(x, q)$ *satisfying condition* (17) *there is a canonical change of variables* $x = x(p, q)$, $y = y(p, q)$, *such that* $\det(\partial x / \partial p) \neq 0$ *and* S *is its generating function.*

Notice that not all canonical changes of variables satisfy condition (17). Here is a simple example: $x = q$, $y = -p$. In this case the method of generating functions may be modified slightly. Suppose, say, that the Jacobian $\det(\partial y / \partial p) \neq 0$. Such canonical transformations are known as *free*. The function $S_*(y, q) = F(p(y, q), q)$ serves as a generating function, meaning that the formulas

$$p = S'_{*q}, \qquad x = -S'_{*y}$$

give the free canonical transformation in question. Proposition 10 is valid also for free canonical transformations. These remarks can be generalized.

Proposition 11. *Let g be a canonical transformation given by $2n$ functions $x=x(p,q)$, $y=y(p,q)$. Then one can always take as independent local coordinates one of the 2^n sets of functions (x_i, y_j, q), $i\in I$, $j\in\{1,\dots,n\}\setminus I$, such that*

$$\frac{\partial(x_i, y_j)}{\partial(p_i, p_j)}\neq 0.$$

Moreover, for such a choice, transformation g is locally obtained from the generating function

$$\tilde{S}(x_i, y_j, q)=\sum_{i\in I} x_i y_i+\int(p\,dq-x\,dy)$$

through the relations

$$p=\frac{\partial\tilde{S}}{\partial q},\qquad y_i=\frac{\partial\tilde{S}}{\partial x_i},\qquad x_j=-\frac{\partial\tilde{S}}{\partial y_j}.$$

3.3. Symplectic Structure of the Cotangent Bundle. Let N^n be a smooth manifold and T_q^*N the cotangent space to N at the point q, i.e., the space of all 1-forms on the tangent space T_qN. The union $\bigcup_{q\in N} T_q^*N=M$ has a natural structure of $2n$-dimensional smooth manifold. It is called the (total) space of the cotangent bundle to N and is denoted by T^*N. If $q=(q_1,\dots,q_n)$ are local coordinates on N, then every 1-form is specified by its n components $p=(p_1,\dots,p_n)$ in the basis $dq_1,\dots,dq_n$. The sets of numbers $(p_1,\dots,p_n, q_1,\dots,q_n)$ serve as local coordinates on M.

The symplectic structure of the cotangent bundle T^*N is determined exclusively by the smooth structure of the manifold N. First we define the remarkable canonical 1-form $\omega^1=p\,dq$ on T^*N: the value of the covector $p\in T_q^*N$ on the tangent vector $\dot{q}\in T_qN$. In coordinates p_i, q_i $(1\leqq i\leqq n)$, ω^1 has the expression $\sum p_i\,dq_i$. The symplectic structure on M is given by the 2-form $\omega^2=d\omega^1$, which is clearly closed and nondegenerate.

Using this symplectic structure one can represent Lagrange's equations, defined on TN, as Hamilton's equations on T^*N. First we examine this transition locally. Let $L(\dot{q}, q, t)$ be a Lagrangian such that

$$\det(L''_{\dot{q}\dot{q}})\neq 0.$$

We set $p=L'_{\dot{q}}$ and regard p as an element of the dual space T_q^*N. This equation can be solved locally with respect to the velocity $\dot{q}$. We consider the function

$$H(p,q,t)=p\cdot\dot{q}-L|_{\dot{q}\to p}$$

and call it a "local" *Hamiltonian.* For fixed q and t function $H(p)$ is called the *Legendre transformation* of the function $L(\dot{q})$.[10] It is readily verified that

[10] As a matter of fact, the "Legendre transformation" already appears in Euler's work.

$\det(H''_{pp}) \neq 0$, and that $\dot{q} = H'_p$, $L(\dot{q}, q, t) = \dot{q} \cdot p - H|_{p \to \dot{q}}$. Therefore Legendre's transformation is involutive.

Theorem 5. *Let $q(t)$ be a solution of Lagrange's equations $[L]_{q(t)} = 0$. Then the functions $q(t)$ and $p(t) = L'_{\dot{q}}|_t$ satisfy Hamilton's equations $\dot{p} = -H'_q$, $\dot{q} = H'_p$.*

Momenta were first used instead of velocities in the works of Lagrange and Poisson.

To pass from Lagrange's to Hamilton's equations globally we shall assume that the smooth function $L: TN \times R \to R$ is convex in velocities, i.e., the matrix $(L''_{\dot{q}\dot{q}})$ is positive definite for all $\dot{q}$, q, and t. We define the "global" Hamiltonian by the formula

$$H(p, q, t) = \sup_{\dot{q}} (p \cdot \dot{q} - L(\dot{q}, q, t)). \tag{18}$$

If in the neighborhood of the point $\dot{q} = 0$ the Lagrangian L satisfies the supplementary condition $L(\dot{q}, q, t) \geqq |\dot{q}|^2$, where $|\cdot|$ designates (the length in) some Riemann metric on N, then a local Hamiltonian is defined for all p and coincides everywhere with the global Hamiltonian. Consequently, H is a smooth function on $M \times R$. Moreover, H is convex in p, and

$$L(\dot{q}, q, t) = \sup_{p} (\dot{q} \cdot p - H(p, q, t)). \tag{19}$$

From formulas (18) and (19) we obtain the convexity inequality $p \cdot \dot{q} \leqq H(p) + L(\dot{q})$.

Let us apply these considerations to natural mechanical systems. Suppose $|\cdot|$ is a Riemann metric on the configuration space N. The Lagrangian is $L = T - U$, where $T = |\dot{q}|^2/2$ and $U(q)$ are the kinetic and the potential energy of the system. If $|\dot{q}| = A(q)\dot{q} \cdot \dot{q}$, then $p = A\dot{q}$ and hence

$$H(p, q) = (T + U)|_{\dot{q} \to p} = \tfrac{1}{2}|p|^2 + U,$$

where $|p|^2 = A^{-1}p \cdot p$. Function H is equal to the total energy of the system. In a more general case the Lagrangian also contains linear terms, i.e.,

$$L = \tfrac{1}{2}|\dot{q}|^2 + \langle v(q), \dot{q} \rangle - U(q),$$

where v is a smooth vector field on N. Then

$$H = \tfrac{1}{2}|p|^2 - p \cdot v(q) + \tfrac{1}{2}|v(q)|^2 + U(q).$$

3.4. The Problem of *n* Point Vortices. One should not conclude that Hamilton's equations arise in mechanics only by applying the Legendre transformation to Lagrange's equations. Thus, consider a planar steady flow of an ideal incompressible fluid. Let $v = (a(x, y), b(x, y))$ be the velocity field of fluid particles in Cartesian coordinates x, y. From the incompressibility condition $\operatorname{div} v = 0$ it follows that the 1-form $a\,dy - b\,dx$ is the differential of a function $\Psi(x, y)$. Then the equation of motion of a fluid particle can be written in Hamilton's form

$$\dot{x} = \Psi'_y, \qquad \dot{y} = -\Psi'_x \tag{20}$$

with Hamiltonian Ψ. Since $\Psi=$const along streamlines, i.e., along trajectories of particles, Ψ is called the stream function.

In hydromechanics an important role is played by the flow with stream function

$$\Psi=\frac{\Gamma}{2\pi}\ln r,\qquad r=\sqrt{(x-x_0)^2+(y-y_0)^2}.$$

The corresponding velocity field is

$$v=\frac{\Gamma}{2\pi r}\left(-\frac{y-y_0}{2},\frac{x-x_0}{2}\right).$$

In this case we say that the flow generates a *vortex* of intensity Γ at the point (x_0, y_0). It is readily seen that the intensity of the vortex is equal to the integral

$$\oint(a\,dx+b\,dy)$$

taken over any circle centered at (x_0, y_0).

If we are given n point vortices in the plane, with intensities Γ_s and coordinates (x_s, y_s), $s=1, \ldots, n$, the natural choice for a stream function is

$$\Psi=-\frac{1}{2\pi}\sum_{s=1}^{n}\Gamma_s\ln\sqrt{(x-x_s)^2+(y-y_s)^2}.$$

The motion of a fluid particle with coordinates (x, y) is described by equations (20). According to *Thomson's theorem* (see §3.7) the vortices are "frozen" in the ideal fluid and their intensity does not change in time. It is therefore[11] natural to describe the dynamics of the vortices themselves by the system of differential equations

$$\dot{x}_s=\frac{\partial\tilde{\Psi}}{\partial y_s},\qquad \dot{y}_s=-\frac{\partial\tilde{\Psi}}{\partial x_s};$$
$$\tilde{\Psi}=-\frac{1}{2\pi}\sum_{k\neq s}\Gamma_k\ln\sqrt{(x_s-x_k)^2+(y_s-y_k)_2}. \tag{21}$$

Upon introduction the function

$$H=-\frac{1}{2\pi}\sum_{s\neq k}\Gamma_s\Gamma_k\ln\sqrt{(x_s-x_k)^2+(y_s-y_k)^2},$$

we can write equations (21) of the system of point vortices in the form

$$\Gamma_s\dot{x}_s=H'_{y_s},\quad \Gamma_s\dot{y}_s=-H'_{x_s}\qquad(1\leqq s\leqq n).$$

These equations are Hamiltonian relative to the symplectic structure in $R^{2n}\{x, y\}$ given by the Poisson bracket

$$\{f, g\}=\sum_s\frac{1}{\Gamma_s}\left(\frac{\partial f}{\partial y_s}\frac{\partial g}{\partial x_s}-\frac{\partial f}{\partial x_s}\frac{\partial g}{\partial y_s}\right).$$

[11] This argument is heuristic.

3.5. The Action Functional in Phase Space. Again, let $M=T^*N$ and let $H: M\times\Delta\to R$ be a smooth Hamiltonian. The "energy-momentum" 1-form $p\,dq-H\,dt$ is correctly defined on the extended phase space $M\times\Delta$. Consider a smooth path $\omega: [0,1]\to M\times\Delta$ whose trajectory in the extended phase space is represented by the equations $p=p(t)$, $q=q(t)$, $t_1\leqq t\leqq t_2$. We denote the set of all such paths by $\tilde{\Omega}$. A variation (with varying extremities) of the path ω is, by definition, a map $\hat{\alpha}: (-\varepsilon,\varepsilon)\to\tilde{\Omega}$ such that

1) $\hat{\alpha}(0)=\omega$,

2) the map $\alpha: (-\varepsilon,\varepsilon)\times[0,1]\to M\times\Delta$ given by $\alpha(u,s)=\hat{\alpha}(u)(s)$, $-\varepsilon<u<\varepsilon$, $0\leqq s\leqq 1$, is smooth.

We call the functional $F: \tilde{\Omega}\to R$, defined by the formula

$$F(\omega)=\int_\omega p\,dq-H\,dt, \tag{22}$$

the *action in phase space* (along the path ω). The action functional is differentiable and its differential (variation) can be calculated by the formula

$$\delta F=\left.\frac{dF(\hat{\alpha}(u))}{du}\right|_{u=0}=(pq'-Ht')\Big|_{t_1}^{t_2}+\int_{t_1}^{t_2}(p'\dot{q}-q'\dot{p}-H'+t'\dot{H})\,dt. \tag{23}$$

Here the prime designates the derivative with respect to u at $u=0$.

The subset of $\tilde{\Omega}$ consisting of paths with fixed extremities will be denoted by Ω (as in §2). The restriction $F: \Omega\to R$ is also differentiable, and its variation is given by formula (23) with the first term removed.

Theorem 6. *The path ω is a critical point of the functional $F: \Omega\to R$ if and only if its trajectory is a solution of Hamilton's equations with Hamiltonian H.*

The principle of stationary action in phase space was formulated by Poincaré ([33]; see also [38]).

An alternative approach to Poincaré's variational problem is to write the action as the integral

$$F=\int_{t_1}^{t_2}(p\cdot\dot{q}-H)\,dt$$

and regard the integrand as a Lagrangian $L: TM\to R$. Then the Euler-Lagrange equations of the variational problem $\delta F=0$ actually become Hamilton's equations; indeed,

$$(L'_{\dot{p}})^{\cdot}=0=L'_p=\dot{q}-H'_p,$$
$$(L'_{\dot{q}})^{\cdot}=\dot{p}=L'_q=-H'_q.$$

If M is an arbitrary symplectic manifold, the action (22) is defined only locally (within the domain of a canonical chart on M). Under canonical changes of local coordinates the action functional $F: \Omega \to R$ may change only by a constant, and in this sense it is "correctly" defined.

3.6. Integral Invariants. Let ω be a smooth closed path in the extended phase space. The points on the trajectory of ω may be taken as initial conditions for the solutions of Hamilton's equations. The set of all solutions with initial conditions on ω is a smooth surface Γ in $M \times \Delta$, called a *tube of trajectories.* Let $\hat{\alpha}(u)$, $0 \leqq u \leqq 1$, be a smooth family of closed paths in the extended phase space, lying on Γ and such that $\hat{\alpha}(0) = \omega$.

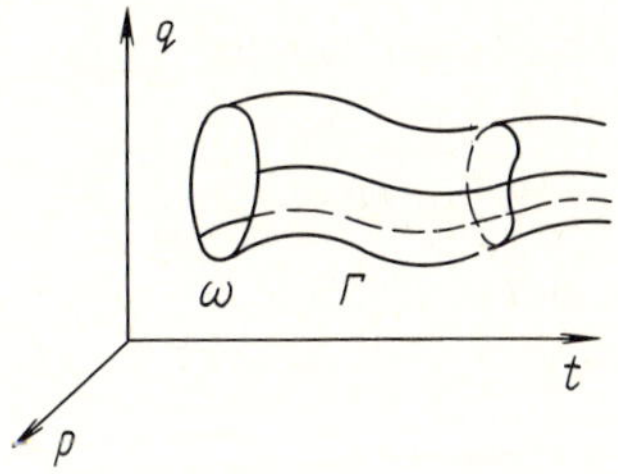

Fig. 4. Tube of trajectories

Theorem 7. *The value of the integral*

$$\oint_{\hat{\alpha}(u)} p\,dq - H\,dt \tag{24}$$

does not depend on u.

The integral (24) is known as the Poincaré-Cartan *integral invariant* [33, 13].

◁ Let $s (\mathrm{mod}\, 1)$ be an angular variable parametrizing the closed paths of family $\hat{\alpha}$. Consider the action

$$F(\gamma(s)) = \int_{\gamma(s)} p\,dq - H\,dt$$

along those paths on surface Γ which are solutions of Hamilton's equations that start at points of $\hat{\alpha}(0) = \omega$ and terminate at points of $\hat{\alpha}(u)$. By the first variation formula

$$\frac{dF(\gamma(s))}{ds} = \left(p \frac{\partial q}{\partial s} - H \frac{\partial t}{\partial s} \right) \Bigg|_0^u .$$

Integrating this expression with respect to s we get

$$\int_0^1 \frac{dF(\gamma(s))}{ds} ds = \oint_{\hat{\alpha}(u)} (p\,dq - H\,dt) - \oint_{\hat{\alpha}(0)} (p\,dq - H\,dt) = 0. \ \triangleright$$

Corollary 1. *Let g^t be the phase flow of Hamilton's equations. Let γ be a closed curve in the phase space M. Then the value of the integral*

$$\int_{g^t\gamma} p\,dq \tag{25}$$

does not depend on t.

Corollary 2. *The mapping $g^t\colon M \to M$ is canonical for all values of t.*

A meaning can be attached to the integral (25) even in the case where the symplectic differential form ω^2 is not exact (i.e., the 1-form ω^1 is not uniquely defined on the entire phase space M). Indeed, by *Stoke's formula*

$$\oint_{\gamma} p\,dq = \iint_{\sigma} dp \wedge dq = \iint_{\sigma} \omega^2$$

for every surface σ in M with boundary γ.

Corollary 3. *The value of the integral*

$$\iint_{g^t\sigma} \omega^2$$

does not depend on t.

Since surface $\sigma \subset M$ is arbitrary, this clearly implies that the differential of the mapping $g^t\colon M \to M$ preserves the symplectic structure ω^2. Consider the exterior powers of the 2-form ω^2:

$$\omega^4 = \omega^2 \wedge \omega^2, \ \ldots, \ \omega^{2n} = \omega^2 \wedge \omega^2 \wedge \ldots \wedge \omega^2.$$

Since the tangent map $dg^t\colon TM \to TM$ preserves the differential forms ω^{2k}, the integrals

$$\int \ldots \int_{g^t\sigma^{2k}} \omega^{2k}, \tag{26}$$

calculated over "moving" $2k$-dimensional surfaces $\sigma^{2k} \subset M$, are clearly independent of t.

The form ω^{2n}, written in canonical coordinates p, q, is proportional to

$$dp_1 \wedge \ldots \wedge dp_n \wedge dq_1 \wedge \ldots \wedge dq_n. \tag{27}$$

Therefore, for $k = n$ the integral (26) may be naturally termed the *volume* of the domain σ^{2n}. The phase volume element has form (27) to within a constant factor in any system of canonical coordinates.

Corollary 4. *The phase flow preserves volume in phase space.*

This important statement (known as "*Liouville's theorem* on the conservation of phase volume") permits us to apply to Hamiltonian mechanics the results of ergodic theory (*Poincaré's recurrence theorem, Birkhoff's theorem on mean values*, etc.). It is useful to keep in mind the following remark when such applications are envisaged: If the given Hamiltonian system has the first integrals $F_1, \ldots, F_m$ (one of which might coincide with the Hamiltonian

H), then the restriction of the phase flow g^t to a nonsingular invariant manifold $M_c=\{p, q: F_1=c_1, \ldots, F_m=c_m\}$ preserves a certain measure having a smooth positive density. More precisely, one can show that $g^t|_{M_c}$ preserves the value of the integral

$$\int_D \frac{d\sigma}{V_m}$$

taken over a "moving" $(2n-m)$-dimensional domain D in M_c; here $d\sigma$ is the volume element that M_c possesses as an embedded submanifold of $R^{2n}\{p, q\}$ and V_m is the m-dimensional volume of the parallellipiped with edges grad $F_1, \ldots$, grad F_m. We note that the form $d\sigma/V_m$ is in fact completely determined by the symplectic structure and does not depend on the choice of a metric on R^{2n}.

3.7. Applications to the Dynamics of Ideal Fluids. Euler's equation governing the flow of an ideal fluid in a conservative force field has the form

$$a=-\frac{1}{\rho}\operatorname{grad} p+\operatorname{grad} U. \tag{28}$$

Here a is the acceleration of fluid particles, ρ the density, p the pressure, and U the potential of mass forces. In the case of barotropic flows, p and ρ are connected by a relation $p=p(\rho)$, and hence one can determine the pressure function

$$P=\int \frac{dp}{\rho}.$$

It follows from equation (28) that every fluid particle behaves like a unit mass point placed in a conservative force field with potential $U-P$. In this case the Poincaré-Cartan integral invariant takes the form

$$\oint \langle v, dr\rangle - E\,dt,$$

where r is the position vector of the particle, v its velocity, and $E=v^2/2+P-U$ is the Bernoulli function. In the case of a steady flow function E is constant along streamlines.

In particular, if one takes a "fluid" contour made of particles observed at the same moment of time, the integral

$$\oint \langle v, dr\rangle$$

called *circulation*, assumes constant values. This assertion is the well known theorem of Thomson on the conservation of circulation, from which one can deduce the basic results of the dynamics of ideal fluids. We mention here two of these results. The first is *Lagrange's theorem* on the conservation of the potential (irrotational) nature of the flow: if at the initial time curl $v\equiv 0$, then this equality holds perpetually. The second is *Helmholtz's theorem*, asserting that vortex lines (i.e., the integral curves of the vector field curl v) are "frozen": if the particles of fluid form a vortex line at some instant, they form such a line during the whole motion.

3.8. Principle of Stationary Isoenergetic Action. Let (M, L) be a Lagrangian system with Lagrangian $L=L_2+L_1+L_0$, where L_k are smooth functions on TM which are homogeneous of degree k in the velocities. We shall assume that the quadratic form L_2 is positive definite, so that L_2, the kinetic energy of the system, defines a Riemann metric on M. Function $L_0: TM \to R$ can be regarded as a force function $V: M \to R$.

Lagrange's equation $[L]=0$ possesses the energy first integral $H=L_2-L_0$. For a fixed value $H=h$, the motion can occur only in the domain $B_h = \{x \in M: V+h \geqq 0\}$, called the region of possible motions. If $h > \bar{h} = \sup_M(-V)$, the set B_h coincides with the configuration space M. If $h < \bar{h}$, then the boundary $\partial B_h \neq \emptyset$. In the typical situation, where h is a regular value of the function $H: TM \to R$, B_h is a smooth manifold with boundary $\partial B_h = \Sigma_h$; $\dim \Sigma_h = \dim B_h - 1$.

To simplify formulas, suppose that $h=0$ (if $h \neq 0$ one can replace L_0 by L_0+h), and let $B=B_0$, $\Sigma=\Sigma_0$. We assume that $B \setminus \Sigma \neq \emptyset$.

Definition. The functional

$$F^* = \int_{t_1}^{t_2} (2\sqrt{L_0 L_2} + L_1)\, dt = F - \int_{t_1}^{t_2} (\sqrt{L_2} - \sqrt{L_0})^2\, dt,$$

defined on smooth paths $x: [t_1, t_2] \to B$, is called the *isoenergetic action* or the *Maupertuis action*.

The integrand in F^* is a homogeneous function of degree one in velocities. Hence, the value of the isoenergetic action does not depend on the parametrization of the integration path.

Proposition 12. *A smooth path $x: [t_1, t_2] \to B \setminus \Sigma$ such that $H(x(t))=0$ for all $t \in [t_1, t_2]$ is a solution of Lagrange's equation $[L]=0$ if and only if it is a critical point of the functional F^*.*[12]

◁ Let $[L]_{x(t)}=0$ and $L_2(\dot{x}(t)) \equiv L_0(x(t))$. Then

$$\delta F^* = \delta F - 2 \int_{t_1}^{t_2} (\sqrt{L_2} - \sqrt{L_0})\, \delta(\sqrt{L_2} - \sqrt{L_0})\, dt = 0.$$

Conversely, let $x: [s_1, s_2] \to B \setminus \Sigma$ be a smooth path which is a stationary point of the functional F^*. Set

$$t = \int_{s_1}^{s} \frac{L_2(x(\tau))}{L_0(x(\tau))}\, d\tau.$$

[12] Historically, the "principle of Maupertuis" (Proposition 12) preceded the simpler principle of stationary action, discovered by Hamilton. "The actual content of this principle was not quite clear to Maupertuis. The precise formulation given in the text is due to Jacobi and to his predecessors, Euler and Lagrange" ([42]).

Then obviously the smooth path $x(s(t))\colon [t_1, t_2] \to B \setminus \Sigma$ satisfies the equation $L_2 = L_0$. From the formula for δF^* it follows that $\delta F = 0$. ▷

We define in region B a Riemann metric $\langle\,,\rangle$ setting $\langle \dot{x}, \dot{x} \rangle = 4V(x)\,L_2(\dot{x})$, $\dot{x} \in TB$. This is known as the *Maupertuis metric*. For natural systems ($L_1 \equiv 0$) Proposition 12 says that in region $B \setminus \Sigma$ the trajectories of motions with zero total energy coincide with the geodesics of the Maupertuis metric.

When $h > \bar{h}$, $B = M$ and $(B, \langle\,,\rangle)$ is an ordinary Riemannian manifold. This observation permits us to apply topological theorems on Riemannian manifolds to the study of mechanical systems. Thus, for example, consider the torus T^2 with some Riemann metric. Among the family of all closed curves on T^2 which wind around the torus m times in parallels' direction and n times in meridians' direction, there is a curve of minimal length. It is a closed geodesic. On the other hand, T^2 is the configuration space of a planar double pendulum. It follows that for all nonnegative integers m and n there is a periodic motion of the double pendulum such that the first link makes m rotations over a period of time in which the second link makes n rotations. Moreover, such periodic motions exist for every sufficiently large value of the energy constant. For the variational theory of closed geodesics we refer the reader to [106, 138].

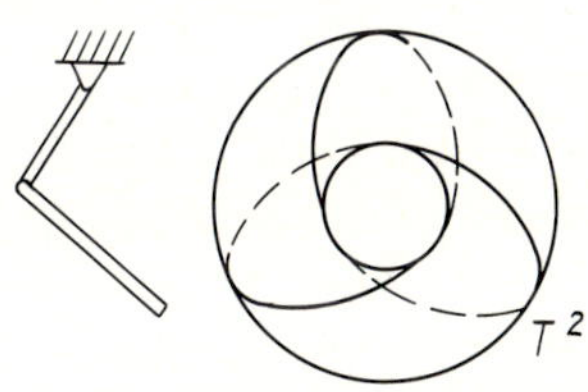

Fig. 5. Periodic oscillations of a double pendulum

If $h < \bar{h}$, the boundary Σ of the region B is not empty and the Maupertuis metric becomes singular: the curves on Σ have length zero. In this case the geometry of the region of possible motions is not similar to the familiar Riemannian geometry of closed manifolds (see [110]). Problems concerning the existence of closed trajectories for $h < \bar{h}$ are considered in [63, 65, 177].

§4. Vakonomic Mechanics

In §2.5 we considered constrained Lagrangian systems, whose motion obeys the d'Alembert-Lagrange principle (equivalent to the Gauss and Hölder principles). In this section we offer a different mathematical model of the motion of constrained systems which is based on a certain natural generalization of Hamilton's principle of stationary action; for this model we coined the name *vakonomic mechanics* (mechanics of **v**ariational **a**xiomatic **k**ind). As it turns out, in the case of completely integrable constraints vakonomic mechanics is identical with the usual mechanics of holonomic systems. If, however, the constraints are not integrable, then the principles of d'Alembert-Lagrange and of Hamilton, applied to the same Lagrangian system, yield different equations of motion.

4.1. Lagrange's Problem. Let M be a smooth manifold and $L: TM \times R \to R$ a smooth function. Let $f_k: TM \times R \to R$ $(1 \leq k \leq n)$ be a set of smooth functions such that the covectors $f'_{1\dot{q}}, \ldots, f'_{m\dot{q}}$ are linearly independent. The *problem of Lagrange* is the following variational problem: find the stationary points of the action functional

$$F = \int_{t_1}^{t_2} L\,dt$$

in the class of curves with fixed extremities which satisfy the equations

$$f_1 = \ldots = f_m = 0. \tag{29}$$

In contrast to the variations used in Hölder's principle (see §2.5), the variations of admissible paths in Lagrange's problem must also satisfy equations (29). This statement should not however be taken too literally since this may lead to serious difficulties.

Example 7 (Carathéodory). Suppose a constraint is imposed through the equation

$$\dot{x}_2 = \sqrt{1 + \dot{x}_1^2}, \qquad (x_1, x_2) \in R^2. \tag{30}$$

If we fix the values of the coordinates $(x_1, x_2) = x$ at time t_1, then a smooth curve $x: [t_1, t_2] \to R^2$ satisfying equation (30) is uniquely determined by its projection $x_1(\cdot)$; moreover, the difference of the x_2-coordinates at the extremities of the curve $x(\cdot)$ is equal to the length of the x_1-projection. If, in particular, x_1 is a linear function of time, then the admissible curve $x(\cdot)$ has the property that its endpoint $x(t_2)$ cannot be joined to the point $x(t_1)$ by any different admissible curve. △

This difficulty, connected with the "rigidity" of constraints, can be circumvented by slightly modifying the definition of the notion of a variation.

Definition. A *variation* of the admissible path $\omega: [t_1, t_2] \to M$ is any smooth family of paths $\hat{\alpha}(u): [t_1, t_2] \to M$, $u \in (-\varepsilon, \varepsilon)$, such that

1) $\hat{\alpha}(0) = \omega$;
2) the values $\alpha(u, t_i)$, $i = 1, 2$, do not depend on u;
3) the path $\alpha(u, t)$ satisfies equations (29) up to a quantity $o(u)$.

Lemma 4. *A smooth vector field $W(t)$ along the admissible path ω is a variation vector field if and only if:*

1) $W(t_1)=0,\ W(t_2)=0,$

2) $(a_k \cdot W)^{\cdot} = b_k W$ *for all* $t_1 < t < t_2$ *and all* $1 \leqq k \leqq m$, *where* $a_k = f'_{k\dot{q}}|_\omega$, $b_k = [f_k]|_\omega$.

Corollary. $\int_{t_1}^{t_2} b_k W dt = 0 \ (1 \leqq k \leqq m)$.

The variation of the action F is defined in the usual manner as

$$\delta F(W) = \left.\frac{dF(\hat{\alpha}(u))}{du}\right|_{u=0}, \qquad W = \left.\frac{d\hat{\alpha}(u)}{du}\right|_{u=0}.$$

A criterion for the stationarity of the action functional is provided by

Theorem 8. *The admissible path* $\omega: [t_1, t_2] \to M$ *is a conditional extremal of action* F *if and only if there exist* m *smooth functions* $\lambda_k: [t_1, t_2] \to R$ *such that the following equality holds along* ω:

$$[L] = \sum_k \lambda_k [f_k] + \sum_k \dot{\lambda}_k f'_{k\dot{q}}. \tag{31}$$

This equation together with the constraint equations (29) form a "closed" system from which the solutions of Lagrange's problem may be found. Equation (31) can be obtained by *the method of Lagrange multipliers.* Thus, upon introducing the new Lagrangian $\mathscr{L} = L - \sum \lambda_k f_k$ and regarding $\lambda_1, \ldots, \lambda_m$ as additional coordinates we reduce Lagrange's problem to a variational problem without constraints. If now in this new problem we ignore the constraint equations, the Euler-Lagrange equations take the form

$$(\mathscr{L}'_{\dot{q}})^{\cdot} = \mathscr{L}'_q, \qquad (\mathscr{L}'_{\dot{\lambda}})^{\cdot} = \mathscr{L}'_\lambda.$$

The first of these equations is in fact (31), whereas the second is identical with the set of constraint equations (29). A rigorous proof of Theorem 8 is based on Lemma 4 (see [58]).

Remark. Theorem 8 is not valid for the "classical" variant of Lagrange's problem, in which the variation $\hat{\alpha}(u)$ satisfies the constraint equations exactly for all values of u. In this case the equations for the extremals are equations (31) with the "amended" Lagrangian $\mathscr{L} = \lambda_0 L - \sum \lambda_k f_k$, where λ_0 is some constant (which may be zero), and the multipliers $\lambda_0, \lambda_1, \ldots, \lambda_m$ are not simultaneously zero. In Example 8 the constant λ_0 is indeed equal to zero.

4.2. Vakonomic Mechanics. The d'Alembert-Lagrange principle is not the only rational way of defining the motion of constrained Lagrangian systems. Instead, we may use Hamilton's principle, whereby the motions of the system are extremals of the variational problem of Lagrange (in the sense of the definition given in §4.1), and the equations of motion are (29) and (31). The mathematical model of the motion of constrained Lagrangian systems based on this extension of Hamilton's principle, will be referred to hereafter simply as vakonomic mechanics.[13]

The Lagrange equation (31) of vakonomic mechanics differ from the nonholonomic equations

$$[L]=\sum_k \mu_k f'_{k\dot q}, \quad f_1=\ldots=f_m=0 \tag{32}$$

by the presence of the term $\sum \lambda_k [f_k]$. If this sum vanishes identically (as a consequence of systems (29) and (31)), then equations (31) and (32) are identical. For example, suppose that a system is subjected to a set of integrable constraints $g_k(q, t)=0$ $(1 \leqq k \leqq m)$. We may replace these by the equivalent equations $\dot g_k=0$ $(1 \leqq k \leqq m)$. Since $[\dot g_k]\equiv 0$, it follows that in the case of integrable constraints vakonomic mechanics reduces to ordinary holonomic mechanics. We note that in contrast to nonholonomic mechanics, the motion of a vakonomic Lagrangian system is determined by the restriction of the Lagrangian to the submanifold of TM specified by the constraint equations (29).

Lagrange's equations (29), (31) can be put in Hamiltonian form. To this end we introduce the canonical momenta

$$p=\mathcal{L}'_{\dot q}=L'_{\dot q}+\sum \lambda_k f'_{k\dot q}, \tag{33}$$

where the multipliers λ_k are not determined for the present. We also consider the constraint equations (29), and solve system (29), (33) with respect to $\dot q$ and λ. The condition of local solvability is

$$\det \left[\begin{array}{c|c} A & f'_{1\dot q} \cdots f'_{m\dot q} \\ \hline (f'_{1\dot q})^* & \\ & 0 \\ (f'_{m\dot q})^* & \end{array}\right] \neq 0, \tag{34}$$

where $A=(\mathcal{L}''_{\dot q\dot q})$, and $(f'_{k\dot q})^*$ are the covectors $f'_{k\dot q}$, written as rows. If the matrix A is nonsingular, we can reexpress inequality (34) in the form

$$\det (f'_{i\dot q} \cdot A^{-1} f'_{j\dot q}) \neq 0.$$

If, in particular, the constraints are linear in velocities, then $\mathcal{L}''_{\dot q\dot q}=L''_{\dot q\dot q}$, and condition (34) becomes inequality (13); this guarantees the deterministic behavior of a constrained Lagrangian system obeying the d'Alembert-Lagrange principle (see §2).

Thus, if condition (34) is fulfilled, then by the implicit function theorem $\dot q=\dot q(p, q, t)$ and $\lambda=\lambda(p, q, t)$. Following the usual rule we introduce the Hamiltonian

$$H(p, q, t)=\dot q(p, q, t)-L(\dot q(p, q, t), q, t).$$

[13] The study of vakonomic mechanics is developed in [116]. We notice that equation (31) appears in works of Hertz, Hölder, G.K. Suslov, and other authors, in connection with the investigation of the applicability of the principle of stationary action (in the spirit of Lagrange's problem) to nonholonomic mechanics (see [38]). As it turned out, in the case of nonintegrable constraints the principles of d'Alembert-Lagrange and Hamilton are not equivalent.

Proposition 13. *A smooth path $q\colon \Delta \to M$ is a motion of the vakonomic system with Lagrangian L and constraints $f_1 = \ldots = f_m = 0$ if and only if function $q(\cdot)$ together with a "canonically conjugate" function $p(\cdot)$ satisfies Hamilton's equations*

$$\dot{q} = H'_p, \qquad \dot{p} = -H'_q. \tag{35}$$

The Hamiltonian H is degenerate in the momenta: the rank of the Hessian matrix H''_{pp} drops by m. In the case of natural systems (when the Lagrangian is a positive definite quadratic form in velocities) the mapping $p \mapsto \dot{q}$ defined through equations (29) and (33) is a degenerate linear mapping. Since every finite dimensional linear space is canonically identified with its bidual, equations (35) may be regarded as Hamilton's equations on T^*M.

Example 8. Consider a skate on an inclined plane (Example 2 in §2). The Lagrangian of this system and the constraint equation are $L = (\dot{x}^2 + \dot{y}^2 + \dot{\varphi}^2)/2 + x$ and $\dot{x} \sin\varphi - \dot{y}\cos\varphi = 0$, respectively. Hence, the canonical momenta are given by the formulas

$$\begin{gathered} p_x = \dot{x} - \lambda \sin\varphi, \qquad p_y = \dot{y} + \lambda\cos\varphi, \qquad p_\varphi = \dot{\varphi}; \\ \lambda = p_y \cos\varphi - p_x \sin\varphi. \end{gathered} \tag{36}$$

The corresponding Hamiltonian is

$$H = \tfrac{1}{2}\left[(p_x \cos\varphi + p_y \sin\varphi)^2 + p_\varphi^2\right] - x. \tag{37}$$

Suppose that at the initial instant the angle φ and the momentum p_y are equal to zero. Since p_y is a first integral of Hamilton's equations, $p_y \equiv 0$. It follows that in this case the Hamiltonian takes the simple form

$$H = \tfrac{1}{2}(p_x^2 \cos^2\varphi + p_\varphi^2) - x.$$

The canonical equations with this Hamiltonian are probably not integrable. Nevertheless, qualitative conclusions on the sliding motion of the vakonomic skate may be drawn. Thus, since $\dot{p}_x = -H'_x = 1$, the momentum p_x equals t modulo an additive constant. From the Hamilton's equations $\dot{\varphi} = p_\varphi$, $\dot{p}_\varphi = p_x^2 \sin\varphi\cos\varphi$ it follows that

$$\ddot{\varphi} = t^2 \sin\varphi\cos\varphi. \tag{38}$$

Next, from relations (36) we obtain equations describing the motion of the point of contact in Cartesian coordinates:

$$\dot{x} = t\cos^2\varphi, \qquad \dot{y} = t \sin\varphi\cos\varphi.$$

The first of these equations shows that the skate slides monotonically down the inclined plane. One can further show that almost all solutions of equation (38) tend to one of the points $\pi/2 + k\pi$ $(k \in Z)$ as $t \to \infty$ (see [115]). Moreover, there exist the limits

$$\lim_{t\to\infty} y(t) \quad \text{and} \quad \lim_{t\to\infty} \frac{1}{t}\int_0^t \dot{x}(s)\,ds > 0.$$

Therefore, asymptotically the skate slides down along a straight line with nonzero mean velocity and tends to turn sideways to its mean motion. It is interesting to compare this kind of motion with that of the nonholonomic skate, which for the same initial conditions moves sideways along a cycloid.

4.3. The Principle of Determinacy. Consider the motion of a vakonomic natural system with the Lagrangian $L=(A\dot{q},\dot{q})/2+V(q)$ and time-independent linear constraints $f_k=a_k(q)\cdot\dot{q}=0$ $(1\leqq k\leqq m)$. The linear map $\Psi_q\colon T_q^*M\to T_qM$ defined by equations (29) and (33) is degenerate; its m-dimensional kernel consists of the linear combinations $\sum\mu_k a_k$. Let $\Gamma_q(\dot{q})$ denote the pre-image of the point $\dot{q}\in T_qM$ under Ψ_q; $\Gamma_q(\dot{q})$ is an m-dimensional plane in T_q^*M. Pick an initial state $(q_0,\dot{q}_0)$, allowed by the constraints, and consider the family of solutions of Hamilton's equations (35) with initial conditions $q(0)=q_0$, $p(0)\in\Gamma_{q_0}(\dot{q}_0)$. If the constraints are completely integrable the function $q(t,q(0),p(0))$ (that is, the motion of the given vakonomic system) does not depend on the choice of the initial momentum in the plane $\Gamma_{q_0}(\dot{q}_0)$. The converse of this simple remark is also true:

Proposition 14. *If for every admissible state $(q_0,\dot{q}_0)$ the motion $q(t,q(0),p(0))$ does not depend on $p(0)\in\Gamma_{q_0}(\dot{q}(0))$, then the linear constraints in question are completely integrable.*

Corollary. *The principle of determinacy is not valid for vakonomic systems with nonintegrable constraints.*

Example 9. In the problem of a skate sliding on an inclined plane (Problem 8), the initial value of the constant momentum p_y does not affect the skate's initial configuration if $\varphi(0)=0$. However, the solutions of Hamilton's equations with Hamiltonian (37) depend essentially on p_y. △

Remark. The principle of determinacy, though not valid globally, may nevertheless hold for isolated states. Thus, one can show that the following "generalized" principle of determinacy is valid in vakonomic mechanics: the motion of the system on some time interval uniquely determines its past and future motion. △

Let $\Psi\colon T^*M\to TM$ be the map which coincides on each fiber T_q^*M with the linear map Ψ_q. We say that function $F\colon T^*M\to R$ is *observable* if there is a function $G\colon TM\to R$ such that the following diagram commutes:

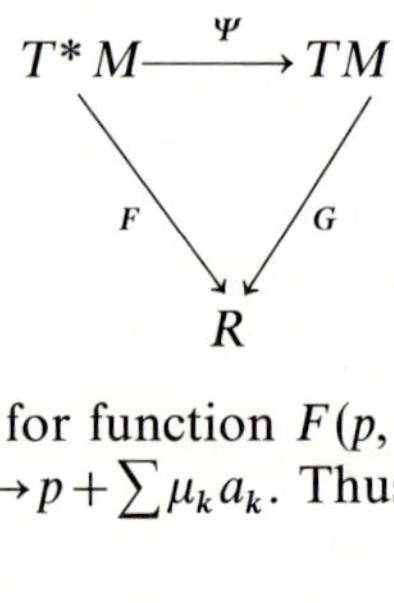

The observability condition for function $F(p,q)$ is that it be invariant under the family of translations $p\mapsto p+\sum\mu_k a_k$. Thus, for example, the total energy of the system is observable, whereas the Lagrange multipliers λ_k are not. Notice that the observability property does not depend on whether the constraints are integrable or not.

Proposition 15. *The linear space of all observable functions is closed under the Poisson bracket (defined by the standard symplectic structure $dp \wedge dq$) if and only if the constraints are completely integrable.*

In conclusion, we remark that the problem of "hidden" parameters – unobservable quantities intervening in the description of the dynamics of a system – has been discussed repeatedly in quantum mechanics, exactly in connection with the principle of determinacy (see [62]).

4.4. Hamilton's Equations in Redundant Coordinates. In the case of completely integrable constraints equations (35) are Hamilton's equations for a holonomic system, written in *redundant* coordinates. Consider, for example, the motion of a point mass (m, r) in Euclidean space E^3 on a smooth regular surface Σ, given by the equation $f(r)=0$. Suppose that a conservative force field with potential $U(r)$ acts on the point. Set (according to (33))

$$p=m\dot{r}+\lambda f_r', \qquad \lambda=\frac{\langle p, f_r'\rangle}{\langle f_r', f_r'\rangle}. \tag{39}$$

The motion of the point is described by Hamilton's equations

$$\dot{r}=H_p', \quad \dot{p}=-H_r'; \qquad H=\frac{m}{2}\langle \dot{r}, \dot{r}\rangle+U=\frac{1}{2m}(p\times n)^2+U, \tag{40}$$

where n is the unit vector normal to surface Σ. Therefore, equations (40) are determined by surface Σ itself, and do not depend on the particular form of the equation $f=0$ by which this surface is given.

Equations (40) possess the energy integral H and the "geometric" integral $f(r)$. It is clear that in the standard symplectic structure $dp \wedge dr$ the Poisson bracket $\{H, F\}=0$. Let $g(\dot{r}, r)$ be a first integral of the equations of motion

$$m\ddot{r}=-U_r'+\lambda f_r', \qquad f(r)=0,$$

and denote by G the function g expressed in canonical variables. Then obviously $\{H, G\}=0$, and the involutiveness of function G and F is readily verified.

The relative Poincaré invariant

$$\oint_\Gamma \langle p, dr\rangle,$$

evaluated under the assumption that the contour Γ lies on the hypersurface $\{f=0\} \subset R^6\{p, r\}$, equals

$$\oint_\gamma \langle \dot{r}, dr\rangle \tag{41}$$

to within a constant factor m; here the closed contour γ in $R^6\{r, \dot{r}\}$ is the image of Γ under mapping (39). In hydrodynamics an integral of form (41) is called circulation of the velocity along the contour γ.

§ 5. Hamiltonian Formalism with Constraints

5.1. Dirac's Problem. Let (M, ω^2) be a symplectic manifold, $H: M \to R$ a smooth function, and N a submanifold of M. We call the quadruplet (M, Ω^2, H, N) a *Hamiltonian system with constraints* or a *constrained Hamiltonian system*. The restriction of the form Ω^2 and of the function H to N will be denoted by ω^2 and F, respectively. The form ω^2 is obviously closed, but it might be degenerate (for example, when the dimension of N is odd).

Definition. The smooth path $x: \Delta \to M$ is called a *motion* of the Hamiltonian system (M, Ω^2, H, N) if $\omega^2(\cdot, \dot{x}(t)) = dF(x(t))$ for all $t \in \Delta$.

Our next objective is to describe the set of all motions of a constrained Hamiltonian system.[14]

If $N = M$ the constrained Hamiltonian system reduces to an ordinary Hamiltonian system (see §3) and its motions are simply the solutions of Hamilton's equations on M. There is another case in which Dirac's problem reduces to solving Hamiltonian equations: when the form ω^2 is nondegenerate (N, ω^2) is a symplectic submanifold and the motions of the system (M, Ω^2, H, N) are solutions of Hamilton's equations on N with Hamiltonian F. Moreover, to each initial state $x_0 \in N$ there corresponds a unique motion of the constrained system. In the degenerate case two alternatives are possible: in Dirac's problem there may be either several distinct motions with initial state x_0 or none. As we shall see, these two situations do indeed occur.

Suppose the form Ω^2 is exact on M, that is $\Omega^2 = d\Omega^1$. Then $\omega^2 = d\omega^1$, where ω^1 is the restriction of the 1-form Ω^1 to N. In the general case (where Ω^2 is not exact) these relations hold on M only locally.

Lemma 5. *The smooth path* $x: [t_1, t_2] \to N$ *is a motion of the system* (M, Ω^2, H, N) *if and only if* $x(\cdot)$ *is a critical point of the action functional*

$$\int_{t_1}^{t_2} (\omega^1(\dot{x}) - F)\, dt$$

on the space of smooth paths on N *with fixed extremities.*

This assertion reduces Dirac's problem to the investigation of Lagrange's variational problem (see §4.1) with Lagrangian $L(\dot{x}) = \Omega^1(\dot{x}) - H$ in which the integrable constraints are given by the submanifold N.

---

[14] This problem was first considered by Dirac in 1950 for purposes of quantum mechanics (see [16]).

Let us write a number of explicit formulas that will be used below. Let $x=(p, q)$ be local symplectic coordinates on M and suppose that submanifold N is given by the system of equations

$$\Phi_1(p, q)=\ldots=\Phi_m(p, q)=0, \tag{42}$$

and that functions $\Phi_1, \ldots, \Phi_m$ are functionally independent at all points of N. The equations of the extremals of Lagrange's problem with Lagrangian $L=p\cdot\dot{q}-H(p, q)$ and constraints (42) can be written as the equations with multipliers $[L]=\sum\lambda_k\Phi'_{kx}$ or, explicitly,

$$\dot{q}=H'_p+\sum\lambda_k\Phi'_{kp}, \qquad \dot{p}=-H'_q-\sum\lambda_k\Phi'_{kq}, \tag{43}$$

to which we must adjoin equations (42). Since the function L is degenerate in the velocities (it does not depend on $\dot{p}$ at all), the method of §4 is not applicable to equations (43).

From equations (42) and (43) we derive the "compatibility conditions"

$$\dot{\Phi}_i=\{\Phi_i, H\}+\sum\lambda_k\{\Phi_i, \Phi_k\}=0 \tag{44}$$

on N (i.e., where all $\Phi_k=0$). If the matrix of Poisson brackets $(\{\Phi_i, \Phi_j\})$ is nonsingular, equations (44) uniquely define λ_k as functions of p and q. In this case m is necessarily even and N is a symplectic submanifold of M. The symplectic structure on N is specified by the Poisson bracket

$$\{F_1, F_2\}'=\{F_1, F_2\}+\sum_{i,j=1}^{m}\{\Phi_i, F_1\}\,c_{ij}\{\Phi_j, F_2\},$$

where (c_{ij}) is the inverse of the matrix $(\{\Phi_i, \Phi_j\})$. One can show that the restriction of the bracket $\{F_1, F_2\}'$ to N depends only on the restrictions of F_1 and F_2 to N. If some of equations (44) do not contain the multipliers λ_k, we obtain new constraint equations $\Psi_j=\{\Phi_j, H\}=0$, which are usually called *secondary constraints*. In the most general case the secondary constraints are algebraic conditions for the solvability of equations (44) with respect to λ_k. The functions Ψ_j must be adjoined to Φ_k; if the elements of the resulting collection are functionally independent, one can repeat the analysis of the compatibility conditions. In the end we reach either a contradiction (in which case Dirac's problem has no solutions), or the conclusion that system (44) is compatible for an appropriate choice of coefficients λ_k. In the latter situation it may happen that the multipliers λ_k are not uniquely determined, in which case the initial conditions do not determine a unique solution of system (42)–(43).

Example 10. Suppose that $m=1$ and the bracket $\{H, \Phi\}\neq 0$ at all points of N. Then Dirac's problem has no solutions, since the compatibility condition (44) is not fulfilled. Now again let $m=1$ but assume that $\{H, \Phi\}\equiv 0$ on N. In this case the coefficient λ is an arbitrary smooth function on N, and hence through each point $x_0\in N$ there passes, at the same instant, a whole family of distinct motions. Moreover, there exist infinitely many distinct motions that coincide on an entire interval of the time axis. In vakonomic mechanics this does not happen (see §4.3). △

Remark. To solve Dirac's problem it obviously suffices to know the restriction of the Hamiltonian H to the submanifold N.

5.2 Duality. If the Hamiltonian and the constraint equations (42) are known, then one can pass to the Lagrangian L by the usual recipe: $L = \dot{q}\cdot p - H$. Let $\mathscr{H} = H + \sum \lambda_k \Phi_k$. If $\det(\mathscr{H}''_{pp}) \neq 0$ and $\det(\Phi'_{ip}\cdot(\mathscr{H}''_{pp})^{-1}\Phi'_{jp}) \neq 0$, then one can, at least locally, solve the equations

$$\dot{q} = H'_p + \sum \lambda_k \Phi'_{kp}, \qquad \Phi_1 = \ldots = \Phi_m = 0$$

for p as a function of $\dot{q}$ and q. In the final analysis the Lagrangian is a function of states $(q, \dot{q})$, degenerate in velocities. We note that the transition from H to L in Hamiltonian mechanics with constraints is dual to the transition from L to H in vakonomic mechanics (see §4).

Conversely, given a Lagrangian $L(\dot{q}, q)$ degenerate in velocities one can introduce the canonical momenta $p = L'_{\dot{q}}$ and derive from these equations a number of independent relations of the form (42). In quantum mechanics the latter are usually referred to as *primary constraints*. Then one introduces the Hamiltonian $H = p\cdot\dot{q} - L$ which, due to the degeneracy of the Lagrangian, is defined only on the manifold $N = \{\Phi_1 = \ldots = \Phi_m = 0\}$. As we have already seen this limitation is not essential. As a matter of fact, in his construction of the Hamiltonian formalism with constraints, Dirac was originally concerned with a degenerate Lagrangian.

§6. Realization of Constraints

6.1. Various Methods of Realizing Constraints. We begin with a simple example. Consider the straight-line motion of two bodies of masses M and m, connected by elastic springs with elasticity coefficients k and c (as in Fig. 6). Let x and y be the distances from the "wall" to the points M and m. The motion is described by the following simple system of linear differential equations

$$M\ddot{x} = -kx - c(x-y), \qquad m\ddot{y} = -c(y-x).$$

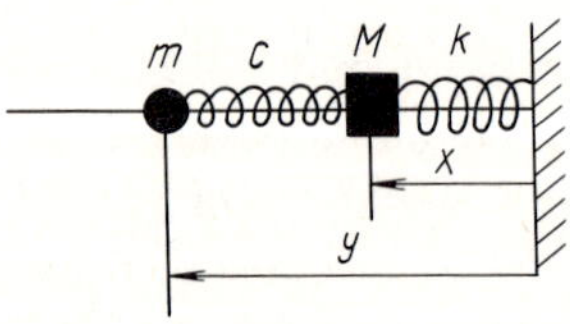

Fig. 6

Now fix the parameters M, m, and c and let $k \to \infty$. Let $x(t, k)$ and $y(t, k)$ be the solution of these equations with initial conditions $x(0, k) = \dot{x}(0, k) = 0$, and $y(0, k)$, $\dot{y}(0, k)$, the latter being independent of k. Then clearly

$$\lim_{k \to \infty} x(t, k) = 0,$$

whereas the limiting motion

$$\hat{y}(t) = \lim_{k \to \infty} y(t, k)$$

is a harmonic oscillation with frequency $\omega = \sqrt{c/m}$. In this case, considering a spring of "infinite" stiffness is equivalent to subjecting the system to the holonomic constraint $x = 0$.

The same constraint may be realized in a different way. To this end it suffices to let the mass $M \to \infty$ and assume again that $x(0) = \dot{x}(0) = 0$.

There is yet another, physically obvious way of realizing this constraint, which resorts to forces of *viscous friction.* Thus, let us assume that the body of mass M is also acted upon by the resistance force $F = -\alpha \dot{x}$ with $\alpha = \text{const} > 0$. Then we must add this term to the right-hand side of the first equation of motion. If we now let $\alpha \to \infty$ then again $x(t) \to 0$. Here it is no longer necessary to take $x(0)$ and $\dot{x}(0)$ equal to zero. The equality

$$\lim_{\alpha \to \infty} x(t) = 0$$

will hold for all $t > 0$.

We may consider a more general case, when both M and α tend to infinity but their ratio α/M tends to a finite value $\mu > 0$. In this limit, the equations of motion take the simple form

$$\ddot{x} = -\mu x, \qquad m\ddot{y} = -c(y - x).$$

They admit again the solutions $(x, y) = (0, y_0)$ such that $\ddot{y}_0 + \omega^2 y_0 = 0$, where $\omega^2 = c/m$.

Finally, we consider the case where the second mass m is small. Then in the limit it has no effect on the motion of the body M (which will perform harmonic oscillations with frequency $\sqrt{k/M}$). If $c > 0$, then m will obviously replicate the motion of M: $y(t) \equiv x(t)$.

If, however, the value of c also tends to zero in such a way that $c/m \to \mu > 0$, then in the limit we get a "restricted" two-body problem: M performs harmonic oscillations following the law $x_0(t)$, whereas m performs forced oscillations governed by the equation

$$\ddot{y} + \mu y = \mu x_0(t).$$

These simple observations can be generalized.

6.2. Holonomic Constraints. Consider a holonomic natural system with configuration space M, kinetic energy $T(\dot{q}, q)$, and potential energy $U(q)$. Let $f: M \to R$ be a smooth function such that $df \neq 0$ on the set $\Lambda = \{f = 0\} \subset M$.

Consider a new system with potential energy $U+Nf^2$ depending on a parameter N (subsequently, we will let $N\to\infty$). We write $q(t, N)$ to denote its motion with initial conditions on Λ:

$$q(0)=q_0\in\Lambda, \qquad \dot q(0)=\dot q_0\in T_{q_0}\Lambda,$$

which do not depend on N.

Theorem 9. *On every bounded time interval* $0\leqq t\leqq t_0$ *there exists the limit*

$$\lim_{N\to\infty} q(t, N)=\hat q(t)\in\Lambda.$$

The limiting function $\hat q\colon [0, t]\to\Lambda$ *satisfies Lagrange's equations with the Lagrangian*

$$L=(T-U)|_{T\Lambda}.$$

A detailed proof of this theorem can be found, for example, in [171, 184]. The key step if to observe that, as a consequence of conservation of energy, the point $q(t, N)$ cannot move away from Λ at a distance exceeding $c/\sqrt{N}$, which tends to zero as $N\to\infty$.

Theorem 9 may be used to construct arbitrary holonomic Lagrangian systems from systems of free points masses.

6.3. Anisotropic Friction. We begin with by defining forces of viscous friction. We say that the Lagrangian system with Lagrangian $L=(A(q)\dot q\cdot\dot q)/2-U(q)$ is subjected to forces of viscous friction if its motion is described by the equation

$$[L]=-F'_{\dot q}, \tag{45}$$

where F is a nonnegative definite quadratic form in velocities, called the *dissipation function* or *Rayleigh's function.* The derivative of the total energy of the system along the solutions of equation (45) equals $-F$. If the form F is positive definite (in which case the forces of friction are called *totally dissipative*) then the energy decreases monotonically along all motions, except for equilibrium positions. Here we shall consider friction forces with Rayleigh function of the form $F_N=-N(a(q)\cdot\dot q)^2/2$, where a is a covector field and $N=\text{const}>0$. It is readily seen that the quadratic form F_N is degenerate and that the only motions $q(\cdot)$ along which the total energy does not decrease are those which satisfy the equation

$$a(q)\cdot\dot q=0. \tag{46}$$

Of course, motions of this kind, which are not equilibria, do not always exist. A friction with dissipation function F_N is also called *anisotropic friction.*

Let $q(t, N)$ be a solution of equation (45) with an initial condition which does not depend on N.

Theorem 10. *The limit*

$$\lim_{N\to\infty} q(t, N)=\hat q(t) \tag{47}$$

exists on every bounded time interval $0<t\leqq t_0$. *The limiting function satisfies the system of nonholonomic equations*

$$[L]=\lambda a, \qquad a(q)\cdot\dot{q}=0.$$

In particular, $\hat{q}(t)$ *satisfies the linear constraint equation* (46).

If the initial state $(q_0, \dot{q}_0)$ belongs to the set of solutions of equation $a(q)\cdot\dot{q}=0$, then the limit (47) exists for $t=0$ too, and the convergence is uniform on every bounded time interval. In the general case this convergence is not uniform on the interval $0<t\leqq t_0$.

Theorem 10 follows from a known theorem of A.N. Tikhonov on singularly-perturbed systems (see [67, 101]). The idea of realizing constraints which are linear in velocities by means of forces of viscous friction belongs to Carathéodory, who also obtained the first results in this direction [72].

Consider from this point of view a problem of nonholonomic mechanics mentioned in §2, namely, the motion of a homogeneous ball inside a vertical tube of large radius. Here we assume that the ball may slip. Let $v\neq 0$ be the velocity of the point of contact. We introduce the force of viscous friction $-kv$, where $k=\text{const}>0$, applied at the point of contact. For sufficiently large values of k the motion of such a ball will closely resemble the motion of the nonholonomic ball and hence, at least during the first moments of time, we may observe that the ball with friction moves up the tube.

6.4. Adjoining Masses. Consider the motion of a natural system whose Lagrangian

$$L_N=\frac{1}{2}(A(q)\dot{q}\cdot\dot{q})+\frac{N}{2}(a(q)\cdot\dot{q})^2-U(q)$$

depends on a parameter $N\geqq 0$. Here $a(q)$ is again a nonnull covector field on the configuration space.

Let $q(t, N)$ denote the motion with initial condition $q_0, \dot{q}_0$ satisfying $a(q_0)\cdot\dot{q}_0=0$.

Theorem 11 (see [115]). *The limit*

$$\lim_{N\to\infty} q(t, N)=\hat{q}(t)$$

exists on every bounded time interval $0\leqq t\leqq t_0$.

The limiting function is an extremal of Lagrange's variational problem on the stationarity of the functional

$$\int_{t_1}^{t_2} L_0\,dt, \qquad L_0=\tfrac{1}{2}A\dot{q}\cdot\dot{q}-U,$$

under the linear constraint $a\cdot\dot{q}=0$.

Therefore, the limiting function $\hat{q}(\cdot)$ is a motion of the vakonomic system with Lagrangian L_0 and constraint $a\cdot\dot{q}=0$.

Let us examine this passage to limit more closely. For $N \geqq 0$, we introduce in the usual manner the canonical momenta

$$p = A\dot{q} + N(a \cdot \dot{q})\,a.$$

Solving this equation for the velocity:

$$A\dot{q} = p - \frac{A^{-1}p \cdot a}{A^{-1}a \cdot a}\,a + \frac{1}{1 + N(A^{-1}a \cdot a)}\,\frac{A^{-1}p \cdot a}{A^{-1}a \cdot a}\,a,$$

we see that for $N \to \infty$ it becomes the familiar equality

$$A\dot{q} = p - \frac{A^{-1}p \cdot a}{A^{-1}a \cdot a},$$

used in vakonomic mechanics to introduce the momenta.

For $N > 0$ let us consider the motion of the holonomic system with initial conditions

$$q(0) = q_0, \quad p_\alpha(0) = p_0 + \alpha a, \qquad \alpha \in R,$$

where $p_0 = A\dot{q}_0$ and $a(q_0) \cdot \dot{q}_0 = 0$. For $\alpha = 0$ we obtain the initial conditions considered in Theorem 11. For each fixed value of α the initial conditions $q(0)$ and $\dot{q}_\alpha(0) = A^{-1}(q_0)\,p_\alpha(0)$ satisfy the equation $a \cdot \dot{q} = 0$ up to $1/N$.

The Hamiltonian of the holonomic system with Lagrangian L_N equals $H_N = H_0 + O(1/N)$, where H_0 is the vakonomic Hamiltonian (see §4). Consequently, for fixed α and $N \to \infty$ the following limit exists:

$$\lim_{N \to \infty} q_\alpha(t, N) = \hat{q}_a(t), \tag{48}$$

and represents one of the motions of the vakonomic system with Lagrangian L_0 and constraint $a \cdot \dot{q} = 0$.

For $N \to \infty$ the initial state $q(0)$, $\dot{q}_\alpha(0)$ does not depend on α. However, in the case of nonintegrable constraints, limit (48) depends essentially on the parameter α (see §4). Thus, when N is large, errors of order $1/N$ in the initial conditions may lead to finite deviations over times $t \sim 1$. This is one of the qualitative explanations of the nondeterministic behavior of vakonomic systems.

Example 11. Let us show how we can physically realize the motion of the vakonomic skate on an inclined plane, as studied in §4. To this end we consider the motion of an elongated weightless elliptic plate, to which points of positive mass are rigidly attached, in a boundless ideal fluid (see Fig. 7). We assume that the only force acting on the points is that of gravity. The symmetry of the problem allows motions in which the x-axis is horizontal, while the y- and z-axes lie invariably in some vertical plane.

Let ω be the projection of the angular velocity of the body on the x-axis, and let u and v denote the projections of the velocity of its barycenter on

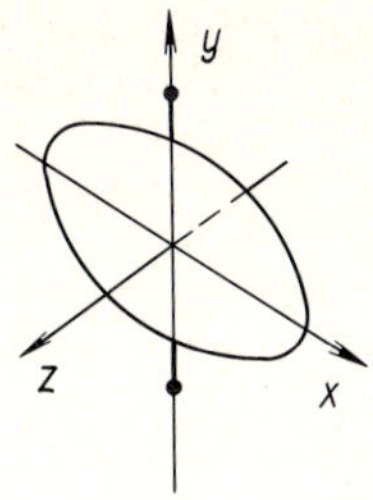

Fig. 7

the y- and z-axes, respectively. Consider first the motion of the ellipsoid

$$\frac{x^2}{a^2}+\frac{y^2}{b^2}+\frac{z^2}{c^2}=1$$

in a homogeneous fluid of density ρ. The kinetic energy of the fluid is equal to

$$\tfrac{1}{2}(A\omega^2+Bu^2+Cv^2), \tag{49}$$

where

$$A=\frac{1}{5}\frac{(b^2-c^2)(\gamma_0-\beta_0)}{2(b^2-c^2)+(b^2+c^2)(\beta_0-\gamma_0)}\frac{4}{3}\pi\rho abc,$$

$$B=\frac{\beta_0}{2-\beta_0}\frac{4}{3}\pi\rho abc, \qquad C=\frac{\gamma_0}{2-\gamma_0}\frac{4}{3}\pi\rho abc,$$

with

$$\beta_0=abc\int_0^\infty\frac{d\lambda}{(b^2+\lambda)D}, \qquad \gamma_0=abc\int_0^\infty\frac{d\lambda}{(c^2+\lambda)D},$$

and

$$D=\{(a^2+\lambda)(b^2+\lambda)(c^2+\lambda)\}^{1/2}.$$

These formulas may be found, for example, in Lamb's book [121].

We now let $c\to 0$ and set $b=\varepsilon$, $a=\varepsilon^{-\alpha}$. For small ε the last relations yield the asymptotic formulas

$$A\sim\tfrac{2}{15}\pi\rho\varepsilon^{4-\alpha}, \qquad B=0, \qquad C\sim\tfrac{4}{3}\pi\rho\varepsilon^{2-\alpha}.$$

Thus, if $2<\alpha<4$, then $A\to 0$ and $C\to\infty$ as $\varepsilon\to 0$.

The kinetic energy of the "body + fluid" system is again of form (49), with the difference that to A one must add the moment of inertia of the body relative to the x-axis, while adding the mass of the body to B and C. Following these modifications the quantities A and B tend to finite limits whereas $C\to\infty$ as $\varepsilon\to 0$. Therefore, the conditions of the theorem are met: as $\varepsilon\to 0$ the motion of the elliptic plate with initial velocity $v(0)=0$ tends to a motion of the limiting vakonomic system.

6.5. Adjoining Masses and Anisotropic Friction. Consider a multidimensional natural mechanical system with Lagrangian

$$L_N=\frac{1}{2}(A(q)\dot{q}\cdot\dot{q})+\frac{\alpha N}{2}(a(q)\cdot\dot{q})^2+V(q).$$

Suppose that, in addition to the conservative forces with force function V, the system is acted upon by forces of anisotropic friction with Rayleigh's dissipation function

$$F_N=\frac{\beta N}{2}(a(q)\cdot\dot{q})^2.$$

The equation of motion takes the form of Lagrange's equation

$$[L_N]=-F'_{N\dot{q}}.$$

We fix the values of the parameters $\alpha>0$, $\beta\geqq 0$, and let $N\to\infty$.

Let $q(t, N)$ denote the solutions of the equation of motion with an initial condition independent of N which satisfies the equation $a\cdot\dot{q}=0$.

Theorem 12 (see [116]). *The limit*

$$\lim_{N\to\infty} q(t, N)=\hat{q}(t)$$

exists on every bounded time interval.

The limiting motion $\hat{q}(\cdot)$, together with a "canonically conjugate" function $p(\cdot)$, satisfies the differential equations

$$\dot{p}=-\frac{\partial H}{\partial q}-\mu\frac{A^{-1}p\cdot a}{A^{-1}a\cdot a}a,\quad \dot{q}=\frac{\partial H}{\partial p};\qquad \mu=\frac{\beta}{\alpha},\tag{50}$$

where $H(p, q)$ is the Hamiltonian of the vakonomic system with Lagrangian L_0 and constraint $a\cdot\dot{q}=0$.

It follows from the second of equations (50) that the limiting motion $\hat{q}(\cdot)$ satisfies the constraint equation $a\cdot\dot{q}=0$.

For $\beta=0$ Theorem 12 coincides with Theorem 11. In a different limiting case, where the ratio $\mu=\beta/\alpha$ is large, equations (50) become singular. According to the general method for studying such equations, when $\alpha=0$ we obtain from (50) the "simplified" equation

$$\lambda=\frac{A^{-1}p\cdot a}{A^{-1}a\cdot a}=0.$$

Differentiating the equation $A\dot{q}=p-\lambda a$ with respect to time and taking into account the condition $\lambda=0$ we get

$$(L'_{0\dot{q}})^{\cdot}=(A\dot{q})^{\cdot}=\dot{p}-\dot{\lambda}a-\lambda\dot{a}=-H'_q-\dot{\lambda}a=L'_{0q}-\dot{\lambda}a.$$

This relation and the constraint equation forms a closed system of nonholonomic equations. Using Tikhonov's theorem alluded to above, one can show that for $\mu \to +\infty$ the solutions of equations (50) converge indeed to solutions of the nonholonomic equations.

For each fixed value of the parameter μ equations (50) may be regarded as the equations of motion of the mechanical system with Lagrangian L_0 and constraint $a \cdot \dot{q} = 0$. Thus, we have an entire family of consistent mathematical models of motion. Each of them is a synthesis of traditional nonholonomic mechanics, based on the d'Alembert-Lagrange principle, and vakonomic mechanics, at the basis of which we laid Hamilton's principle. The choice of a model in each concrete problem can be decided only on empirical bases. A discussion of these problems can be found in [116].

6.6. Small Masses. We conclude this section by a critical discussion of the validity of Dirac's generalized Hamiltonian formalism. As we already mentioned (see §5), constraints in phase space arise, for example, when the Lagrangian is degenerate in the velocities. In this connection, consider the holonomic system with the Lagrangian

$$L_\varepsilon - L_0(\dot{q}, q, Q) + \frac{\varepsilon \dot{Q}^2}{2} + \varepsilon L_1(\dot{q}, q, Q, \varepsilon), \qquad q \in R^n, \ Q \in R,$$

where ε is a small parameter. We assume that the function L_0 is nondegenerate in $\dot{q}$.

For $\varepsilon = 0$ we get a degenerate system. The equation $P = 0$, where $P = L'_{0\dot{Q}}$, serves as a primary constraint (in Dirac's sense). The compatibility condition gives us the secondary constraint

$$\{P, H_0\} = -H'_{0Q} = 0, \tag{51}$$

in which $H_0(p, q, Q) = p \cdot \dot{q} - L_0|_{\dot{q} \to p}$.

Let $Q = f(p, q)$ be a solution of equation (51). Then one can represent the secondary constraint as the equation $\Psi = Q - f(p, q) = 0$; here $\{P, \Psi\} = -1 \neq 0$.

Dirac's Hamiltonian $\mathscr{H}$ is the sum $H_0 + \lambda P + \mu(Q - f)$; the coefficients λ and μ are uniquely determined by the compatibility conditions

$$\{P, \mathscr{H}\} = \{P, H_0\} - \mu = 0, \qquad \{Q - f, \mathscr{H}\} = -\{f, H_0\} - \lambda = 0.$$

These yield $\mu = -H'_{0Q}$ and $\lambda = \{H_0, f\}$. Hamilton's equations with constraints obviously take the form

$$\dot{p} = -\hat{H}'_{0q}, \qquad \dot{q} = \hat{H}'_{0p}, \qquad P = 0, \qquad Q = f, \tag{52}$$

where $\hat{H}_0(p, q) = H_0(p, q, Q)|_{Q=f}$. The Hamiltonian of the complete system (with $\varepsilon \neq 0$) equals $H_0(p, q, Q) + P^2/2\varepsilon + \varepsilon H_1(p, q, Q, \varepsilon)$. The corresponding canonical equations are

$$\begin{aligned} \dot{p} &= -H'_{0q} - \varepsilon H'_{1q}, \qquad & \dot{q} &= H'_{0p} + \varepsilon H'_{1p}, \\ \dot{P} &= -H'_{0Q} - \varepsilon H'_{1Q}, \qquad & \dot{Q} &= P/\varepsilon. \end{aligned} \tag{53}$$

Proposition 16. *If* $H''_{0QQ}|_{Q=f} \neq 0$, *then equations* (53) *admit a unique solution given by formal power series in* ε:

$$p = p_0(t) + \varepsilon p_1(t) + \ldots, \qquad q = q_0(t) + \varepsilon q_1(t) + \ldots,$$
$$P = \varepsilon P_1(t) + \ldots, \qquad Q = f(p_0(t), q_0(t)) + \varepsilon Q_1(t) + \ldots,$$

where $p_0(t)$, $q_0(t)$ *is some prescribed solution of equations* (52).

Unfortunately, these series do not always converge. In the case where function H_0 does not depend on Q equations (53) are no longer singular: instead of the momentum P we must take the new variable P/ε. Then the solutions of these equations can be expressed as convergent power series, and the initial conditions $Q(0)$ and $\dot{Q}(0)$ may be arbitrary. This is precisely the case in the "restricted" n-body problem, when one of the masses tends to zero.

Thus, Dirac's mechanics may be viewed as a mechanics of small masses. On the contrary, vakonomic mechanics is convenient for describing the dynamics of large masses.

The assertions of this subsection may serve as motivation for our theoretical constructions concerning the dynamics of constrained mechanical systems.

Chapter 2
The n-Body Problem

§1. The Two-Body Problem

1.1. Orbits. Suppose that two points (r_1, m_1) and (r_2, m_2) mutually interact with potential energy $U(|r_1 - r_2|)$, so that the equations of motion have the form

$$m_1 \ddot{r}_1 = -\frac{\partial U}{\partial r_1}, \qquad m_2 \ddot{r}_2 = -\frac{\partial U}{\partial r_2}.$$

Proposition 1. *The relative position vector $r = r_1 - r_2$ in the two-body problem changes according to the same law as the position vector of a point of mass $m = m_1 m_2/(m_1 + m_2)$ moving in a central force field with potential $U(|r|)$.*

If

$$\xi = \frac{m_1 r_1 + m_2 r_2}{m_1 + m_2}$$

is the barycenter of the two points m_1 and m_2, then obviously

$$r_1 = \xi + \frac{m_2}{m_1 + m_2} r, \qquad r_2 = \xi - \frac{m_1}{m_1 + m_2} r.$$

It follows from this formulas that in a barycentric coordinate system the trajectories of the two point masses are similar planar curves (with similarity ratio m_1/m_2). Therefore, the problem reduces to the investigation of the single equation

$$m\ddot{r} = -\frac{\partial U}{\partial r}, \qquad r \in R^3.$$

Let x, y be Cartesian coordinates in the plane of the orbit. Then $K_z = m(x\dot{y} - y\dot{x}) = \text{const}$. In polar coordinates $x = r\cos\varphi$, $y = r\sin\varphi$, we have $K_z = mr^2\dot{\varphi}$. Consequently, $r^2\dot{\varphi} = c = \text{const}$. If $c = 0$, then $\varphi = \text{const}$ (the point moves along a straight line). We shall assume that $c \neq 0$. In this case φ is a monotonic function of t, and hence the inverse function $t = t(\varphi)$ exists. As the point m moves, its position vector sweeps out a curvilinear sector of area

$$S(t) = \tfrac{1}{2} \int_{\varphi(0)}^{\varphi(t)} r^2\, d\varphi = \tfrac{1}{2} \int_0^t r^2 \dot{\varphi}\, dt = \frac{ct}{2}.$$

Therefore, $S=c/2=\text{const}$ (i.e., the "sectorial velocity" is constant). This fact is known as the *area integral* or as *Kepler's second law*; c is called the area constant.

Proposition 2 (Newton). *For a fixed value of the area integral* c

$$m\ddot{r}=-\frac{\partial U_c}{\partial r} \quad \textit{with} \quad U_c=U+\frac{mc^2}{2r^2} \quad (r>0). \tag{1}$$

This equation describes the motion of a point of mass m on the half line $R^+=\{r>0\}$ under the action of the conservative force with potential U_c. It is integrable by quadratures using the energy integral

$$\frac{m\dot{r}^2}{2}+U_c=h.$$

The function U_c is called the *reduced* (*effective*, or *amended*) *potential.*

The energy and area integrals can be used to find the equation of orbits without solving equation (1). In fact, since $\dot{r}=\sqrt{2(h-U_c)/m}$ and $r^2\dot{\varphi}=c$, it follows that

$$\frac{dr}{d\varphi}=\frac{dr}{dt}\frac{dt}{d\varphi}=\frac{r^2}{c}\sqrt{\frac{2(h-U_c)}{m}}.$$

Integrating this equation we get

$$\varphi=\int\frac{c\,dr}{r^2\sqrt{\frac{2(h-U_c)}{m}}}.$$

In calculating orbits it is sometimes useful to keep in mind the following proposition:

Proposition 3 (Clairaut). *Let* $\rho=1/r$ *and write the equation of the orbit as* $\rho=\rho(\varphi)$. *Then*

$$m\frac{d^2\rho}{d\varphi^2}=-\frac{1}{c^2}\frac{d}{d\rho}U_c\left(\frac{1}{\rho}\right).$$

For fixed values h and c the orbit lies in the domain

$$B_{c,h}=\left\{(r,\varphi)\in R^2\colon U+\frac{mc^2}{2r^2}\leqq h\right\},$$

which is a union of several annuli. Now suppose that h is not a critical value of the reduced potential U_c and that $B_{c,h}$ coincides with the annulus $0<r_1\leqq r\leqq r_2<\infty$.

We show that in this cases $r(t)$ is a periodic function of time, and that

$$\min r(t)=r_1, \qquad \max r(t)=r_2.$$

To see this, let

$$u = \frac{\pi}{\tau} \int_{r_1}^{r} \frac{dx}{\sqrt{\frac{2}{m}(h - U_c(x))}}, \qquad \tau = \int_{r_1}^{r_2} \frac{dx}{\sqrt{\frac{2}{m}(h - U_c(x))}}.$$

Then obviously $r(u)$ is a 2π-periodic function of u, and $\dot{u} = \pi/\tau = \text{const}$. The period of the function $r(\cdot)$ is clearly 2τ.

Angle φ varies monotonically (provided, of course, that $c \neq 0$). The points on the orbit lying closest (farthest) to the center are called *pericenters* (respectively, *apocenters*). The orbit is symmetric with respect to the lines passing through the point $r=0$ and its pericenters (apocenters). The angle Φ between the lines joining the center with two adjacent apocenters (or pericenters) is called the *apsidal angle*. The orbit is invariant under rotation through angle Φ. If the apsidal angle

$$\Phi = 2 \int_{r_1}^{r_2} \frac{c\,dr}{r^2 \sqrt{\frac{2}{m}(h - U_c)}}$$

is commensurable with π, the orbit is closed; otherwise it densely fills the annulus $B_{c,h}$. If $r_2 = \infty$ the orbit is unbounded.

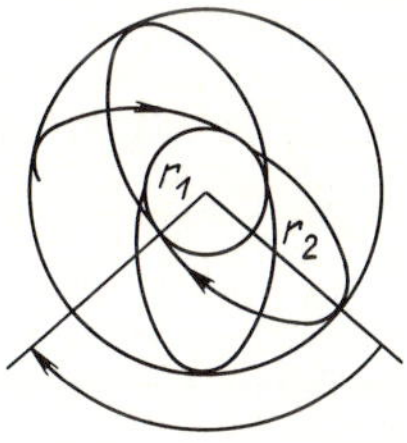

Fig. 8. Orbit in a central field

The motion of the point on a circle $r = r_0$ is called a *relative equilibrium.* It is clear that such a motions are uniform and that the values r_0 are the critical points of the reduced potential U_c. If at the point $r = r_0$ function U_c has a local minimum, then the corresponding circular motion is orbitally stable.

Theorem 1 (J.L.F. Bertrand). *Suppose that for some $c \neq 0$ there exists a stable relative equilibrium and that the potential U_c is analytic for $r > 0$. If all orbits sufficiently close to the given circular orbit are closed, then either $U = \gamma r^2$ or $U = -\gamma/r$ (where $\gamma > 0$).*

In the first case the system is a harmonic oscillator and its orbits are ellipses centered at the point $r=0$. The second case is that of gravitational attraction. The problem of the motion of a point in a conservative field with potential $U = -\gamma/r$ is usually called *Kepler's problem.*

The reduced potential in Kepler's problem is

$$U_c = \frac{c^2}{2r^2} - \frac{\gamma}{r}.$$

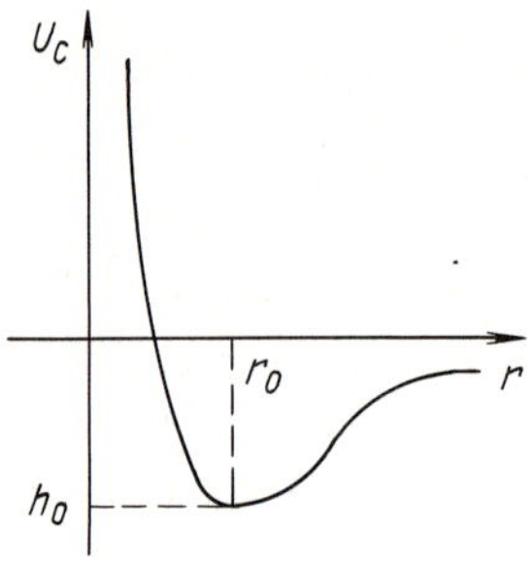

Fig. 9. Reduced potential of Kepler's problem

According to *Clairaut's equation* (Proposition 3), $\frac{d^2\rho}{d\varphi^2} = -\rho + \frac{\gamma}{c^2}$. This linear nonhomogeneous equation is readily solved:

$$\rho = A\cos(\varphi - \varphi_0) + \frac{\gamma}{c^2} = \frac{1}{p}(1 + e\cos(\varphi - \varphi_0)), \tag{2}$$

where e, φ_0 are constants, and $p = c^2/\gamma > 0$. Hence

$$r = \frac{p}{1 + e\cos(\varphi - \varphi_0)},$$

so that the orbits of Kepler's problem are conics with one focus at the center of attraction (this is *Kepler's first law*).

For fixed $c \neq 0$ there exists a unique relative equilibrium $r_0 = c^2/\gamma$. Its energy $h_0 = -\gamma^2/2c^2$ is minimal. Using the simple formula

$$v^2 = \dot{r}^2 + r^2\dot{\varphi}^2 = c^2(\rho^2 + \rho'^2), \qquad \rho' = \frac{d\rho}{d\varphi},$$

we can rewrite the energy integral as

$$\frac{c^2}{2}(\rho'^2 + \rho^2) - \gamma\rho = h.$$

On substituting in this formula the equation of the orbit (2) we obtain the expression of the eccentricity: $e = \sqrt{1 + 2c^2 h/\gamma^2}$. Since $h \geqq h_0 = -\gamma^2/2c^2$, the eccentricity assumes only real values.

If $h = h_0$, then $e = 0$ and the orbit is circular. If $h_0 < h < 0$, then $0 < e < 1$ and the orbit is an ellipse. If $h = 0$, then $e = 1$ and the orbit is a parabola. Finally, if $h > 0$, then $e > 1$ and the point moves along one of the branches of a hyperbola.

Figure 10 shows the *bifurcation set* Σ in the plane of the parameters c, h. It consists of the curve $h=-\gamma^2/2c^2$ and the two coordinate axes $c=0$ and $h=0$. At the points of Σ the topological type of the region of possible motions $B_{c,h}$ changes (in Fig. 10 the regions $B_{c,h}$ are shaded).

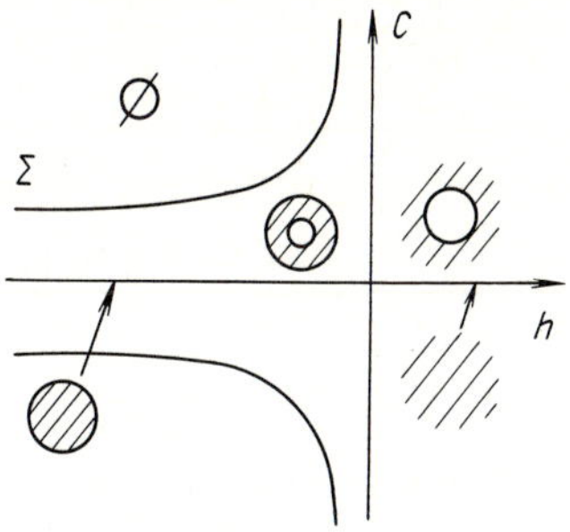

Fig. 10

In the case of the harmonic oscillator the period of revolution (orbital period) does not depend on the initial configuration. In Kepler's problem the picture is different. Thus, for elliptic motions *"Kepler's third law"* holds: $a^3/T^2=\gamma/4\pi^2=\text{const}$, where a is the major semiaxis of the ellipse and T is the period of revolution. Since $a=\dfrac{p^2}{1-e^2}=\dfrac{\gamma}{2|h|}$, T depends only on the constant.

1.2. Anomalies. To solve Kepler's problem completely we must still determine the law governing the motion on the already known orbits. To this end, we take the coordinate axes x and y along the principal axes of the conic representing the given orbit. The equation of this conic can be written in the following parametric form:

$$\begin{array}{llll} x=a(\cos u-e), & y=a\sqrt{1-e^2}\sin u & (0\leqq e<1), & \text{if } h<0, \\ x=a(\cosh u-e), & y=a\sqrt{e^2-1}\sinh u & (e>1), & \text{if } h>0, \\ x=\frac{1}{2}(p-u^2), & y=\sqrt{p}\,u, & & \text{if } h=0. \end{array} \tag{3}$$

In astronomy the auxiliary variable u is known as the *eccentric anomaly*, and the angle φ made by the line joining the center to the pericenter (the x-axis) and the position vector of the point – as the *true anomaly*. These anomalies are related by

$$\tan\frac{\varphi}{2}=\begin{cases} \sqrt{\dfrac{1+e}{1-e}}\tan\dfrac{u}{2}, & \text{if } h<0, \\ \sqrt{\dfrac{e+1}{e-1}}\tanh u, & \text{if } h>0, \\ \dfrac{u}{p}, & \text{if } h=0. \end{cases}$$

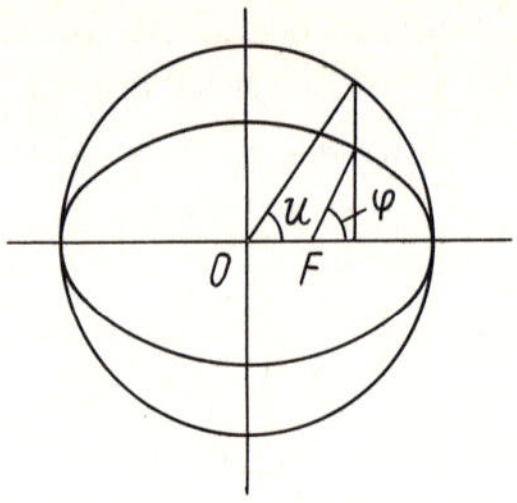

Fig. 11

Substituting expressions (3) in the area integral $x\dot{y}-y\dot{x}=c$ and integrating we obtain the relation between time and eccentric anomaly:

$$u-e\sin u=n(t-t_0),\quad n=\frac{\sqrt{\gamma}}{p^{3/2}},\qquad \text{if } h<0,$$

$$u-e\sinh u=n(t-t_0),\quad n=-\frac{\sqrt{\gamma}}{p^{3/2}},\qquad \text{if } h>0,$$

$$u+\frac{u^3}{3p}=n(t-t_0),\quad n=\frac{2\sqrt{\gamma}}{p},\qquad \text{if } h=0.$$

Here t_0 represents the time of passage at apocenter. These equations (or at least the first of them) are called *Kepler's equations.* The linear function $\zeta=n(t-t_0)$ is usually referred to as the *mean anomaly.*

Thus, in the elliptic case of Kepler's problem we must solve the transcendental equation of Kepler

$$u-e\sin u=\zeta.$$

It is clear that for $0\leqq e<1$ this equation has an analytic solution $u(e,\zeta)$ such that the difference $u(e,\zeta)-\zeta$ is 2π-periodic in the mean anomaly ζ. To represent the function $u(e,\zeta)$ in a form suitable for computations we may proceed in two ways:

1) expand the difference $u-\zeta$ for fixed values of e in a Fourier series in variable ζ with coefficients depending on e, or

2) attempt to write $u(e,\zeta)$ as a power series in the eccentricity e with coefficients depending on ζ.

The first approach gives

$$u=\zeta+2\sum_{m=1}^{\infty}\frac{J_m(me)}{m}\sin m\zeta, \tag{4}$$

where

$$J_m(z)=\frac{1}{2\pi}\int_0^{2\pi}\cos(mx-z\sin x)\,dx=\sum_{k=0}^{\infty}\frac{(-1)^k(z/2)^{m+2k}}{k!\,(m+k)!},\qquad m=0,1,\dots,$$

are the *Bessel function* of order m. These functions, "though usually associated with the name of Bessel, have been used extensively, precisely in this connection (which is that of Bessel), and more than half a century prior to Bessel, by Lagrange and others" (A. Winter, [42]).

The proof of expansion (4) is a straightforward calculation:

$$\frac{du}{d\zeta}=\frac{1}{1-e\cos u}=\frac{1}{2\pi}\int_0^{2\pi}\frac{d\zeta}{1-e\cos u}+\sum_{m=1}^{\infty}\frac{\cos m\zeta}{\pi}\int_0^{2\pi}\frac{\cos m\zeta\, d\zeta}{1-e\cos u}=$$

$$=\frac{1}{2\pi}\int_0^{2\pi} du+\sum_{m=1}^{\infty}\frac{\cos m\zeta}{\pi}\int_0^{2\pi}\cos[m(u-e\sin u)]\,du-$$

$$=1+2\sum_{m=1}^{\infty}J_m(me)\cos m\zeta,$$

and now it remains to integrate this formula with respect to ζ.

In the second approach we get the expansion

$$u(e,\zeta)=\sum_{m=0}^{\infty}c_m(\zeta)\frac{e^m}{m!}, \tag{5}$$

where

$$c_m(\zeta)=\left.\frac{\partial^m u(e,\zeta)}{\partial e^m}\right|_{e=0}.$$

Using Lagrange's well-known formula for the local inversion of holomorphic functions (which he discovered exactly in his attempt to solve Kepler's problem), we obtain the expressions for the coefficients of series (5):

$$c_0(\zeta)=\zeta;\quad c_m(\zeta)=\frac{d^{m-1}}{d\zeta^{m-1}}\sin^m\zeta,\quad m\geqq 1.$$

The functions $c_m(\zeta)$ are trigonometric polynomials in the mean anomaly ζ. Rearranging the terms of series (5) one can obtain expansion (4). As a matter of fact, this is the way in which Lagrange arrived at formula (4).

By the implicit function theorem (and in view of the periodicity of the function $u(e,\zeta)-\zeta$) series (5) converge on the entire real axis $\zeta\in R$ for e small enough. A careful analysis of expansion (4) shows that Lagrange's series converges for $e\leqq 0.6627434\ldots$[15].

1.3. Collisions and Regularization. Thus far we have assumed that the area constant $c\neq 0$. Now let $c=0$. In this case the point moves along a straight line, and we may assume that the latter is the x-axis. If at some instant the velocity $\dot{x}$ is directed toward the center of attraction, then $x(t)\to 0$ and

[15] "In fact, a principal impetus for Cauchy's discoveries in complex function theory was his desire to find a satisfactory treatment for Lagrange's series." (A. Wintner [42]).

$\dot{x}(t) \to \infty$ as t approaches a certain value t_0. Therefore, at time $t = t_0$ a *collision* of the two bodies occurs. It is obvious that for $c=0$ the function $x(t)$, $t \in R$, has necessarily a singularity of the indicated type.

We show that the eccentric anomaly u is a *regularizing variable*, which resolves the singularity of the analytic function $x(t)$. Since $c=0$, $e=1$ in the elliptic and hyperbolic cases, and $p=0$ in the parabolic case. Consequently, formulas (3) take respectively the form

$$x = a(\cos u - 1), \qquad x = a(\cosh u - 1), \qquad x = -\frac{u^2}{2}. \tag{6}$$

It follows from these formulas that if $h<0$ a collision occurs for $u = 2\pi k$, $k \in Z$, whereas if $h \geqq 0$ it occurs only for $u=0$. Actually, in the elliptic case as well it suffices to consider the value $u=0$.

To simplify the discussion we set $t_0 = 0$. From Kepler's equation (for $e=1$) it readily follows that in the neighborhood of the point $u=0$ we can write

$$t = u^3 f(u),$$

where f is a function analytic in the neighborhood of zero, and $f(0) \neq 0$. From (6) we derive an analogous representation

$$x = u^2 g(u)$$

with an analytic function g, $g(0) \neq 0$. Upon eliminating the eccentric anomaly u from the last two formulas we get the Puiseux expansion

$$x(t) = (\sqrt[3]{t})^2 \sum_{n=0}^{\infty} c_n (\sqrt[3]{t})^n.$$

Here the coefficients c_n with n odd obviously vanish, and $c_0 \neq 0$. Therefore, $x(t)$ is an even function of time, i.e., the moving point is reflected back from the center of attraction after collision. If x and t are regarded as complex variables, then $t=0$ is an algebraic branching point of the analytic function $x(t)$. At the collision point $t=0$ the three sheets of the Riemann surface of $x(t)$ merge; also, the real values of $x(t)$ for $t>0$ and $t<0$ lie on the same sheet. Hence, the function $x(t)$ admits a unique real analytic continuation.[16]

In conclusion we mention a result of K. Bohlin concerning the *regularization* of the two-body problem in the general elliptic case ($h<0$).

Introducing the complex variable $z = x + iy$ we rewrite the equation of Kepler's problem and the energy integral as

$$\ddot{z} = -\frac{\gamma z}{|z|^3}$$

and

$$\frac{|\dot{z}|^2}{2} = \frac{\gamma}{|z|} + h, \tag{7}$$

[16] The regularization of collisions in the two-body problem goes back to Euler.

respectively. Next, we change the independent variable z to w and the time variable t to τ according to the rules

$$z=w^2, \qquad t'=\frac{dt}{d\tau}=4\,|w^2|=4\,|z|. \tag{8}$$

In the new variables w, τ, equation (7) becomes

$$\frac{|w'|^2}{2}=4\gamma+4h\,|w^2|. \tag{9}$$

This further becomes $w''+8\,|h|\,w=0$, i.e., the equation describing the vibrations of a harmonic oscillator. Thus, the nonlinear mapping (8) takes the orbits of Kepler's problem with energy constant $h<0$ into orbits of the harmonic oscillator lying on the energy level (9). This conclusion is an interesting supplement to Bertrand's theorem.

The regularizing variable τ depends linearly on the eccentric anomaly u. In fact, since

$$|z|=r=a(1-e\cos u) \quad \text{and} \quad nt=u-e\sin u,$$

it follows that

$$\frac{du}{d\tau}=\frac{n}{1-e\cos u}=\frac{na}{r},$$

whence $u=4na\tau$.

1.4. Geometry of the Kepler Problem. J. Moser observed that by suitably changing the time variable one can transform the phase flow of Kepler's problem into the geodesic flow on a surface of constant curvature. In our presentation of this result we follow Yu. S. Osipov (Usp. Mat. Nauk **27**, No. 2, 161 (1972)).

Lemma 1. *Let $x(t)$ be a solution of the Hamiltonian system with Hamiltonian $H(x)$ which lies on the level surface $H=0$. Change the time variable t to τ along this trajectory according to the formula $d\tau/dt=1/G(x(t))\neq 0$. Then $x(\tau)=x(t(\tau))$ is a solution of the Hamiltonian (with respect to the same symplectic structure) system with Hamiltonian $\tilde{H}=HG$. If $G=2(H+\alpha)$ with* $\alpha=\text{const}$, *one can also put $\tilde{H}=(H+\alpha)^2$.*

Let us write the Hamiltonian of Kepler's problem in the notations of §3.1: $H=|p|^2/2-\gamma/|z|$, with $p=\dot{z}$. Now perform the change of time $\dot{\tau}=|z|^{-1}$ on the submanifold $H=h$ (cf. formula (8)). According to Lemma 1, this leads to the new Hamiltonian $|z|(H-h)=|z|(|p|^2-2h)/2-\gamma$. Then again change the time τ to $'\tau$ by $d'\tau/d\tau=2\,|z|((H-h)+\gamma)$ on the same level surface $H=h$. In the end we get the Hamiltonian system with Hamiltonian

$$\tilde{H}=|z|^2(p^2-2h)^2/4.$$

Finally, apply Legendre's transformation, regarding p as a coordinate and z as the canonically conjugate momentum. We thus obtain a natural Lagrangian system with the Lagrangian

$$L=\frac{|p'|^2}{(2h-|p|^2)^2}\,. \tag{10}$$

This (length) function defines a Riemann metric of constant Gaussian curvature (positive for $h<0$ and negative for $h>0$). In the case $h<0$ the geodesics of the metric (10) (which is defined for all $p\in R^2$) are images of great circles on the sphere under stereographic projection, whereas in the case $h>0$ (in which the metric is defined in the disc $|p|^2<2h$) the geodesics are the straight lines of Lobachevsky's geometry (in the Poincaré model).

Remark (A.B. Givental'). Suppose $\{(\dot{x}, \dot{y})\}$ is the configuration plane of the Kepler problem with the Lagrangian $L=(\dot{x}^2+\dot{y}^2)/2+1/\sqrt{x^2+y^2}$. In the space $\{(x, y, z)\}$ consider the right circular cone $z^2=x^2+y^2$ and the family of inscribed paraboloids of revolution $z=(x^2+y^2)/4\alpha+\alpha$, where α is a parameter. By "projection" we mean the projection of the space $\{(x, y, z)\}$ onto the plane $\{(x, y)\}$ parallel to the z-axis. Then one can show that

1) the trajectories of Kepler's problem are projections of plane sections of the cone (in particular, the vertex of the cone is a focus of these projections);

2) trajectories with the same value of the total energy are projections of sections of the cone by planes tangent to one and the same paraboloid;

3) trajectories with the same value of the angular momentum are projections of sections of the cone by planes passing through one and the same point of the z-axis.

§2. Collisions and Regularization

2.1. Necessary Conditions for Stability. We now turn to the general *n-body problem* in which n point masses $(m_1, r_1), \ldots, (m_n, r_n)$ attract one another according to the law of universal attraction. The kinetic energy of this system is

$$T=\tfrac{1}{2}\sum m_i\dot{r}_i^2,$$

and the force function

$$V=\sum_{j<k}\frac{m_j m_k}{|r_{jk}|}, \qquad r_{jk}=r_j-r_k,$$

is everywhere positive. Consider an inertial reference frame with origin at the center of mass and let r_i be the position vectors of the points in this new frame. The equations of the n-body problem are Lagrange's equations with the Lagrangian $L=T+V$.

We say that the motion $r_k(t)$ $(1\leqq k\leqq n)$ is *stable* if the following two conditions are fulfilled:

a) $r_{ij}(t)\neq 0$ for all values of t and all $i\neq j$ (absence of collisions).

2) $|r_{ij}(t)|\leqq c$ $(c=\text{const})$.

Theorem 2 (C. Jacobi). *If a motion is stable, then its total energy* $h=T-V$ *is negative.*

◁ We apply Lagrange's formula

$$\ddot{I}=2V+4h, \tag{11}$$

where $I=\sum m_i r_i^2$ is the polar moment of inertia. If $h\geqq 0$, then the function $I(t)$, $t\in R$, is upward convex and hence cannot be simultaneously bounded from below and from above. To complete the proof it remains to use Lagrange's identity:

$$I\sum m_i=\sum_{j<k} m_j m_k r_{jk}^2+\left(\sum m_i r_i\right)^2. \ \triangleright$$

Under the additional assumption that mutual distances are bounded from below ($|r_{ij}(t)|\geqq c>0$), it follows from the energy integral and Lagrange's formula (11) that along a stable motion the mean values

$$\lim_{S\to\infty}\frac{1}{S}\int_0^S V(t)\,dt \quad \text{and} \quad \lim_{S\to\infty}\frac{1}{S}\int_0^S 2T(t)\,dt$$

exist and are equal to $-2h>0$.

The necessary condition for stability $h<0$ is not sufficient for $n>2$.

2.2. Simultaneous Collisions. If the position vectors $r_i(t)$ of all points tend to the same limit r_0 as $t\to t_0$, then we say that a simultaneous (or general) collision occurs at time t_0. The point r_0 must obviously coincide with the center of mass, i.e., $r_0=0$. A simultaneous collision takes place if and only if the polar moment of inertia $I(t)$ tends to zero as $t\to t_0$.

Theorem 3. *If* $I(t)\to 0$ *as* $t\to t_0$, *then the constant angular momentum vector*

$$K=\sum m_i(r_i\times\dot{r}_i)$$

is equal to zero.

For $n=3$ this fact was already known to Weierstrass.

◁ Since $V(t)\to+\infty$ as $t\to t_0$, it follows from the equation $\ddot{I}=2V+4h$ that $\ddot{I}(t)>0$ for t close to T_0. Hence, prior to collision $I(t)$ decreases monotonically.

We now use the inequality $K^2\leqq 2IT$ (given in §1.1) which, in view of Lagrange's formula, is equivalent to

$$\ddot{I}\geqq\frac{K^2}{I}+2h.$$

Multiplying this inequality by the positive function $-2\dot{I}$ and integrating the result over the interval (t_1, t), where $t<t_0$, we get

$$\dot{I}^2(t_1)-\dot{I}^2(t)\geqq 2K^2\ln\frac{I(t_1)}{I(t)}+4h(I(t_1)-I(t)).$$

This clearly gives the inequality

$$2K^2 \ln \frac{I(t_1)}{I(t)} \leqq I^2(t_1) + 4|h|\, I(t_1),$$

from which we conclude that $I(t)$ has a positive lower bound in the interval (t_1, t_0) provided $K^2 \neq 0$. ▷

2.3. Binary Collisions. We say that a *binary collision* occurs at time t_0 if the distance between two points, say m_1 and m_n, tends to zero as $t \to t_0$, whereas the mutual distances between the remaining points are bounded from below by a positive constant for t close to t_0. For such values of t the influence of the points $m_2, \ldots, m_{n-1}$ upon the motion of m_1 and m_n is clearly negligible in comparison with the interaction between m_1 and m_n. It is therefore natural to expect that close to the collision time t_0 the vector $r_{1n}(t)$ will behave roughly as in the case of collisions in the two-body problem (§1). In that problem a local uniformizing variable was the true anomaly $u(t)$, proportional to the integral of the inverse of the distance between the points. Hence, in the case of a binary collision it is natural to attempt to regularize the solution by means of the variable

$$u(t) = \int_{t_0}^{t} \frac{ds}{|r_{1n}(s)|}. \tag{12}$$

One can show that this is indeed a good choice: the functions $r_k(u)$ are regular in a neighborhood of the point $u=0$ (corresponding to the binary collision); furthermore, $t(u) - t_0 = u^3\, p(u)$, where $p(\cdot)$ is holomorphic in the neighborhood of $u=0$, and $p(0) \neq 0$. Therefore, in the case of a binary collision, as in the two-body problem, the coordinates of the points r_k are holomorphic functions of the variable $\sqrt[3]{t-t_0}$ and hence each admits a unique real analytic continuation for $t > t_0$. One can further show that functions $r_2(t), \ldots, r_{n-1}(t)$ are actually holomorphic in a neighborhood of the point t_0.

To ensure that the uniformizing variable $u(t)$ will be suitable for any pair of points and any time of binary collision, we must replace (12) by the formula

$$u(t) = \int_0^t V(s)\, ds = \int_0^t \sum_{j<k} \frac{m_j m_k}{|r_{jk}(s)|}\, ds.$$

If in the three-body problem the polar angular momentum is different from zero, then the only possible singularities are binary collisions. K.F. Sundman showed that the functions $r_k(u)$, $1 \leqq k \leqq 3$, are holomorphic in a strip $|\operatorname{Im} u| < \delta$ of the complex plane $u \in C$ which contains the real axis. Now let us map this strip conformally onto the unit disc $|\omega| < 1$ by the transformation

$$\omega = \left(e^{\frac{\pi u}{2\delta}} - 1\right)\left(e^{\frac{\pi u}{2\delta}} + 1\right)^{-1},$$

which takes the real axis $-\infty < u < +\infty$ into the segment $-1 < \omega < 1$. Then the coordinates of the points r_k become functions holomorphic in the disc $|\omega| < 1$, and they can be expressed as convergent power series in the new variable ω. These series describe the motion of the three bodies for all values of the time t, from $-\infty$ through $+\infty$.[17]

This result was obtained by Sundman (1913). It followed earlier investigations of Poincaré and Weierstrass, in which convergent power series expansions in the auxiliary variable ω were obtained for the solutions of the n-body problem in the absence of collisions. As for collisions, in the three-body problem they are extremely rare. In fact, using the theorem on simultaneous collisions and also regularization of binary collisions one can show that in the 12-dimensional configuration space of the three-body problem (with the center of mass fixed) the collision trajectories lie on certain special analytic 10-dimensional surfaces. Their measure is, of course, zero. However, it is not known whether these surfaces may densely fill whole domains in the state space.

In conclusion we give, for illustrative purposes, the results of numerical calculations for the "Pythagorean" variant of the three-body problem, in which three bodies of masses 3, 4, and 5 are at rest at the initial time at the point of the (x, y)-plane with coordinates $(1, 3)$, $(-2, -1)$, and $(1, -1)$, respectively.

Calculations of the Pythagorean three-body problem were initiated by C. Burrau as early as 1913 and have been continued in our time by V. Szebehely with the aid of fast electronic computers. In Figures 12–14 one can

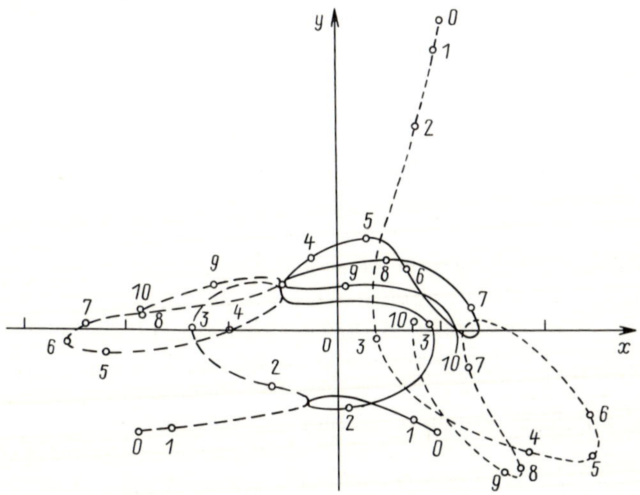

Fig. 12. Motion of the mutually attracting masses in the Pythagorean three-body problem over the time interval $0 \leqq t \leqq 10$

[17] These power series in ω are however of no use in practical calculations because of their extremely low rate of convergence.

respectively see the points moving very close (near encounters of the points), binary collisions, and the dispersal of the triple system. Figure 15 shows a "final" motion: the point of mass $m=5$ moves along a straight line away from the "double star" constituted by the points $m=3$ and $m=4$, which collide periodically. It is worth noting that although in this case the angular momentum is zero, no triple collisions occur.

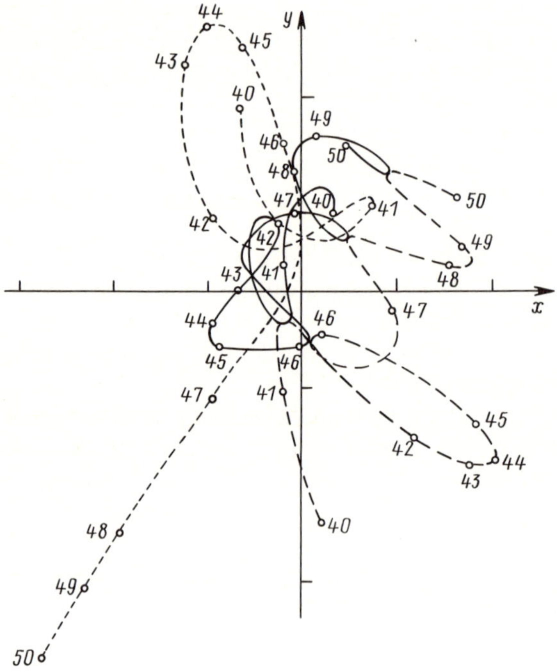

Fig. 13. Aspect of orbits in the Pythagorean three-body problem over the time interval $40 \leqq t \leqq 50$

2.4. Singularities of Solutions in the *n*-Body Problem. In the case of multiple collisions, that is, when $k \geqq 3$ point masses interacting gravitationally collide simultaneously, the singularities of the coordinates r_k of the masses as functions of time have a rather complicated analytic structure. Generally, these singularities are not algebraic; moreover, the functions r_j $(1 \leqq j \leqq n)$ do not admit a real analytic continuation beyond the collision time.

This can be seen even in examples of simultaneous collisions in the three-body problem. It turns out that for arbitrary values of the masses m_1, m_2, m_3 there exist solutions of the form

$$r_j(t) = t^{2/3} \sum_{m=0}^{\infty} a_{jm} t^{\alpha m}. \tag{13}$$

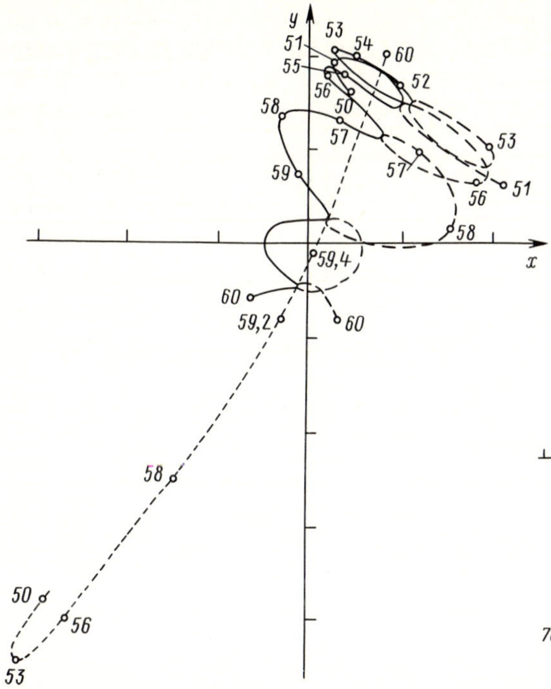

Fig. 14

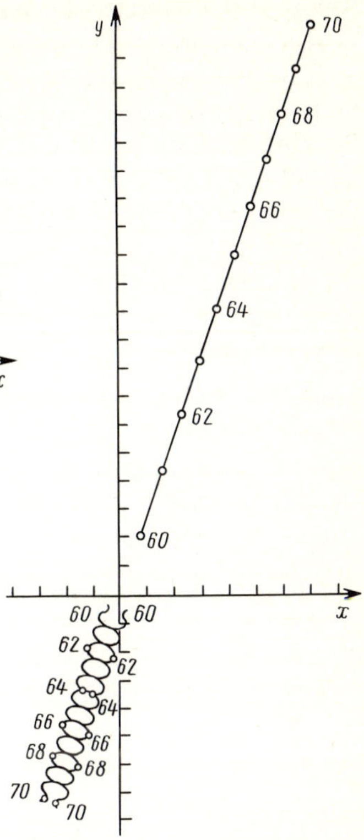

Fig. 15

Fig. 14. Evolution of orbits in the Pythagorean three-body problem over the time interval $50 \leqq t \leqq 60$

Fig. 15. Formation of a double star in the Pythagorean three-body problem ($60 \leqq t \leqq 70$)

Here the positive number α is a nonconstant algebraic function of the masses m_1, m_2, m_3, and the coefficients a_{jm} are not simultaneously zero. At time $t=0$ a triple collision occurs. In the typical case, where α is irrational, series (13) has an isolated logarithmic singularity at $t=0$. In particular, this solution, which is real for $t>0$, has infinitely many distinct analytic branches for $t<0$, which, however, are all complex.

Solution (13) was discovered by H. Block (1909) and J.F. Chazy (1918) by means of the following device: for arbitrary values of the masses, the equations of the three-body problem admit a "homographic" solution, for which the triangle formed by the bodies always remains similar to itself. This solution is analytically given by the formula

$$r_j(t)=a_{j0}\, t^{3/2} \qquad (1 \leqq j \leqq 3). \tag{14}$$

Among the characteristic numbers of the variational equation (i.e., of the system of equations linearized about this solution), there is a negative number $-\alpha$. According to well-known results of Lyapunov and Poincaré, the equations of motion have a solution (13) which is asymptotic to solution (14). We remark that the method of Block and Chazy has been in fact used earlier (in 1894) by Lyapunov to prove that the solutions of the equations describing the rotation of a heavy rigid body around a fixed point are not single-valued as functions of complex time.

Consider a solution $r_k(t)$ of the three-body problem. Suppose that at the initial time t_0 all $r_{jk} \neq 0$ $(j \neq k)$. Let us track this solution for $t > t_0$. Three cases are possible:

(a) there are no collisions for any $t > t_0$; then the motion proceeds without singularities till $t = +\infty$;

(b) at some instant $t_1 > t_0$ a "continuable" collision, i.e., one admitting analytic continuation, occurs;

(c) at some instant a "non-continuable" singularity, i.e., one for which analytic continuation is not possible, occurs.

Consider case (b). Then for $t > t_1$ one of the three variants (a)–(c) is again possible. Continuing this process, we may either reach case (a) or (c) after a finite number of steps, or observe infinitely many continuable collisions at times $t_1, t_2, \ldots, t_k, \ldots$. One can show that for $n = 3$ in the latter case

$$\lim_{k \to \infty} t_k = +\infty .$$

In the n-body problem with $n \geqq 4$, however, a principially different type of singularity is possible. Already in the 4-body problem on a line there exist motions in which infinitely many binary collisions occur over a finite time interval $[0, t_0]$. Moreover, for $t \to t_0$ three of the bodies diverge at infinity: one of them in a certain direction and the other two in the opposite direction, as in the Pythagorean three-body problem. However, in contrast to the three-body problem, the colliding bodies unboundedly approach one another, and this provides the kinetic energy which causes the solution to become unbounded in finite time. The fourth body bounces back and forth between them. When the fourth body passes close to the two bodies which are approaching one another, then some kind of an almost triple collision occurs. The existence of such a motion was demonstrated by J. Mather using the McGehee regularization of the simultaneous collision in the three-body problem (see [136]).

§3. Particular Solutions

In the n-body problem only a few exact solutions have been found. Basically, all these solutions were already known to Euler and Lagrange.

3.1. Central Configurations. We shall say that in an barycentric reference frame n point masses (m_i, r_i) form a *central configuration* if

$$\frac{\partial V}{\partial r_i}=\sigma\frac{\partial I}{\partial r_i}, \quad 1\leqq i\leqq n, \tag{15}$$

where

$$V=\sum_{j<k}\frac{m_k m_j}{|r_{kj}|}$$

is the gravitational force function $I=\sum m_i r_i^2$ is the polar moment of inertia, and the scalar function σ does not depend on the index i. It follows from Euler's formula for homogeneous functions that $\sigma=-V/2I$. Thus, we can

$$I\frac{\partial V}{\partial r_i}=-\frac{1}{2}V\frac{\partial I}{\partial r_i}.$$

That is to say, to central configurations there correspond critical points of the function IV^2. Since this function is homogeneous, it follows that if (m_i, r_i) is a central configuration, then so is $(m_i, \alpha r_i)$ for every $\alpha\neq 0$; we shall not distinguish between such central configurations.

Finding all central configurations for an arbitrary number n of points is a difficult problem, which is still open. Leaving aside the trivial case $n=2$, we list the result known presently in this direction.

For $n=3$ the only noncollinear central configuration is formed by the vertices of an equilateral triangle (Lagrange). For $n=4$ the unique noncoplanar (nonflat) configuration is given by the vertices of a regular tetrahedron. The collinear central configurations are described by the following *theorem of Moulton* [42]: each enumeration of the points uniquely determines a central configuration in which the points lie collinearly in the given order. Therefore, there are exactly $n!/2$ distinct collinear central configurations. For $n=3$ there are exactly three such configurations, discovered by Euler.

The notion of central configuration proves useful in the analysis of simultaneous collisions: it turns out that the configuration of gravitating points at the instant of a simultaneous collision is (in an asymptotic sense) central. From formula (15) it follows that if at the initial time the points form a central configuration and are at rest, then their configuration obviously does not change until the simultaneous collision occurs.

3.2. Homographic Solutions. A given solution of the n-body problem is called *homographic* if in an inertial barycentric reference frame the configuration formed by the n bodies remains similar to itself at all moments of time. If, in this case, the configuration does not rotate, then the solution will be referred to as *homothetic*. As an example we offer the configurations mentioned at the end of §3.1. If, however, the configuration remains congruent to itself (in the sense of Euclidean geometry, i.e., it rotates without dilation) then the solution is called a *relative equilibrium*.

It is not hard to show that

a) an homographic solution is homothetic if and only if the polar angular momentum is equal to zero;

b) an homographic solution is a relative equilibrium if and only if it is planar and the corresponding configuration rotates with constant nonzero angular velocity.

The next two facts are less readily established:

c) if an homographic solution is not coplanar, then it is homothetic;

d) if an homographic solution is coplanar, then it is planar.

In particular, every homographic solution is either planar or homothetic. In the three-body problem, for every homographic solution, the three bodies in a barycentric reference frame remain in an invariable plane which contains the center of mass (Lagrange).

Proposition 4. *In an homographic solution the bodies form a central configuration at any moment of time.*

This result provides a method for constructing homographic solutions. As an example we give here the well-known *theorem of Lagrange* (1772).

Theorem 4. *For arbitrary values of the masses the three-body problem admits an exact solution with the following properties:*

1) *the plane which contains the point masses is fixed in a barycentric reference frame;*

2) *the resultant of the Newtonian forces of attraction applied to each of the three masses passes through the common barycenter;*

3) *the triangle formed by the three bodies is equilateral;*

4) *the trajectories of the bodies are similar conics with one focus at the common barycenter.*

In the particular case of equal masses these conics are congruent and differ from each other by a rotation through 120°. This observation is readily generalized: the problem of n bodies of equal masses has solutions in which each body describes a conic with a focus at the common center of mass; their trajectories are congruent and differ one from another by a rotation through the angle $2\pi/n$.

3.3. The Amended Potential and Relative Equilibria

Proposition 5. *The relative equilibria configurations with polar angular momentum K coincide with the critical points of the function*

$$U_K = U + \frac{K^2}{2I}.$$

The function U_K is called the *amended* (*reduced*, or *effective*) *potential.* It was already used in §1, Ch. 1, to describe the regions of possible motions in the planar n-body problem, and also in §2.1 for determining the trajectories in the two-body problem.

◁ Suppose that a relative equilibrium configuration rotates around the center of mass with constant angular velocity ω. Then clearly $K=I\omega$. Let us pass to a reference frame with coordinates u, v, which rotates with angular velocity ω. In the new frame the given relative equilibrium configuration is fixed. The Lagrangian here is

$$L=T+V=\tfrac{1}{2}\sum m_i(\dot{u}_i^2+\dot{v}_i^2)+\omega\sum m_i(u_i\dot{v}_i-\dot{u}_i v_i)+V_\omega,$$

where $V_\omega=V+I\omega^2/2$. The corresponding equations of motion are

$$\begin{aligned} m_i\ddot{u}_i&=2m_i\omega\dot{v}_i+\frac{\partial V_\omega}{\partial u_i},\\ m_i\ddot{v}_i&=-2m_i\omega\dot{u}_i+\frac{\partial V_\omega}{\partial v_i}. \end{aligned} \tag{16}$$

Proposition 2 follows easily from these equations upon observing that the functions U_K and V_ω have the same critical points because at these points $K=I\omega$. ▷

§4. Final Motions in the Three-Body Problem

4.1. Classification of Final Motions According to Chazy

Theorem 5 (Chazy, 1922). *Every solution $r_k(t)$, $k=1, 2, 3$, of the three-body problem belongs to one of the following seven classes:*

1°. *H (hyperbolic motions): $|r_k|\to\infty$, $|\dot{r}_k|\to c_k>0$ as $t\to+\infty$;*

2°. *HP_k (hyperbolic-parabolic motions): $|r_i|\to\infty$, $|\dot{r}_k|\to 0$, $|\dot{r}_i|\to c_i>0$ $(i\neq k)$;*

3°. *HE_k (hyperbolic-elliptic motions): $|r_i|\to\infty$, $|\dot{r}_i|\to c_i>0$ $(i\neq k)$, $\sup\limits_{t\geqq t_0}|r_k|<\infty$;*

4°. *PE_k (parabolic-elliptic motions): $|r_i|\to\infty$, $|\dot{r}_i|\to 0$ $(i\neq k)$, $\sup\limits_{t\geqq t_0}|r_k|<\infty$.*

5°. *P (parabolic motions): $|r_i|\to\infty$, $|\dot{r}_i|\to 0$.*

6°. *B (bounded motions): $\sup\limits_{t\geqq t_0}|r_k|<\infty$.*

7°. *OS (oscillatory motions): $\overline{\lim\limits_{t\to+\infty}}\,\sup\limits_{k}|r_k|=\infty$, $\underline{\lim\limits_{t\to+\infty}}\,|r_k|<\infty$.*

Examples of motions of the first six types were known to Chazy. The existence of oscillatory motions was proved by K.A. Sitnikov in 1959.

It is only natural to attach to these seven types of final motions subsets of the 12-dimensional phase space M^{12} of the three-body problem with fixed center of mass: each such subset consists of all the phase trajectories that correspond to motions of a given type. The qualitative picture of the partition of M^{12} into classes of final motions is shown in Fig. 16. The sets H and HP_k lie in the domain in which the total energy h is positive, P lies on the hypersurface $h=0$, and the sets B, PE_k, and OS in the domain $h<0$; motions of class HE_k are possible for all values of h. It is known that H and HE_k are open in M^{12}, that HP_k consist of analytic manifolds of codimension 1, and that P consists of three connected manifolds of codimension 2 (represented by the three points in Fig. 16) and a manifold of codimension 3 (which is not shown in Fig. 16). The topology of the other classes of motions is not sufficiently studied.

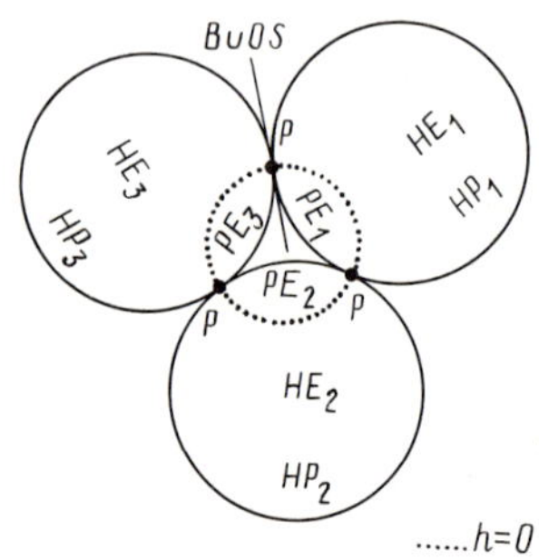

Fig. 16

4.2. Symmetry of Past and Future. By a theorem of Chazy one can introduce seven analogous *final classes* of motions when t tends to $-\infty$ instead of $+\infty$. To distinguish the classes corresponding to the cases $t\to\pm\infty$ we shall use superscripts $(+)$ and $(-)$ to write H^+, HE_3^-, and so on. In one of his works (1929) Chazy mistakenly asserted that for each given solution of the three-body problem the final types of behavior for $t\to+\infty$ and $t\to-\infty$ coincide. There has been a longheld belief in this "symmetry" of past and future, despite a numerical counterexample constructed by L. Becker (1920) which indicated the possibility of *"exchange"*: $HE_1^-\cap HE_2^+\neq\emptyset$. Becker's example was attributed to errors in the numerical integration. In 1947 O.Yu. Schmidt indicated an example of *"capture"* in the three-body problem: $H^-\cap HE^+\neq\emptyset$. This example, also constructed by numerical calculations, was proposed by Schmidt to support his well-known cosmological hypothesis. A rigorous proof of the possibility of capture was offered by K.A. Sitnikov in 1953.

The present state of our knowledge of final motions in the three-body problem is sketchily shown in Tables 1 and 2, which are taken from the work of V.M. Alekseev [3]. To each cell there corresponds one of the logically possible combinations of final types of past and future motion. The Lebesgue measure of the corresponding subsets of M^{12} is indicated (where it is known).

Table 1

<table>
<tr><td colspan="2" rowspan="2">$h>0$</td><td colspan="2">$t \to +\infty$</td></tr>
<tr><td>H^+</td><td>HE_i^+</td></tr>
<tr><td rowspan="3">$t \to -\infty$</td><td>H^-</td><td>Lagrange 1772
(isolated examples);
Chazy 1922
Measure >0</td><td>PARTIAL CAPTURE
Measure >0
Schmidt (numerical example), 1947
Sitnikov (qualitative methods), 1953</td></tr>
<tr><td rowspan="2">HE_j^-</td><td rowspan="2">COMPLETE DISPERSAL
Measure >0</td><td>$i=j$ Measure >0
Birkhoff 1927</td></tr>
<tr><td>$i \neq j$ EXCHANGE Measure >0
Becker (numerical examples), 1920
Alekseev (qualitative methods), 1956</td></tr>
</table>

Table 2

<table>
<tr><td colspan="2" rowspan="2">$h<0$</td><td colspan="3">$t \to +\infty$</td></tr>
<tr><td>HE_i^+</td><td>B^+</td><td>OS^+</td></tr>
<tr><td rowspan="4">$t \to -\infty$</td><td rowspan="2">HE_j^-</td><td>$i=j$ Measure >0
Birkhoff, 1927</td><td rowspan="2">COMPLETE CAPTURE
{ Measure $=0$
Chazy, 1920
Merman 1954 }
Littlewood, 1952
Alekseev, 1968
$\neq \varnothing$</td><td rowspan="2">{ Measure $=0$
Chazy, 1929
Merman, 1954 }
Alekseev, 1968
$\neq \varnothing$</td></tr>
<tr><td>EXCHANGE
$i \neq j$ Measure >0
Becker, 1920
(numerical examples)
Alekseev, 1956
(qualitative methods)</td></tr>
<tr><td>B^-</td><td>PARTIAL
DISPERSAL
$\neq \varnothing$
Measure $=0$</td><td>Euler, 1772
Lagrange, 1772
Poincaré, 1892
(isolated examples);
Measure >0
Arnol'd, 1963</td><td>Littlewood, 1952
Measure $=0$;
Alekseev, 1968
$\neq \varnothing$</td></tr>
<tr><td>OS^-</td><td>Measure $=0$, $\neq \varnothing$</td><td>Measure $=0$, $\neq \varnothing$</td><td>Sitnikov, 1959
$\neq \varnothing$
Measure $=$?</td></tr>
</table>

§5. The Restricted Three-Body Problem

5.1. Equations of Motion. The Jacobi Integral. Suppose that the Sun S and Jupiter J revolve around their center of mass in circular orbits. We choose units of length, time and mass so that the angular velocity of rotation, the sum of masses of S and J, and the gravitational constant are all equal to one. It is readily verified that for this choice the distance between S and J is also equal to one.

Consider the motion of an asteroid A in the plane of the orbits of S and J. We assume that the mass of A is considerably smaller than the masses of the Sun and Jupiter, and accordingly we neglect the influence of A on the motion of the two larger bodies.

It is convenient to pass to a moving reference frame which rotates with unit angular velocity around the center of mass of the S, J system; in this frame S and J are at rest. We choose coordinates x, y in the moving frame so that the points S and J lie invariably on the x-axis and their center of mass is the origin of the coordinate system. Then the equations governing the motion of the asteroid can be written in the form (see (16)):

$$\ddot{x}=2\dot{y}+\frac{\partial V}{\partial x}, \quad \ddot{y}=-2\dot{x}+\frac{\partial V}{\partial y}, \qquad V=\frac{x^2+y^2}{2}+\frac{1-\mu}{\rho_1}+\frac{\mu}{\rho_2}, \tag{17}$$

where μ is Jupiter's mass and ρ_1, ρ_2 are the distances from the asteroid A to S and J. Since S and J have coordinates $(-\mu, 0)$ and respectively $(1-\mu, 0)$,

$$\rho_1^2=(x+\mu)^2+y^2, \quad \rho_2^2=(x-1+\mu)^2+y^2.$$

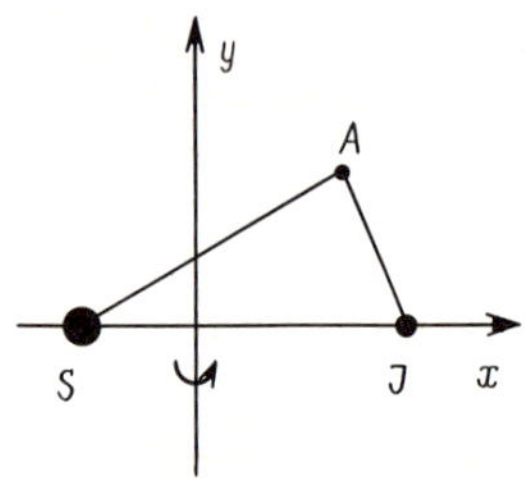

Fig. 17. The restricted three-body problem

Equations (17) have first integral

$$\frac{\dot{x}^2+\dot{y}^2}{2}-V(x, y)=h,$$

known as the *Jacobi integral*. It expresses the conservation of energy in the relative motion of the asteroid.

For fixed h the motion of the asteroid takes place in the *Hill region*

$$\{(x, y)\in R^2 \colon V(x, y)+h\geqq 0\}.$$

5.2. Relative Equilibria and the Hill Region. The shape of the Hill region depends on the position of the critical points of function $V(x, y)$. To each such critical point (x_0, y_0) there corresponds an "equilibrium" solution $x(t) \equiv x_0$, $y(t) \equiv y_0$, which may naturally be termed a relative equilibrium. Let us show that for each value $\mu \in (0, 1)$ there are exactly five such points.

We calculate

$$V'_y = yf, \quad f = 1 - \frac{1-\mu}{\rho_1^3} - \frac{\mu}{\rho_2^3},$$

$$V'_x = xf - \mu(1-\mu)\left(\frac{1}{\rho_1^3} - \frac{1}{\rho_2^3}\right),$$

and now solve the system of algebraic equations $V'_x = V'_y = 0$. Suppose first that $y \neq 0$. Then $f = 0$, hence $\rho_1 = \rho_2 = \rho$. From the equation $f = 0$ we find that $\rho = 1$. Thus, in this case the points S, J, and A are the vertices of an equilateral triangle. There are exactly two such equilibria, called *equilateral* or *triangular libration points*. They should be thought of as a particular case of Lagrange's solutions of the general, "unrestricted" three-body problem (see §3). Lagrange himself regarded these solutions as a "pure curiosity" and considered them to be useless in astronomy. However, 1907 brought the discovery of an asteroid, Achilles, which moves practically on Jupiter's orbit being constantly ahead of it by 60°. Close to Achilles there are nine other asteroids (the "Greeks"), and five other (the "Trojans") were discovered on the opposite side, which also form an equilateral triangle with the Sun and Jupiter.

Now let us turn to the relative equilibria lying on the x-axis. They are the critical points of the function

$$g(x) = \frac{x^2}{2} + \frac{1-\mu}{|x+\mu|} + \frac{\mu}{|x-1+\mu|}.$$

Since $g(x) > 0$ and $g(x) \to +\infty$ as $x \to \pm\infty$, $x \to -\mu$, and $x \to 1-\mu$, it follows that in the intervals $(-\infty, -\mu)$, $(-\mu, 1-\mu)$, $(1-\mu, +\infty)$ into which the points S and J divide the x-axis there exist three local minima of g. Since $g''(x) > 0$, these are the only critical points of g. They were found by Euler and are called *collinear libration points*.

One can show that the collinear libration points (we denote them by L_1, L_2, L_3) are of hyperbolic type, whereas the equilateral libration points (L_4 and L_5) are nondegenerate minima of function V. Figure 18 shows how the topology of Hill's region changes as the Jacobi constant h varies from $-\infty$ to $+\infty$; the mass of Jupiter is taken to be smaller than that of the Sun. If h is larger than the negative number

$$-\tfrac{1}{2}(3 - \mu + \mu^2),$$

then Hill's region coincides with the entire plane $R^2 = \{(x, y)\}$. For $\mu = 1/2$ the Hill region is symmetric not only about the x-axis, but also about the y-axis.

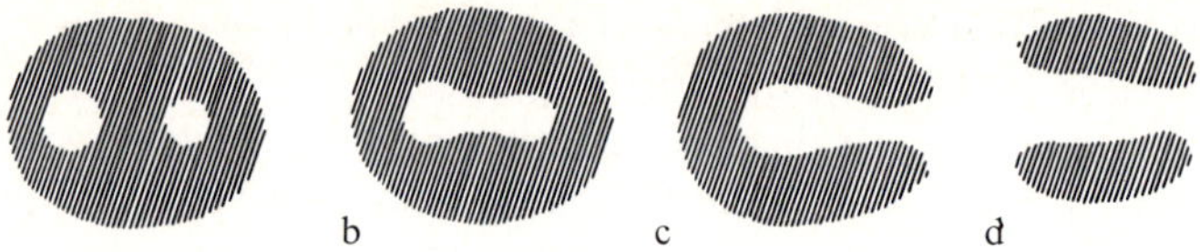

Fig. 18

The collinear librations points are always unstable: among the roots of the characteristic equation of the variational equations (i.e., the linearized equations of motion) there are some having positive real parts. For the case of equilateral libration points these roots are purely imaginary and distinct only when

$$27\mu(1-\mu)<1. \tag{18}$$

If this condition is fulfilled, then the equilateral relative equilibria are stable in the first approximation. The investigation of the Lyapunov stability of these points turns out to be a considerably more difficult task; we postone it until Chapter 7. We remark in conclusion that for the real Sun-Jupiter system condition (18) is clearly fulfilled.

5.3. Hill's Problem. Let us take as the origin of the rotating coordinate system the point where the body of the mass μ lies. Then the coordinates x, y of the third, small-mass body must be replaced by $x-(1-\mu)$, y. Preserving the notation x, y for these new variables, we see that the equations of motion have the same form (17), with the only difference being that the potential is now the function

$$V=(1-\mu)x+\tfrac{1}{2}(x^2+y^2)+(1-\mu)(1+2x+x^2+y^2)^{-1/2}+\mu(x^2+y^2)^{-1/2}. \tag{19}$$

We follow G.W. Hill and introduce a further simplification of the problem, borrowed from astronomy. Suppose again that the body of mass $1-\mu$ represents the Sun, μ is the Earth, and that a third body, the Moon, of negligible mass, moves in the vicinity of the point (0, 0) where the Earth is at rest. Let us neglect in equations (17) all the terms of order two or higher in x, y. Equivalently, we ignore in (19) the terms of order at least three in x, y. Modulo such terms, function V is given by

$$V=\frac{\mu}{2}(x^2+y^2)+\frac{3}{2}(1-\mu)x^2+\mu(x^2+y^2)^{-1/2}.$$

Since the Earth's mass μ is considerably smaller than the Sun's mass $1-\mu$, we can also neglect the first term in this expression of V.

It is convenient to change the units of mass and length slightly as follows:

$$x\to\alpha x,\qquad y\to\alpha y,\qquad \mu\to\beta\mu,\qquad 1-\mu\to\beta(1-\mu),$$

where

$$\alpha=\left(\frac{\mu}{1-\mu}\right)^{1/3},\qquad \beta=(1-\mu)^{-1}.$$

Following these transformations, the equations governing the motion of the Moon take the form

$$\ddot{x}-2\dot{y}=\frac{\partial V}{\partial x}, \quad \ddot{y}+2\dot{x}=\frac{\partial V}{\partial y}, \quad V=\tfrac{3}{2}x^2+(x^2+y^2)^{-1/2}. \tag{20}$$

They possess the Jacobi first integral

$$\tfrac{1}{2}(\dot{x}^2+\dot{y}^2)-V(x, y)=h.$$

It is readily verified that on passing from the restricted three-body problem to its limit version just described, known as the *Hill problem*, two of the equilateral libration points and one of the collinear ones disappear. In fact, the system of equations $V_x'=V_y'=0$ has only the two solutions $(x, y)=(\pm 3^{-1/3}, 0)$. Hill's region

$$\{V(x, y)+h \geqq 0\}$$

is symmetric relative to the x- and y-axes for every value of h. If $h \geqq 0$, Hill's region coincides with the entire plane. For $h<0$ its boundary has two asymptotes parallel to the y-axis: $x=+(-2/3\,h)^{1/2}$. The shape of these regions depends on whether the constant $(-h)$ is larger than, equal to, or smaller than the unique critical value of V, which is $3^{1/3}\cdot 3/2$. These three cases are depicted in Fig. 19. The only case of interest in astronomical applications is (a), and there only a domain surrounding the origin of coordinates.

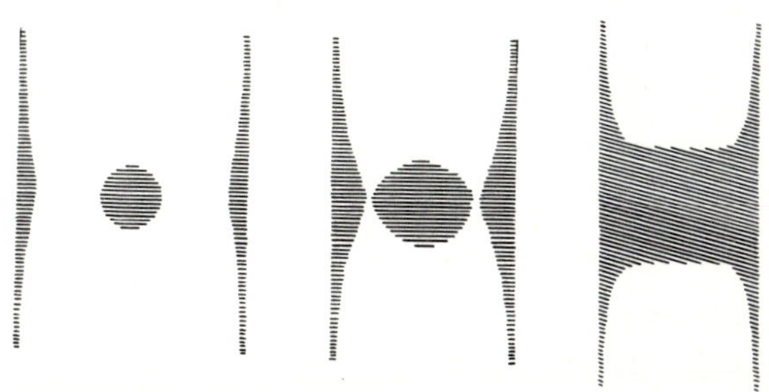

a b c

Fig. 19

We now turn to various questions connected with the regularization of Hill's problem. To this end, we shift to new, parabolic coordinates through the formulas $x=\xi^2-\eta^2$, $y=2\xi\eta$, and also change the time variable t along the trajectories to τ according to the rule

$$\frac{dt}{d\tau}=4(\xi^2+\eta^2).$$

Denoting differentiation with respect to τ by a prime, we write the equations of motion in the new variables as

$$\xi''-8(\xi^2+\eta^2)\eta'=\hat{V}_\xi', \quad \eta''+8(\xi^2+\eta^2)\xi'=\hat{V}_\eta',$$

where

$$\hat{V}=4+4(\xi^2+\eta^2)h+6(\xi^2-\eta^2)(\xi^2+\eta^2).$$

The energy integral takes the form

$$\tfrac{1}{2}(\xi'^2+\eta'^2)-\hat{V}(\xi,\eta,h)=0.$$

This regularization of Hill's problem, proposed by Birkhoff, permits a straightforward investigation of the analytic singularities of the solutions corresponding to Moon-Earth collisions. Suppose that such a collision occurs at the time $t=0$ and let $\tau(0)=0$. Then obviously

$$\xi=(\sqrt{8}\sin\alpha)\tau+\ldots,\quad \eta=(\sqrt{8}\cos\alpha)\tau+\ldots,\quad \text{and}\quad t=\tfrac{32}{2}\tau^3+\ldots,$$

where α is an integration constant. Thus, the new time τ is a uniformizing variable and, as in the case of binary collisions in the general three-body problem, the solution $\xi(t), \eta(t)$ admits a unique real analytic continuation beyond the collision time.

As we have already mentioned, in astronomy one is interested only in motions occuring near the point $\xi=\eta=0$. For negative h of large absolute value it is convenient to use the new variables

$$\varphi=2\xi\{-2h-3(\xi^2-\eta^2)^2\}^{1/2},\quad \psi=2\eta\{-2h-3(\xi^2-\eta^2)^2\}^{1/2}.$$

Then the energy integral takes the very simple form

$$\xi'^2+\eta'^2+\varphi^2+\psi^2=8.$$

This is the equation of a three-dimensional sphere in the four-dimensional phase space of variables $\xi', \eta', \varphi, \psi$. Since to the points (ξ,η) and $(-\xi,-\eta)$ there corresponds one and the same point in the (x,y)-plane, Moon's states $(\xi',\eta',\varphi,\psi)$ and $(-\xi',-\eta',-\varphi,-\psi)$ must be identified. We thus conclude that for negative h of large absolute value the component of the three-dimensional energy level which is of interest to us is diffeomorphic to the three-dimensional projective space. This is of course valid for all $h<-\tfrac{3}{2}3^{1/3}$.

We conclude this section by a discussion of certain periodic solutions of Hill's problem, which have important astronomical applications. Specifically, we are interest in periodic solutions $x(t), y(t)$ of small period τ, near the Earth (the point $x=y=0$), whose orbits are symmetric relative to the x- and y-axes. These symmetry requirements are expressed by the equalities

$$x(-t)=x(t)=-x\left(t+\frac{\tau}{2}\right),\quad y(-t)=-y(t)=y\left(t+\frac{\tau}{2}\right).$$

Accordingly, we seek these solutions as trigonometric series

$$x(t)=\sum_{n=-\infty}^{\infty}a_n(m)\cos(2n+1)\frac{t}{m},\quad y(t)=\sum_{n=-\infty}^{\infty}a_n(m)\sin(2n+1)\frac{t}{m},$$

where

$$m=\frac{\tau}{2\pi}.$$

Substituting these series in the equations of motion (20) we obtain an infinite nonlinear system of algebraic equations for infinitely many unknown coefficients $a_n(m)$. Hill showed (1878) that this system admits a unique solution, at least for small values of m (see [36, 42]). The value $m_0=0.08084\ldots$ for the real Moon falls in this admissible range. The convergence of Hill's series was established by Lyapunov in 1895.

One can show that the following asymptotic expansions hold for the coefficients $a_n(m)$:

$$a_0=m^{2/3}\left(1-\frac{2}{3}m+\frac{7}{18}m^2-\ldots\right),$$

$$\frac{a_1}{a_0}=\frac{3}{16}m^2+\frac{1}{2}m^3+\ldots, \qquad \frac{a_{-1}}{a_0}=-\frac{19}{16}m^2-\frac{5}{3}m^3-\ldots,$$

$$\frac{a_2}{a_0}=\frac{25}{256}m^4+\ldots, \qquad \frac{a_{-2}}{a_0}=0\cdot m^4+\ldots, \ldots.$$

This shows that for small values of m the main contribution in Hill's periodic solutions comes from the terms

$$x_0(t)=m^{2/3}\cos\frac{t}{m}, \qquad y_0(t)=m^{2/3}\sin\frac{t}{m},$$

which represent the law governing the motion of the Moon when the influence of the Sun is neglected. The presence of the coefficient $m^{2/3}$ is a consequence of Kepler's third law. For small values of the parameter m the Sun, while perturbing the Earth-Moon system, does not destroy the circular periodic orbits, merely deforming them slightly.

§6. Ergodic Theorems in Celestial Mechanics

6.1. Stability in the Sense of Poisson. Let (M, S, μ) be a complete measure space; here S is a σ-algebra of subsets of M, and μ is a countably-additive measure on S. Consider a measure-preserving automorphism g of M. We call the sets

$$\Gamma_p=\bigcup_{n\in Z} g^n(p), \qquad \text{where } g^0(p)=p,$$

and

$$\Gamma_p^+=\bigcup_{n\geq 0} g^n(p)$$

the *trajectory* and the *positive* (or *forward*) *semitrajectory*, respectively, of the point $p\in M$.

Poincaré's Recurrence Theorem. *Let $\mu(M)<\infty$. Then for every measurable set $V \in S$ of positive measure there is a set $W \subset V$ with $\mu(W)=\mu(V)$ such that for every $p \in W$ the set $\Gamma_p^+ \cap W$ is infinite.*

We follow Poincaré and apply this result to the restricted three-body problem. In the notations of the preceding section, the equations governing the motion of the asteroid are Lagrange's equations with the Lagrangian

$$L=\frac{1}{2}(\dot{x}^2+\dot{y}^2)+x\dot{y}-y\dot{x}+V, \qquad V=\frac{1}{2}(x^2+y^2)+\frac{1-\mu}{\rho_1}+\frac{\mu}{\rho_2}.$$

These can be written as Hamilton's equations with the Hamiltonian

$$H=\frac{1}{2}(X^2+Y^2)+yX-xY-G, \qquad G=\frac{\mu}{\rho_1}+\frac{1-\mu}{\rho_2},$$

where $X=\dot{x}-y$ and $Y=\dot{y}+x$ are the momenta canonically conjugate to the coordinates x and y. According to Liouville's theorem, the phase flow of this system, g^t, preserves the usual Lebesgue measure in $R^4=\{(X, Y, x, y)\}$.

Consider the set of all phase points at which $c_1<-H<c_2$, where c_1 and c_2 are sufficiently large positive constants. As we have seen in §5, if this double inequality holds, then the point (x, y) lies in one of the three connected subregions of the Hill region $\{G \leqq c_1\}$. Pick one of the two regions containing the Sun and Jupiter. The corresponding connected region in phase space is obviously invariant under the action of g^t. From it we must eliminate the collision trajectories; the union of these trajectories has measure zero. Denoting the resulting set by M, we show that its measure is finite. In fact, the coordinates (x, y) of the points of M belong to a bounded subset of the plane $R^2=\{(x, y)\}$. The admissible momenta X and Y are subject to the inequalities

$$2(G-c_2)<(Y+x)^2+(Y-y)^2<2(G-c_1),$$

which are consequences of the Jacobi integral. These inequalities determine a circular annulus in the plane $R^2=\{(X, Y)\}$ whose area does not exceed $2\pi(c_2-c_1)$. It follows from these remarks that the measure $\mu(M)$ is finite, and hence that Poincaré's recurrence theorem is applicable: for almost every $p \in M$ the semitrajectory $\{g^t(p), t \geqq 0\}$, intersects any neighborhood of the point p for arbitrarily large values of t. Poincaré called the motions with this property *stable in the sense of Poisson* (or *Poisson-stable*).

6.2. Probability of Capture. Again, let V be a measurable set of positive measure. For $n \in N$ we let V^n denote the set of points $p \in V$ such that $g^k(p) \in V$ for all $0 \leqq k \leqq n$. Clearly, $V^{n_1} \supset V^{n_2}$ whenever $n_1<n_2$. Set

$$B=\bigcap_{n \geqq 0} V^n.$$

The set B is measurable and $\mu(B)<\infty$. If $p \in B$, then obviously $\Gamma_p^+ \subset V$.

Put $B^n = g^n(B)$. The B^n are measurable, and again $B^{n_1} \supset B^{n_2}$ if $n_1 < n_2$. The set

$$D = \bigcap_{n \geqq 0} B^n \subset B$$

is also measurable, and obviously $\Gamma_p \subset V$ for all $p \in D$.

Proposition 6. $\mu(B \setminus D) = 0$.

In order that this assertion be useful we need first to verify that $\mu(B) > 0$. However, in concrete problems proving this might in fact be very difficult. Proposition 6, which is attributed to K. Schwarzschild, is of course valid also for the case of continuous time n.

Suppose, for example, that the Sun-Jupiter system has "captured" from the surrounding space asteroids (the "Greeks" and the "Trojans") in the vicinity of the equilateral libration points. Proposition 6 immediately shows that this event has probability zero. Thus, "capture" phenomena in celestial mechanics should be considered only in mathematical models which account for dissipation of energy.

A more interesting example is provided by the following reasoning of J. Littlewood. Consider the n-body problem with the center of mass at rest. The motion of the n points is described by a Hamiltonian system, whose Hamiltonian is regular in the domain in which the mutual distances $|r_{kl}|$ are different from zero. For arbitrary $c > 1$ we consider the open subset $A(c)$ of phase space defined by the inequalities

$$c^{-1} < |r_{kl}| < c \quad (1 \leqq k < l \leqq n), \qquad -c < H < c.$$

Since $A(c)$ is bounded, $\mu(A(c)) < \infty$. Therefore, by Proposition 6, the set $B(c)$ of points which remain in $A(c)$ for all $t \geqq 0$ differs only by a set of measure zero from the set $D(c)$ of those points which remain in $A(c)$ for all $t \in R$.

If $c_1 < c_2$ then obviously $A(c_1) \subset A(c_2)$, $B(c_1) \subset B(c_2)$, and $D(c_1) \subset D(c_2)$. Hence, a corresponding assertion holds true for the sets

$$A = \bigcup_{c>1} A(c), \qquad B = \bigcup_{c>1} B(c), \qquad D = \bigcup_{c>1} D(c).$$

Thus, for the points $p \in B$ the mutual distances $|r_{kl}|$ remain bounded from above and from below for all $t \geqq 0$ by some positive constants which do not depend on p. For points $p \in D$ this property holds for all $t \in R$. Almost all points of D belong to B.

Suppose, for example, that it is known that the planetary system is stable "in the past". If it captures a new body, say, a speck of dust arriving from "infinity", then the newly formed system of bodies looses its stability: with probability one, either a collision occurs, or one of the bodies escapes again at infinity. Moreover, it is by no means necessary that the speck of dust is the one that leaves the Solar system: Jupiter, or even the Sun may equally well leave!

Chapter 3
Symmetry Groups and Reduction (Lowering the Order)

§1. Symmetries and Linear First Integrals

1.1. E. Noether's Theorem. Let (M, L) be a Lagrangian system and v a smooth vector field on M. The field v generates a one-parameter group of diffeomorphisms $g^\alpha: M \to M$, where $g^\alpha(x)$ is the solution of the differential equation

$$\frac{d}{d\alpha} g^\alpha(x) = v(g^\alpha(x)) \tag{1}$$

with initial condition $g^0(x) = x$.

Definition. The Lagrangian system (M, L) *admits the group* g^α if the Lagrangian L is invariant under the maps $g^\alpha_*: TM \to TM$. The group g^α and the vector field v are naturally called a *symmetry group* and a *symmetry vector field*, respectively.

Let $\gamma: \Delta \to M$ be a motion of the Lagrangian system (M, L). Then $g^\alpha \circ \gamma: \Delta \to M$ is also a motion, for every value of α.

In the nonautonomous case L is a smooth function on the tangent bundle of the *extended configuration space* $'M = M \times R$. We call the one-parameter group of diffeomorphisms $'g^\alpha: {}'M \to {}'M$ a symmetry group of the system $('M, L)$ if $'g^\alpha(x, t) = (y, t)$ for all $(x, t) \in M \times R$ (i.e., $'g^\alpha$ preserves time) and the maps $'g^\alpha_*$ preserve L. To the group $'g^\alpha$ there corresponds the smooth vector field

$$'v(x, t) = \frac{d}{d\alpha}('g^\alpha(x, t))|_{\alpha=0}$$

on $'M$. Obviously, $'v(x, t) = (v(x, t), 0) \in T_{(x,t)}(M \times R)$, and $v(x, t)$ may be regarded as a vector field on M which depends smoothly on t.

Lemma 1. *System* (M, L) *admits the symmetry group* g^α *if and only if*

$$(p \cdot v)^\cdot = [L] \cdot v. \tag{2}$$

◁ This is a consequence of the identity

$$\left.\frac{d}{d\alpha}\right|_{\alpha=0} L(g_*^\alpha x)=(L'_{\dot{x}}\cdot v)^{\cdot}-[L]\cdot v. \ \triangleright \tag{3}$$

Lemma 1 is valid also in the nonautonomous case. From equality (2) we obtain

Theorem 1. *If system (M, L) admits the group g^α, then $\mathscr{I}=p\cdot v$ is a first integral of the equations of motion.*[18]

Let $(M, \langle\, ,\rangle, V)$ be a natural mechanical system. The Lagrangian $L=\frac{1}{2}\langle\dot{x}, \dot{x}\rangle+V(x)$ is invariant under the action of the group g^α if and only if this property is shared by the Riemannian metric $\langle\, ,\rangle$ and the force function V. For such a system the first integral $\mathscr{I}$ is obviously equal to $\langle v, \dot{x}\rangle$, and hence it is linear in the velocity.

Example 1. Suppose that in some (local) system of coordinates $x_1, \ldots, x_n$ on M the Lagrangian L does not depend on x_1. Then the system (M,L) admits locally the symmetry group g^α: $x_1 \to x_1+\alpha$, $x_k \to x_k$ $(k \geqq 2)$. The vector field corresponding to this group is $v=\partial/\partial x_1$. Theorem 1 is asserts that the quantity $\mathscr{I}=p\cdot v=p_1=L'_{\dot{x}_1}$ is a first integral. In mechanics x_1 is called a *cyclic coordinate*, and $\mathscr{I}$ a *cyclic integral.* In particular, the energy integral is a cyclic integral of a suitable extended Lagrangian system. To show this, we introduce a new time variable τ by the rule $t=t(\tau)$, and define the function $'L\colon T'M \to R$ $('M=M\times R)$ by the formula

$$'L(x', t', x, t)=L(x'/t', x, t)\,t', \qquad (\cdot)'=\frac{d}{d\tau}(\cdot).$$

It follows from Hamilton's variational principle and the equality

$$\int_{\tau_1}^{\tau_2} 'L\, d\tau=\int_{t_1}^{t_2} L\, dt$$

that if $x\colon [t_1, t_2]\to M$ is a motion of the system (M, L), then $(x, t)\colon [\tau_1, \tau_2] \to 'M$ is a motion of the extended Lagrangian system $('M, 'L)$. In the autonomous case time t is a cyclic coordinate and the cyclic integral

$$\frac{\partial' L}{\partial t'}=L-\frac{\partial L}{\partial \dot{x}}\,\dot{x}=\text{const.}$$

coincides with the energy integral. △

Theorem 2. *If $v(x_0)\neq 0$, then in a small neighborhood of the point x_0 there exist local coordinates $x_1, \ldots, x_n$ such that $\mathscr{I}=p\cdot v=p_1$.*

[18] Theorem 1 was first formulated in this form by E. Noether in 1918. Lagrange and Jacobi were already aware of the connection between the conservation of momentum and angular momentum, and the groups of translations and rotations, respectively. Theorem 1 for natural systems was published by M. Lévy in 1878.

This is a consequence of the *theorem on the straightening* of vector fields.[19]

Theorem 3. *Suppose that the equation of motion* $[L]=0$ *has the first integral* $\mathscr{I}=p\cdot v$. *Then the phase flow of equation* (1) *is a symmetry group of the Lagrangian system* (M, L).

Theorems 2 and 3 have the following:

Corollary. *Every first integral, linear in the velocities, of a natural system is locally a cyclic integral.*

If k symmetry fields $v_1, \ldots, v_k$ exist ($k \geqq 1$), then the equations of motion admit an equal number of first integrals $\mathscr{I}_1=p\cdot v_1, \ldots, \mathscr{I}_k=p\cdot v_k$. Assuming that the Lagrangian system (M, L) is natural, we use the Legendre transformation to pass to Hamilton's equations on T^*M. The functions $\mathscr{I}_1, \ldots, \mathscr{I}_k$: $T^*M \to R$ are independent and in involution (in the standard symplectic structure on T^*M) if and only if the fields $v_1, \ldots, v_k$ on M are linearly independent and commute. The existence of linear integrals imposes constraints not only on the Riemannian metric and the potential of the force field, but also on the topology of the configuration space.

Theorem 4. *Let* M *be a connected, compact, orientable, even-dimensional manifold. If a Hamiltonian natural system on* T^*M *has* k *independent linear first integrals in involution with* $k \geqq (\dim M)/2$, *then the Euler-Poincaré characteristic* $\chi(M) \geqq 0$.[20]

Corollary. *Suppose* $\dim M=2$. *If a natural system on* T^*M *has a first integral linear in the velocities, then* M *is diffeomorphic to either a sphere or a torus.*

In the nonorientable case we must also include the projective plane and the Klein bottle.

◁ Let us prove the Corollary. If $\chi(M)<0$, then the symmetry field v has singular points. Since the phase flow of the equation $\dot{x}=v(x)$ is an isometry group of the two-dimensional Riemannian manifold $(M, \langle\, , \rangle)$, it follows that the singular points x_j of v are isolated and of elliptic type. By Poincaré's formula, $\chi(M)=\sum_j \operatorname{ind}(x_j)>0$; we have thus reached a contradiction. ▷

Consider the more general situation in which an arbitrary Lie group G acts (from the left) on M. Let $\mathscr{G}$ be the Lie algebra of G and $\mathscr{G}^*$ the linear space dual to $\mathscr{G}$. We now indicate a natural mapping $\mathscr{I}_G$: $TM \to \mathscr{G}^*$ which sends each point $x \in TM$ into a linear function on $\mathscr{G}$.

To each vector $X \in \mathscr{G}$ there corresponds a one-parameter subgroup g_X of G whose action on M determines a tangent vector field v_X. The mapping

[19] See §2, Ch. 1, Vol. I of this Encyclopedia.

[20] This result was obtained by S.V. Bolotin and D.L. Abrarov.

$X \mapsto v_X$ is a homomorphism of the Lie algebra $\mathcal{G}$ into the Lie algebra of all vector fields on M. We set $\mathcal{I}_G(x) = L'_{\dot{x}} \cdot v_X$; this function is linear in $X \in \mathcal{G}$.

Definition. The mapping $\mathcal{I}_G\colon TM \to \mathcal{G}^*$ is called the *(angular) momentum mapping* of the Lagrangian system (M, L) *relative to the group* G (or, simply, the momentum mapping when there is no danger of confusion).

In addition to the momentum mapping $\mathcal{I}_G\colon TM \to \mathcal{G}^*$ one can define the mapping $P_G\colon T^*M \to \mathcal{G}^*$ by the rule $P_G(p) = p \cdot v_X$. The momentum $\mathcal{I}_G$ is the composition of P_G and the Legendre transformation.

Example 2. Consider n free point masses (r_k, m_k) in three-dimensional Euclidean space. Let $SO(2)$ be the group of rotations around the axis with unit vector e. This group acts in the configuration space $R^3\{r_1\} \times \ldots \times R^3\{R_n\}$, and the corresponding vector field is $(e \times (r_1 - 'r_1), \ldots, e \times (r_n - 'r_n))$, where $'r_k$ designates the position vector of the k-th point with the tail at some point of the rotation axis. Here

$$L = \tfrac{1}{2} \sum m_k \langle \dot{r}_k, \dot{r}_k \rangle + V(r_1, \ldots, r_n),$$

so that

$$\mathcal{I}_{SO(2)} = \sum m_k \langle \dot{r}_k, e \times (r_k - 'r_k) \rangle = \langle e, \sum m_k (r_k - 'r_k) \times \dot{r}_k \rangle$$

is identical with the usual angular momentum of the system relative to the given axis.

Now let $G = SO(3)$ be the group of rotations around some point o. The dual space $\mathcal{G}^* = (so(3))^*$ can be canonically identified with the algebra of vectors of oriented three-dimensional Euclidean space in which the commutator is given by the usual cross product. Then, obviously, $\mathcal{I}_{SO(3)}$ corresponds to the angular momentum of the system relative to the point o. △

Definition. The Lie group G is called a *symmetry group* of the Lagrangian system $(M\ L)$ if $L(g_* \dot{x}) = L(\dot{x})$ for all $\dot{x} \in TM$ and $g \in G$.

Theorem 5. *Suppose that the system* (M, L) *admits the symmetry group* G. *Then the momentum mapping* $\mathcal{I}_G$ *is a first integral* (i.e., $\mathcal{I}_G$ *takes constant values on the motions of the Lagrangian system* (M, L)).

This is a consequence of Theorem 1.

Example 3. We have already seen in Chapter 1 that the equations of the n-body problem admit the group of Galilean transformations. However, the Lagrangian

$$L = \tfrac{1}{2} \sum m_k (\dot{x}_k^2 + \dot{y}_k^2 + \dot{z}_k^2) + \sum_{i<j} \frac{\gamma m_i m_j}{(x_i - x_j)^2 + (y_i - y_j)^2 + (z_i - z_j)^2}$$

is not invariant under the full Galilean group. It admits translations along the time axis, and also isometries of three-dimensional Euclidean space. Corresponding to translations of time is conservation of the total energy, to translations of Euclidean space is conservation of the momentum, and to the group of rotations, conservation of the angular momentum. Consider also the group of homotheties

$$(x, y, z) \mapsto (\alpha x, \alpha y, \alpha z), \qquad \alpha > 0. \tag{4}$$

It is generated by the vector field

$$v = \sum \left(x_k \frac{\partial}{\partial x_k} + y_k \frac{\partial}{\partial y_k} + z_k \frac{\partial}{\partial z_k} \right).$$

For $\alpha = 1$ we have the identity mapping. The Lagrangian of the n-body problem does not admit the group of homotheties. We may, however, use identity (3) for $\alpha = 1$. Since under transformation (4), $T \to \alpha^2 T$ and $V \to \alpha^{-1} V$, equality (3) yields the already known Lagrange's identity

$$\left.\frac{dL}{d\alpha}\right|_{\alpha=1} = (p \cdot v)^{\cdot} \Leftrightarrow 2T - V = \sum m_k (x_k \dot{x}_k + y_k \dot{y}_k + z_k \dot{z}_k)^{\cdot} = \tfrac{1}{2} \ddot{I},$$

where

$$I = \sum m_k (x_k^2 + y_k^2 + z_k^2). \quad \triangle$$

1.2. Symmetries in Nonholonomic Mechanics. Suppose that nonconservative forces $F(x, \dot{x})\colon T_x M \to T_x^* M$ act on the nonholonomic system (M, S, L). The motions of the system are then defined by the d'Alembert-Lagrange principle: $([L] - F) \cdot \xi = 0$ for all virtual velocities ξ.

Definition. The Lie group G is called a *symmetry group* of the nonholonomic system (M, S, L) if

1) G preserves L,

2) the vector fields v_X, $X \in \mathcal{G}$, are fields of virtual velocities.

We call the mapping $\Phi_G\colon TM \to \mathcal{G}^*$, defined by the formula $\Phi_G(x) = F \cdot v_X$, the *moment of the force F relative to the group G*.

Theorem 6. *If (M, S, L) admits the symmetry group G, then $(\mathcal{I}_G)^{\cdot} = \Phi_G$.*

Corollary. *If $F \equiv 0$, then under the assumptions of Theorem 6 the momentum $\mathcal{I}_G$ is conserved.*

Theorem 6 follows from the d'Alembert-Lagrange principle upon using identity (3).

We now apply these general considerations to the dynamics of a system of point masses in oriented three-dimensional Euclidean space. We shall assume that a force F acts on the point (r, m). Consider the group of translations along a moving straight line with direction vector $e(t)$: $r \mapsto r + \alpha e$, $\alpha \in R$.

Theorem 7 ([117]). *Suppose the following conditions are satisfied:*

1) *the vectors $\xi_k = e$ $(1 \leq k \leq n)$ are virtual velocities;*

2) *$\langle P, \dot{e} \rangle = 0$, where $P = \sum m\dot{r}$ is the total momentum.*

Then $\langle P, e \rangle^{\cdot} = \langle \sum F, e \rangle$.

Corollary. *Suppose that at each instant the vectors* $\xi_k=\dot{\eta}=(\sum m\dot{r}/\sum m)$ $(1\leqq k\leqq n)$ *are virtual velocities. If the system moves by virtue of its own inertia* $(F\equiv 0)$, *then the velocity* $\dot{\eta}$ *of its center of mass is constant in time.*

Example 4. Consider a balanced skate sliding on a horizontal plane, and also a homogeneous disc which rolls so that its plane is always vertical. By the Corollary to Theorem 7, in both cases the velocity of the center of mass is constant. △

Consider also the group of rotations of Euclidean space around a moving line l with unit direction vector $e(t)$, which passes through the point with position vector $r_0(t)$. Let K be the angular momentum of a given system of point masses relative to the origin of a fixed reference frame, and let K_l (respectively M_l) be the angular momentum (respectively, the moment of forces) relative to the moving axis.

Theorem 8 ([117]). *Suppose that the following conditions are satisfied:*

1) *when the system rotates as a rigid body around axis* l, *the velocity vectors of the mass points are virtual velocities.*

2) $\langle P, (r_0\times e)^{\cdot}\rangle+\langle K, \dot{e}\rangle=0$.

Then $\dot{K}_l=M_l$.

In particular, if the direction of l in space is constant ($\dot{e}(t)=$const), then condition 2) takes the form of S.A. Chaplygin's condition (1897): $\langle e, \dot{r}_0\times\dot{\eta}\rangle=0$, where $\dot{\eta}$ is the velocity of the center of mass. In case that $r_0=\eta$, condition 2) can be simplified: $\langle K+R_0\times P, \dot{e}\rangle=0$. It is obviously fulfilled under the additional assumption that $e(t)=$const. For example, a balanced skate rotates around the vertical with constant angular velocity.

Example 5. Let us consider *Chaplygin's problem* concerning a dynamically asymmetric ball, whose center of mass coincides with its geometric center, rolling on a horizontal plane. Let o be the point of contact of the ball with the plane, and let K_0 be its angular momentum relative to o. Chaplygin's problem admits the group $SO(3)$ of rotations around the point o. The momentum mapping $\mathscr{I}_{SO(3)}$ is, of course, equal to K_0, whereas the moment of forces is $\Phi_G=0$. Therefore, by Theorem 6, $K_0=$const. This remark enables us to write a closed system of equations governing the rolling of the ball. Let k_0 be the angular momentum in a reference frame attached to the ball, ω the ball's angular velocity, and γ the vertical unit vector. The constancy of the vectors k_0 and γ in the fixed reference frame is expressed by the equations

$$\dot{k}_0+\omega\times k_0=0, \qquad \dot{\gamma}+\omega\times\gamma=0. \tag{5}$$

Let A be the inertia tensor of the ball relative to its center of mass. Let m and a denote the mass and the radius of the ball. Then $k_0=A\omega+ma^2\gamma\times(\omega\times\gamma)$. This relation transforms (5) into a closed system of differential equations for ω and γ. Equations (5) possess four independent first integrals:

$F_1=\langle k_0, k_0\rangle$, $F_2=\langle k_0, \gamma\rangle$, $F_3=\langle \gamma, \gamma\rangle=1$, $F_4=\langle k_0, \omega\rangle$. The last expresses the conservation of the kinetic energy of the ball. Using these integrals one can integrate equations (5) by quadratures (Chaplygin, 1903). △

1.3. Symmetries in Vakonomic Mechanics. Let (M, S, L) be a vakonomic system and G a Lie group acting on M.

Definition. G is called a *symmetry group* of the vakonomic system (M, S, L) if:

1) G takes $S \subset TM$ into S;

2) G preserves the restriction of L to S.

The (angular) *momentum mapping* $\mathcal{I}_G$: $T^*M \to \mathcal{G}^*$ of the vakonomic system (M, S, L) relative to G is defined by the rule $\mathcal{I}_G(p)(X)=p\cdot v_X$, $X \in \mathcal{G}$, where p is the vakonomic momentum.

Example 6. Suppose that (M, S, L) is a natural system and the kinetic energy is defined by a Riemannian metric $\langle\ ,\ \rangle$. If the constraint S is given by the equation $\langle a(x), \dot{x}\rangle=0$, then

$$\mathcal{I}=\langle v, \dot{x}\rangle+\langle p, a\rangle(a\cdot v)/\langle a, a\rangle. \ \triangle$$

Theorem 9. *If the vakonomic system (M, S, L) admits the symmetry group G, then $\mathcal{I}_G=$const.*

The function $\mathcal{I}_G$ is, in general, not observable (see Ch. 1, §4.3). If, however, the symmetry fields v_X, $X \in \mathcal{G}$, are fields of virtual velocities, then $\mathcal{I}_G=L_{\dot{x}}\cdot v_X$ and hence $\mathcal{I}_G$ is observable.

Example 7. A skate on a horizontal plane, regarded as a vakonomic system, admits the group of translations, but not the group of rotations around the vertical axis. Consequently, the vakonomic momentum of the skate is conserved. However, this quantity is not observable. The vakonomic angular momentum relative to the group of rotations coincides with the ordinary angular momentum, which is not a first integral of the equations of motion. △

1.4. Symmetries in Hamiltonian Mechanics. Let (M, ω^2) be a symplectic manifold. Suppose the one-parameter group g^s acts on M as a group of symplectic diffeomorphisms. Consider the vector field

$$v=\left.\frac{d}{ds}\right|_{s=0} g^s.$$

This field is locally Hamiltonian, i.e., the 1-form $\omega^2(\cdot, v)$ is closed. Hence, locally $\omega^2(\cdot, v)=dF$. Extension of function F to the entire manifold M leads, as a rule, to a multi-valued Hamiltonian. If, however, the symplectic form ω^2 is exact, that is, $\omega^2=d\omega^1$, then the vector field v has a single-valued Hamiltonian $F=\omega^1(v)$.

Example 8. Let N be a smooth manifold and g^s the group of diffeomorphisms of N generated by a vector field u. Since every diffeomorphism of N takes 1-forms into 1-forms, g^s also acts on the cotangent bundle $M = T^*N$. Recall that M possesses the standard symplectic structure $\omega^2 = dp \wedge dq = d(p \cdot dq)$, where p, q are "canonical" coordinates on M. Since the group g^s preserves the 1-form $p \cdot dq$, it preserves also the 2-form ω^2, and hence it is a group of symplectic diffeomorphisms of M. The action of g^s on M is generated by the single-valued Hamiltonian $F = p \cdot u$. △

Theorem 10. *The group of symplectic diffeomorphisms g^s with single-valued Hamiltonian F preserves function $H: M \to R$ if and only if F is a first integral of the Hamiltonian system with Hamiltonian H.*

◁ The proof is based on the formula

$$\left.\frac{d}{ds}\right|_{s=0} H(g^s(x)) = \{H, F\}(x). \ \triangleright$$

Now suppose that there is given a *symplectic action* of a Lie group G on M such that to each element X of the Lie algebra $\mathcal{G}$ of G corresponds a one-parameter subgroup with single-valued Hamiltonian F_X. The Hamiltonians F_X are defined to within additive constants.

Definition. A symplectic action of G on M is called a *Poisson action* if the correspondence $X \mapsto F_X$ can be selected so that

1) F_X depends linearly on X,

and

2) $\{F_X, F_Y\} = F_{[X,Y]}$ for all $X, Y \in \mathcal{G}$.

Example 9. Let N be a smooth manifold and G a Lie group acting on N by diffeomorphisms. We lift the action of G to T^*N as indicated in Example 8. The result is a Poisson action; this follows from the linearity of the function $p \cdot v_X$ in X and the formula: $\{p \cdot v_X, p \cdot v_Y\} = p \cdot [v_X, v_Y] = p \cdot v_{[X,Y]}$. △

A Poisson action of the group G on M defines a natural mapping $P_G: M \to \mathcal{G}^*$, which sends each point $x \in M$ into the linear function $X \mapsto F_X(x)$ on the algebra $\mathcal{G}$. P_G is called the *momentum mapping of the given Poisson action of G*.

Proposition 1. *If the Lie group G is connected, then the momentum mapping P takes the Poisson action of G on M into the coadjoint action of G on $\mathcal{G}^*$, i.e., the diagram*

$$\begin{array}{ccc} M & \xrightarrow{\ g\ } & M \\ \downarrow{\scriptstyle P} & & \downarrow{\scriptstyle P} \\ \mathcal{G}^* & \xrightarrow{\ Ad^*_{g^{-1}}\ } & \mathcal{G}^* \end{array}$$

commutes for all $g \in G$.

Suppose (N, L) is a Lagrangian system and the Lie group G acts on N. The Lagrangian L defines the Legendre transformation $TN \to T^*N$. Let $P_G: T^*N \to \mathcal{G}^*$ be the momentum mapping of the Poisson action of G on the symplectic manifold T^*N obtained by lifting the action of G on N. Then the composition of P_G and the Legendre transformation is identical with the momentum mapping $\mathcal{I}_G: TN \to \mathcal{G}^*$ on the Lagrangian system (N, L) relative to the group G, defined previously.

If the function $H: M \to R$ is invariant under a Poisson action of G, then, by Theorem 10, the momentum mapping P_G is a first integral of the Hamiltonian system with Hamiltonian H.

We conclude this section by a result on symmetries in Dirac's generalized mechanics. Let (M, ω^2, H, N) be a constrained Hamiltonian system with Hamiltonian $H: M \to R$ and N a submanifold of M (see Ch. 1, §5.1).

Theorem 11. *Suppose given a Poisson action of the Lie group G on the symplectic manifold (M, ω^2), such that G preserves both the function H and the submanifold N. Then the momentum P_G takes constant values on the motions of the constrained Hamiltonian system (M, ω^2, H, N).*

§2. Reduction of Systems with Symmetry

2.1. Lowering the Order (the Lagrangian Aspect). If the Lagrangian system (M, L) admits a symmetry group g^s then, as it turns out, the number of its degrees of freedom can be reduced. To the group g^s there corresponds the first integral $\mathcal{I}_g$ which locally is always cyclic. We first consider *the classical method of E.J. Routh for eliminating cyclic coordinates,* and only then we turn to lowering of the order in the large.

Suppose that the Lagrangian $L(\dot{q}, \dot{\lambda}, q)$ does not contain the coordinate λ. Using the equality $L'_{\dot{\lambda}} = c$ we express the cyclic velocity $\dot{\lambda}$ as a function of $\dot{q}$, q, and c. Following Routh, we consider the function

$$R_c(\dot{q}, q) = L(\dot{q}, \dot{\lambda}, q) - c\dot{\lambda}|_{\dot{q}, q, c}.$$

Theorem 12. *The function $(q(t), \lambda(t))$ is a motion of the Lagrangian system (M, L) with constant cyclic integral $\mathcal{I}_g = c$ if and only if $q(t)$ satisfies Lagrange's equation $[R_c] = 0$.*

If there are several cyclic coordinates $\lambda_1, \ldots, \lambda_k$, then for the *Routh function* we take $R_{c_1 \ldots c_k} = L - \sum c_j \dot{\lambda}_j$.

A small neighborhood U of an arbitrary nonsingular point of the symmetry field v is "regularly" foliated by orbits of the group g (integral curves of the vector field v): the quotient space $N = U/g$ is a smooth manifold with Cartesian coordinates q. It is natural to refer to the pair (N, R_c) as a *(local) reduced Lagrangian system.* An example of lowering the order in the sense of Routh is the elimination of the polar angle in Kepler's problem (see Ch. 2, §1.1).

Cyclic coordinates are not uniquely defined: among the new variables $Q=q$, $\Lambda=\lambda+f(q)$, the coordinate Λ is also cyclic. Let $\hat{L}(\dot{Q}, \dot{\Lambda}, Q)=L(\dot{q}, \dot{\lambda}, q)$, then obviously $\hat{L}'_{\dot{\Lambda}}=L'_{\dot{\lambda}}=c$. The Routh function corresponding to the new cyclic coordinate Λ is $\hat{R}_c(\dot{Q}, Q)=R_c(\dot{q}, q)+cf'_q\cdot\dot{q}$. In view of the identity $[\dot{f}]\equiv 0$, the term $c(f'_q\cdot\dot{q})$ does not affect the form of the equation $[R_c]=0$. But this means that for $c\neq 0$ the Routh functions is not uniquely determined. These observations are useful in the analysis of lowering the order in the large (i.e., globally), which is our next objective. For the sake of definiteness we shall treat the case of natural Lagrangian systems.

Let (M, N, pr, S, G) be a fibre bundle with total space M, base N, projection $pr\colon M\to N$ (the differential pr_* has rank $\dim N$ at all points of M), fiber S, and structure group G. The group G acts freely and transitively on S from the left. This action can be extended to a left action of G on M, for which all the orbits are diffeomorphic to S. In the case of principal bundles the manifold S is diffeomorphic to the group manifold G. The base N may be regarded as the quotient space of M by the equivalence relation determined by the action of G. The vectors v_X, $X\in\mathcal{G}$, tangent to the orbits of the group G are vertical: $pr_*(v_X)=0$.

Suppose now that G is a symmetry group of a natural mechanical system $(M, \langle\,,\rangle, V)$. We introduce in the bundle (M, N, pr, S, G) the "canonical" connection, for which the horizontal tangent vectors to M are those orthogonal in the metric $\langle\,,\rangle$ to all vectors v_X, $X\in\mathcal{G}$. This connection is compatible with the action of the structure group G, that is, the distribution of horizontal vectors is invariant under the action of G. A smooth path $\gamma\colon[t_1, t_2]\to M$ is said to be *horizontal* if the tangent vectors $\dot{\gamma}(t)$ are horizontal for all $t_1\leqq t\leqq t_2$. It is readily verified that for every smooth path $\tilde{\gamma}\colon[t_1, t_2]\to N$ and every point x_1 lying over $\tilde{\gamma}(t_1)$ (i.e., $pr(x_1)=\tilde{\gamma}(t_1)$), there exists a unique horizontal path $\gamma\colon[t_1, t_2]\to M$ which covers $\tilde{\gamma}$.

We endow the manifold $N=M/G$ with the "quotient metric" $\widetilde{\langle\,,\rangle}$, first restricting the given metric to the distribution of horizontal vectors, and then letting it descend to N. Since the force function $V\colon M\to R$ (or the potential $-V$) is constant on the orbits of G, there is a unique smooth function $\tilde{V}\colon N\to R$ such that the following diagram commutes:

$$\begin{array}{ccc} M & \xrightarrow{pr} & N \\ & {\scriptstyle V}\searrow \quad \swarrow{\scriptstyle \tilde{V}} & \\ & R & \end{array}$$

Theorem 13. *The motions of the natural system $(M, \langle\,,\rangle, V)$ with zero value of the momentum mapping $\mathcal{I}_G$ project in a one-to-one manner onto the motions of the reduced system $(N, \widetilde{\langle\,,\rangle}, \tilde{V})$.*

◁ Let $\tilde{\gamma}\colon[t_1, t_2]\to N$ be a motion of the reduced system, and let $\tilde{\gamma}_\alpha$ be a variation of $\tilde{\gamma}$ with fixed extremities. Let $\gamma_\alpha\colon[t_1, t_2]\to M$ be a horizontal lift of the path $\tilde{\gamma}_\alpha$, such that $\gamma_\alpha(t_1)=\gamma_0(t_1)$ for all α. The variation field u of the family of paths γ_α is such that $u(t_1)=0$, while $u(t_2)$ is a vertical vector. If L (respectively, $\tilde{L}$) is the Lagrangian of the original (respectively, reduced) system, then, according to the first variation formula,

$$\delta \int_{t_1}^{t_2} \tilde{L}\,dt=\delta \int_{t_1}^{t_2} L\,dt=\langle \dot{\gamma}_0, u\rangle\big|_{t_1}^{t_2}=0. \quad \triangleright$$

Example 10. Consider the motion of a point mass m in a central force field. In this problem we have the bundle $(R^3\backslash\{0\}, R^+, pr, S^2, SO(3))$; the projection $pr\colon R^3\backslash\{0\}\to R^+$ is given by the formula $(x, y, z)\mapsto\sqrt{x^2+y^2+z^2}$. The Lagrangian $L=m|\dot{r}|^2/2+V(|r|)$ admits the group $SO(3)$ of rotations around the point $x=y=z=0$. If the angular momentum $\mathscr{I}_{SO(3)}=0$, then on $R^+=\{s>0\}$ one obtains the one-dimensional reduced system with the Lagrangian $\tilde{L}=m\dot{s}^2/2+V(s)$. $\triangle$

Consider now reduction in the case $\mathscr{I}_G\neq 0$. We shall assume that the group G is commutative (this is the only case in which Routh's method works). We shall further assume that (M, N, pr, G) is a principal bundle; in particular, the action of G on M is free. In addition to the quotient metric $\widetilde{\langle\, ,\rangle}$ on the base, we shall need the *curvature form* of the canonical connection. We remind the reader its construction. We first introduce the connection 1-form ω on M with values in the Lie algebra $\mathscr{G}$, defined as follows: if $u\in TM$, then $\omega(u)=X$, where $X\in\mathscr{G}$ is such that v_X coincides with the vertical component of the vector u. In the case of a principal bundle, the kernel of the homomorphism $X\mapsto v_X$ of $\mathscr{G}$ into the Lie algebra of vector fields on M is equal to zero, and hence the connection form ω is correctly defined. If, for example, $\dim G=1$, we can put $\omega(u)=\langle u, v\rangle/\langle v, v\rangle$, where v is the symmetry vector field. The *curvature form* is, by definition, the $\mathscr{G}$-valued 2-form Ω such that $\Omega(u_1, u_2)=d\omega(u_1^\perp, u_2^\perp)$, where $u^\perp$ designates the horizontal component of the tangent vector u. Since G is a commutative symmetry group, the form Ω can be lowered to N. Let $\mathscr{I}_G=c\in\mathscr{G}^*$. Since Ω takes values in $\mathscr{G}$, the value of the momentum on the curvature form is meaningful: $\Omega_c=c\cdot\Omega$. Ω_c is a real-valued form on the base N. From *Cartan's structure equation*, $\Omega=d\omega+[\omega, \omega]$, it follows that the forms Ω and Ω_c are closed.

Lemma 2. *Let $c\in\mathscr{G}^*$. Then for every point $x\in M$ there is a unique vertical tangent vector $w_c\in T_xM$ such that $\mathscr{I}_G(w_c)=c$.*

$\triangleleft$ In fact, w_c can be characterized as the unique element of the set $\{w\in T_xM\colon \mathscr{I}_G(w)=c\}$ which minimizes the length in the metric $\langle\, ,\rangle$. This assertion holds true for an arbitrary group G. $\triangleright$

Definition. The *reduced*[21] *force function* of the natural system with symmetry group G corresponding to the constant value $\mathscr{I}_G=c$ of the momentum mapping is the function $V_c\colon M\to R$ equal to $V-\frac{1}{2}\langle w_c, w_c\rangle$.

[21] Translator's note. Other terms used are *amended* or *effective*.

Lemma 3. *The function V_c is invariant under the isotropy subgroup $G_c \subset G$ at the element $c \in \mathscr{G}^*$ under the coadjoint action of on $\mathscr{G}^*$ (see Proposition 1).*

Corollary. *If G is commutative, then the function V_c is constant on the orbits of G.*

This observation permits us to define correctly the reduced potential $\tilde{U}_c = -\tilde{V}_c$ as a function on the base N.

Theorem 14. *The path $\gamma: \Delta \to M$ is a motion of the natural system $(M, \langle\, , \rangle, V)$ with constant value $\mathscr{I}_G = c$ of the momentum mapping if and only if the projection $\mu = pr \circ \gamma: \Delta \to N$ satisfies the differential equation*

$$[L_c]_\mu = F_c(\dot{\mu}), \tag{6}$$

where

$$L_c = \tfrac{1}{2}\widetilde{\langle \dot{\mu}, \dot{\mu} \rangle} + \tilde{V}_c \quad \textit{and} \quad F_c(v) = \omega_c(\cdot, v).$$

Theorem 14 follows, for example, from Theorem 9.

Equation (6) may be regarded as the equation of motion of the natural system $(N, \widetilde{\langle\, , \rangle}, \tilde{V}_c)$ under the action of the supplementary nonconservative forces F_c. Since $F_c(v) \cdot v = \Omega_c(v, v) = 0$, the work of these forces in the real motion is zero. F_c are called *gyroscopic forces.*

Since the form Ω_c is closed, locally $\Omega_c = d\omega_c$. Therefore, (6) is the Lagrange equation $[R_c] = 0$ with $R_c = L_c - \omega_c$. The Routh function is defined globally on TN if and only if the form Ω_c is exact.

Example 11. Consider the rotation of a rigid body fixed at a point in an axially-symmetric force field. The kinetic energy and the potential admit the group $SO(2)$ of rotations around the symmetry axis of the field. In this problem M is diffeomorphic to the underlying manifold of the group $SO(3)$. The reduction $SO(3)/SO(2)$ was first accomplished by Poisson, as follows: let e be the unit vector of the symmetry axis of the field, regarded as a vector in body coordinates (i.e., in the moving reference frame attached to the body). The action of the subgroup $SO(2)$ on $SO(3)$ by right translations preserves e. The set of all positions of e in body coordinates is a two-dimensional sphere S^2, the "Poisson sphere". The points of S^2 "label" the orbits of the rotation group $SO(2)$. Thus, we have the fibre bundle $SO(3)$ with structure group $SO(2)$ and base S^2. The symmetry group $SO(2)$ generates a first integral, namely, the projection of the angular momentum of the body on the axis with direction vector e. Upon setting this first integral equal to a constant, we can pass from the original problem to the investigation of the reduced system with configuration space S^2. Here the Routh function is not globally defined, because the curvature form Ω is not exact: for arbitrary values of the principal moments of inertia

$$\int_{S^2} \Omega = 4\pi \neq 0.$$

Explicit formulas for lowering the order will be indicated below. △

The theory of lowering the order of Lagrangian systems carries over with obvious changes to nonholonomic mechanics. In order to effect the passage from a nonholonomic system to its quotient by a symmetry group it is necessary to assume that the constraints are invariant under the action of this group. As an example, consider Chaplygin's problem on a ball rolling on a horizontal plane (see Example 5). This problem admits the group $SO(2)$ of rotations of the ball around the vertical axis passing through its center. The group $SO(2)$ preserves the constraints, and its infinitesimal generator is a field of virtual velocities. As a matter of fact, we effected the reduction by the rotation group $SO(2)$ in Example 5, using Poisson's method.

In conclusion we mention also the problem of *hidden motions*, or the "problem of *action at distance*", which troubled physicists at the end of the 19th century. Suppose that a natural system with $n+1$ degrees of freedom moves by virtue of its own inertia and that its Lagrangian, given by the kinetic energy, admits a symmetry group with symmetry vector field v. Reducing the order of the system, we see that the Routh function, which is the Lagrangian of the reduced system with n degrees of freedom, contains a term – the reduced potential $\tilde{U}_c=\langle w_c, w_c\rangle/2=c^2/(2\langle v, v\rangle)$ – which does not depend on velocities. This term may be thought of as the potential of certain forces acting on the reduced system. Helmholtz, J.J. Thomson, and Hertz insisted that every mechanical quantity that manifests itself as a "potential energy" is in fact the kinetic energy of hidden motions under which only cyclic (hidden) coordinates vary. A typical example is the rotation of a symmetric top: since the motion of the top around its symmetry axis cannot be detected, we may perceive that the top does not rotate and explain its strange behavior by the fact that it is acted upon by supplementary conservative forces.

Since $U_c=\langle w_c, w_c\rangle/2>0$, Routh's method can yield only positive potentials. However, since the potential is defined to within an additive constant, this limitation is immaterial in the case where the configuration space is compact.

Theorem 15. *Let* $(M, \langle\ ,\ \rangle, V, \Omega)$ *be a mechanical system such that the gyroscopic force form* Ω *is closed. If* M *is compact, then there exists a principal fibre bundle with base* M *and symmetry structure group* T^k, *where* $k\leq \operatorname{rank} H^2(M, R)$, *such that, following Routh's lowering of the order,* $V_c=V+\text{const}$ *and* $\Omega_c=\Omega$, *for some constant value* $\mathcal{I}_{T^k}=c$ *of the momentum mapping.*

This result was communicated to us by S.V. Bolotin (see [65]).

If $\Omega=0$, then we may take as fibre bundle the direct product $M\times S^1\{\varphi \bmod 2\pi\}$, endowed with the metric $\langle \dot{x}, \dot{x}\rangle+\dot{\varphi}^2/U(x)$, where $\langle\ ,\ \rangle$ is the Riemannian metric on M. The coordinate φ is cyclic, with corresponding cyclic first integral $\dot{\varphi}/U=c$. The Routh function is given by $R_c=\langle \dot{x}, \dot{x}\rangle/2-c^2U/2$. For $c=2$ we get a natural system on $(M\times S^1)/S^1\simeq M$ with potential U.

2.2. Lowering the Order (the Hamiltonian Aspect). Let $F: M \to R$ be a first integral of the Hamiltonian system with Hamiltonian H.

Proposition 2. *If $dF(z) \neq 0$, then in some neighborhood of the point $z \in M$ there exist symplectic coordinates $x_1, \ldots, x_n, y_1, \ldots, y_n$ such that $F(x, y) = y_1$.*

This assertion is the Hamiltonian version of the theorem on straightening of trajectories.

In the coordinates x, y function H does not depend on x_1. Hence, if we fix a value $F = y_1 = c$, then

$$\dot{x}_k = H'_{y_k}, \quad \dot{y}_k = -H'_{x_k} \quad (k \geqq 2)$$

is a Hamiltonian system of equations with $n-1$ degrees of freedom. Thus, existence of a first integral F permits us to reduce the dimension of the *phase space* by two units: one unit drops on fixing the value $F = c$, and the second as a result of eliminating the cyclic variable x_1 along the orbits of the action of the symmetry group g_F^α. This observation can be generalized as follows: if a Hamiltonian system has p independent first integrals in involution, then it can be reduced to a system with $n-p$ degrees of freedom. We notice that in order to effectively use the first integral F to lower the number of degrees of freedom we must first find the orbits of the group g_F^α, which in turn is connected with the integration of the Hamiltonian system with Hamiltonian F.

If the algebra of first integrals is noncommutative, then the dimension of the given Hamiltonian system can be lowered by at least twice the maximal dimension of its commutative subalgebras. Sometimes the family of first integrals can be enlarged by considering nonlinear functions of first integrals.

Example 12. The equations of motion in a central field possess an algebra of first integral isomorphic to the Lie algebras $so(3)$. All its commutative subalgebras are one-dimensional. Let M_i designate the projection of the angular momentum on the i-th coordinate axis of an orthogonal system of Cartesian coordinates. It is readily checked that the functions M_1 and $M^2 = \sum M_i^2$ are independent and Poisson-commute. It follows that the problem in question reduces to the investigation of a Hamiltonian system with one degree of freedom. △

This method of lowering the order of Hamiltonian systems was invented by Poincaré, who applied it to various problems of celestial mechanics. It is essentially the Hamiltonian variant of the lowering method of Routh. If the algebra of first integrals is not commutative, Poincaré's method makes only partial use of the known integrals. This shortcoming of Poincaré's method was eliminated by E. Cartan, who studied the general case of an infinite-dimensional Lie algebra of first integrals (see [13]). Specifically, Cartan considers a Hamiltonian system (M, ω^2, H) with first integrals $F_1, \ldots, F_k$ such that $\{F_i, F_j\} = a_{ij}(F_1, \ldots, F_k)$. The set of integrals $F_1, \ldots, F_k$ yields a natural mapping $F: M \to R^k$. In the general case the functions $a_{ij}: R^k \to R$ are nonlinear.

Theorem 16 (S. Lie – E. Cartan). *Suppose that the point $c \in R^k$ is not a critical value of the mapping F and that in its neighborhood the rank of matrix (a_{ij}) is constant. Then in a small neighborhood $U \subset R^k$ of c one can find k independent functions $\varphi_j: U \to R$ such that the functions $\Phi_j = \varphi_j \circ F: N \to R$, where $N = F^{-1}(U)$, satisfy the relations*

$$\{\Phi_1, \Phi_2\} = \ldots = \{\Phi_{2q-1}, \Phi_{2q}\} = 1, \tag{7}$$

whereas the remaining brackets $\{\Phi_i, \Phi_j\}$ vanish. The number $2q$ equals the rank of the matrix (a_{ij}),

The proof can be found in [13]. Using this theorem we can readily lower the order. Suppose the point $c = (c_1, \ldots, c_k)$ satisfies the conditions of Theorem 16. Then, in particular, the level set $M_c = \{x \in M: \Phi_j(x) = c_j,\ 1 \leqq j \leqq k\}$ is a smooth $(2n-k)$-dimensional submanifold of M, where $2n = \dim M$. Since the functions $\Phi_{2q+1}, \ldots, \Phi_k$ commute with all the Φ_j, $1 \leqq j \leqq k$, their Hamiltonian fields are tangent to M_c. Therefore, there is an action of the commutative group R^l, $l = k - 2q$, on M_c, generated by the phase flows of Hamilton's equations with Hamiltonians Φ_j, $j > 2q$. Thanks to the functional independence of the integrals Φ_j, this action of R^l on M_c has no fixed points. If its orbits are compact (in which case they are l-dimensional tori), then the quotient space $M_c/R^l = \tilde{M}_c$ is a smooth manifold, called the *reduced phase space*. Since $\dim \tilde{M}_c = (2n-k) - l = 2(n-k+q)$, $\dim \tilde{M}_c$ is always even.

The reduced phase space is endowed with a natural symplectic structure ω^2, which can be given, for example, by means of a nondegenerate Poisson bracket $\widetilde{\{\,,\,\}}$. Let $A, B: \tilde{M}_c \to R$ be smooth functions. They can be lifted to smooth functions $'A, 'B$ defined on the level set $M_c \subset M$. Let $\hat{A}$ and $\hat{B}$ be arbitrary smooth functions of M whose restrictions to M_c coincide with $'A$ and $'B$, respectively. Finally, set $\widetilde{\{A, B\}} = \{\hat{A}, \hat{B}\}$.

Lemma 4. *The bracket $\widetilde{\{\,,\,\}}$ is correctly defined (i.e., does not depend on the way in which smooth functions are extended from $\tilde{M}_c$ to M) and is a Poisson bracket on M_c.*

Let $'H$ denote the restriction of the Hamiltonian H to the level set M_c of the first integrals. Since $'H$ is constant on the orbits of the group R^l, there is a smooth function $\tilde{H}: M_c/R^l \to R$ such that the diagram

$$\begin{array}{ccc} M_c & \xrightarrow{pr} & \tilde{M}_c \\ & {}_{'H}\searrow \quad \swarrow_{\tilde{H}} & \\ & R & \end{array}$$

commutes.

Definition. The Hamiltonian system $(\tilde{M}_c, \tilde{\omega}^2, \tilde{H})$ is called the *reduced system.*

Theorem 17. *A smooth path $\gamma: \Delta \to M$ with $F(\gamma(t))=c$ is a motion of the Hamiltonian system (M, ω^2, H) if and only if the composition $pr \circ \gamma: \Delta \to \tilde{M}_c$ is a motion of the reduced Hamiltonian system $(\tilde{M}_c, \tilde{\omega}^2, \tilde{H})$.*

◁ This result can be established arguing as follows: formulas (7) shows that functions $\Phi_1, \ldots, \Phi_k$ are part of a symplectic coordinate system in the neighborhood of the submanifold M_c. More precisely, in a neighborhood of every point of M_c one can introduce symplectic coordinates $x_1, \ldots, x_n$, $y_1, \ldots, y_n$, such that $x_i = \Phi_{2i-1}$, $y_i = \Phi_{2i}$ if $i \leqq q$ and $y_i = \Phi_i$, if $i > 2q$. This is a consequence of the known "completion lemma" due to Carathéodory (see [8]). Since Φ_j are first integrals, the Hamiltonian takes the form: $H(y, x) = H(y_{q+1}, \ldots, y_n, x_{k-q+1}, \ldots, x_n)$ in the variables x, y. It remains to fix the values of the cyclic integrals $y_{q+1}, \ldots, y_{k-q}$ and to observe that the variables x_j, y_j $(j > k-q)$ are local coordinates on $\tilde{M}_c$, in which the form $\tilde{\omega}^2$ takes the "canonical" form $\sum_{j>k-q} dx_j \wedge dy_j$. ▷

Remark. Since the $k-q$ first integrals $\Phi_2, \ldots, \Phi_{2q}, \Phi_{2q+1}, \ldots, \Phi_k$ commute, they can be used to reduce the order of the Hamiltonian system by Poincaré's method. The dimension of the local state space of the reduced system will be equal to $2n-2(k-q)$, i.e., to $\dim \tilde{M}_c$. Moreover, by Theorem 16, the reduction methods of Poincaré and Cartan give locally the same result; however, the factorization by Poincaré's method can be accomplished globally only under more restrictive assumptions. △

In degenerate situations the rank of the matrix of Poisson brackets $(\{F_i, F_j\})$ may, of course, be smaller. Reduction by Cartan's scheme can also be accomplished in such situations if it is further assumed that the integrals $F_1, \ldots, F_k$ generate a finite-dimensional algebra (i.e., the functions $a_{ij}: R^k \to R$ are linear). In fact, suppose we are given a Poisson action of G on the symplectic manifold (M, ω^2). Consider the preimage $M_c = P^{-1}(c)$ of a point $c \in \mathscr{G}^*$ under the momentum mapping $P: M \to \mathscr{G}^*$. If c is not a critical value of P, then M_c is a smooth submanifold of M. Since the action of G is a Poisson action, Proposition 1 shows that the elements of G map the integral manifolds M_c into one another. Let G_c be the isotropy (stationary) subgroup at the point $c \in \mathscr{G}^*$, which consists of all $g \in G$ such that $Ad_g^* c = c$. G_c is a Lie group and acts on M_c. If its orbits on M_c are compact, then the reduced phase space $\tilde{M}_c = M_c/G_c$ is a smooth manifold. One can now define the reduced Hamiltonian system $(\tilde{M}_c, \tilde{\omega}_c, \tilde{H})$ by replicating Cartan's reduction method. The relationship between the original and the reduced Hamiltonian systems is again described by Theorem 17. The necessary proofs can be found in the works of J.-M. Souriau [182] and of J. Marsden and A. Weinstein [132].

Example 13. The motion of a unit point mass in a central field can be described by the Hamiltonian system in the phase space $R^6 = R^3\{x\} \times R^3\{y\}$ with standard symplectic structure and the Hamiltonian $H(y, x) = |y|^2/2 + U(|x|)$. Fix a constant angular momentum vector $x \times y = \mu$ $(\mu \neq 0)$. We may assume that $\mu = ce_3$, where $e_3 = (0, 0, 1)$ and $c > 0$. The level set M_c is then given by the equations $x_3 = y_3 = 0$ and $x_1 y_2 - x_2 y_1 = c$. Clearly, the vector μ is invariant under the group $SO(2)$ of rotations around the axis with unit vector e_3. To effect reduction with respect to this group, we introduce in the plane $R^2 = \{x_1, x_2\}$ polar coordinates r, φ and the canonically-conjugate variables p_r, p_φ by

$$x_1 = r\cos\varphi, \qquad y_1 = p_r \cos\varphi - \frac{p_\varphi}{r}\sin\varphi,$$

$$x_2 = r\sin\varphi, \qquad y_2 = p_r \sin\varphi + \frac{p_\varphi}{r}\cos\varphi.$$

It is clear that in the new coordinates M_c is given by the equations $x_3 = y_3 = 0$ and $p_\varphi = c$. Passing to quotients by the group $SO(2)$ amounts to eliminating the angle variable φ. Consequently, the reduced phase space $\tilde{M}_c = M_c/SO(2)$ is diffeomorphic to $R^+\{r\} \times R\{p_r\}$; $\tilde{M}_c$ is endowed with the reduced symplectic structure $\tilde{\omega}^2 = dp_r \wedge dr$. The reduced Hamiltonian is $\tilde{H} = (p_r^2 + c^2 r^{-2})/2 + U(r)$. △

If the element $c \in \mathscr{G}^*$ is in general position (i.e., the matrix (a_{ij}) has maximal rank[22]), the isotropy group G_c is commutative, and reduction by the indicated scheme gives the same result as Cartan's method. If $c = 0$, then the rank of the matrix (a_{ij}) drops to zero, and the integral manifold M_0 has "maximal symmetry": the isotropy subgroup G_0 is identical with the full group G. In this case we have maximal lowering of the order of the given Hamiltonian system: the order drops by $2k = 2\dim G$ units (cf. Theorem 13).

Let $(N, \langle\,,\,\rangle, V)$ be a natural mechanical system, and suppose that it admits a compact commutative symmetry group G (isomorphic to T^k), which acts freely on the configuration space N. We may regard the given system as a Hamiltonian system with symmetry on $M = T^*N$, and apply the already familiar scheme for lowering the order. The action of G on N lifts to a Poisson action on T^*N; since the latter is free, every value $c \in \mathscr{G}^*$ of the momentum mapping is regular. Hence there exists the smooth invariant level manifold M_c (of codimension $k = \dim G$ in M) and the reduced state space $\tilde{M}_c$ ($\dim \tilde{M}_c = \dim M - 2k$). On the other hand, one can define the smooth configuration space $\tilde{N}$ as the orbit space of the action of G on N. Moreover, for the same value $c \in \mathscr{G}^*$ we have the "seminatural" reduced Lagrangian system $(\tilde{N}, \widetilde{\langle\,,\,\rangle}, \tilde{V}_c, \Omega_c)$ (see §2.1, Theorem 14). The function $L\colon TN \to R$, $\tilde{L}(\dot{x}) = \widetilde{\langle \dot{x}, \dot{x}\rangle}/2 + \tilde{V}_c(x)$, is naturally termed the *reduced Lagrangian*.

[22] In the case of a Poisson algebra of integrals it is more appropriate to speak about the rank of the bilinear form $\{F_X, F_Y\}$, $X, Y \in \mathscr{G}$.

Theorem 18. *For every* $c \in \mathscr{G}^*$ *there is a diffeomorphism* $f: \tilde{M}_c \to T^* \tilde{N}$ *such that:*

1) $f^* \tilde{\omega}^2 = \Omega + \Omega_c$, *where* Ω *is the standard symplectic form on* $T^* \tilde{N}$;

2) *the function* $f \circ \tilde{H}: T^* \tilde{N} \to R$ *is the Legendre transform of the reduced Lagrangian given by the metric* $\widetilde{\langle , \rangle}$.

Corollary. *The manifold* $\tilde{M}_0$ *is symplectomorphic (i.e., symplectically diffeomorphic) to* $T^* \tilde{N}$.

If the group G is not commutative, then in general the reduced phase space $\tilde{M}_c$ cannot be represented as the cotangent bundle of a smooth manifold.

Suppose given a free Poisson action of group G on a symplectic manifold (M, ω^2). In this case one can pass to the reduced manifold $(\tilde{M}_c, \tilde{\omega}^2)$ in the following alternate way: consider the quotient manifold $N = M/G$ and on it the bracket $'\{ , \}$ obtained by descent of the original Poisson bracket $\{ , \}$ to N. The bracket $'\{ , \}$ is obviously degenerate. If $P: M \to \mathscr{G}^*$ is the momentum mapping, there is a smooth mapping $\tilde{P}: N \to \mathscr{G}^*$ such that the diagram

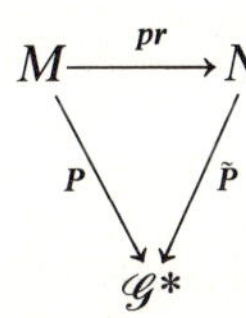

commutes. Since the action of G is free, $c \in \mathscr{G}^*$ is simultaneously a regular or critical value of the maps P and $\tilde{P}$. Assuming that $c \in \mathscr{G}^*$ is regular, consider the smooth manifold $N_c = \tilde{P}^{-1}(c)$ and restrict the bracket $'\{ , \}$ to N_c.

Proposition 3. *The restriction of the bracket* $'\{ , \}$ *to* N_c *defines a symplectic structure* $'\omega^2$; *moreover, the manifolds* $(\tilde{M}_c, \tilde{\omega}^2)$ *and* $(N_c, '\omega^2)$ *are symplectomorphic.*

This observation can be generalized to the case of a noncommutative group G if instead of taking the quotient of M by the entire group G we take the quotient by its center.

Example 14. For the motion of a rigid body with a fixed point, $M = TSO(3) = SO(3) \times R^3$. If the body rotates in an axially-symmetric force field, the system admits the one-parameter symmetry group $SO(2)$. The quotient manifold $M/SO(2)$ is diffeomorphic to $S^2 \times R^3$. The equations of motion on this five-dimensional manifold are usually written as the Euler-Poisson equations

$$\dot{k} + \omega \times k = V' \times e, \qquad \dot{e} + \omega \times e = 0, \qquad (|e| = 1)$$

where $k=A\omega$ is the angular momentum and $V\colon S^2\to R$ the force function (see Ch. 1, §1). The bracket $'\{\,,\}$ on $S^2\times R^3$ is given by the following formulas

$$'\{\omega_1,\omega_2\}=-\frac{A_3}{A_1A_2}\,\omega_3,\ \dots\qquad '\{\omega_1,e_1\}=0,$$
$$'\{\omega_1,e_2\}=-\frac{1}{A_1}\,e_3,\qquad '\{\omega_1,e_3\}=\frac{1}{A_1}\,e_2,\ \dots\qquad '\{e_i,e_j\}=0. \tag{8}$$

The Euler-Poisson equations possess the first integral $\langle k,e\rangle=c$, generated by the symmetry group $SO(2)$. We fix its value c and consider the level surface $N_c=\{\omega,e\colon\langle A\omega,e\rangle=c,\ \langle e,e\rangle=1\}$, which is obviously diffeomorphic to the (co)tangent bundle to the Poisson sphere $S^2=\{e\in R^3\colon\langle e,e\rangle=1\}$. Set $\omega={}'\omega+ce/\langle Ae,e\rangle$; $'\omega$ is a horizontal tangent vector with respect to the canonical connection defined in the principal bundle $(SO(3),\,S^2,\,SO(2))$ by the invariant Riemann metric $\langle A\omega,\omega\rangle/2$. The projection $SO(3)\to S^2$ permits us to regard horizontal vectors $'\omega$ as tangent vectors to the Poisson sphere. Let $\widetilde{\langle\,,\rangle}$ denote the quotient metric on S^2: $\widetilde{\langle 'a,'b\rangle}=\langle 'a,A'b\rangle$. The Lagrangian of the reduced system is obviously equal to

$$\tfrac{1}{2}\langle A\omega,\omega\rangle+V(e)=\tfrac{1}{2}\widetilde{\langle '\omega,'\omega\rangle}+V_c(e),$$

where $\tilde V_c=V-c^2/(2\langle Ae,e\rangle)$ is the reduced (amended) force function. The standard symplectic structure on T^*S^2 is given in the variables $'\omega, e$ by formulas (8). If $c\neq 0$ the reduced structure on T^*S^2 can also be defined using formulas (8), with the difference that in the right-hand sides one must add terms proportional to the constant c. △

2.3. Examples: Free Motion of a Rigid Body and the Three-Body Problem. We consider first Euler's problem on the rotation of a rigid body around a fixed point under its own inertia (i.e., in the absence of external forces, see Ch. 1, §2.4). Here $M=TSO(3)=SO(3)\times R^3$ and the symmetry group is the rotation group $SO(3)$. To G there corresponds a Poisson algebra of first integrals isomorphic to the Lie algebra $so(3)$. Fix a value $c\in\mathscr{G}^*\simeq R^3$ of the angular momentum and consider the integral level set $M_c=P_{SO(3)}^{-1}(c)$. It is readily verified that M_c is a three-dimensional manifold diffeomorphic to the group manifold $SO(3)$ for all values of c. The isotropy subgroup G_c is the one-dimensional group $SO(2)$ of rotations of the body in fixed space around the constant angular momentum vector. The reduced phase space $\tilde M_c=SO(3)/SO(2)$ is diffeomorphic to the two-dimensional sphere.

This reduction can be effected, for example, as follows: Since the Hamiltonian vector field on M admits the group G, it can be lowered to the quotient space $M/G\simeq R^3$. The differential equation on R^3 arising in this manner is Euler's equation

$$\dot k+\omega\times k=0,\qquad \text{where } \omega=A^{-1}k,$$

and it can be put in Hamiltonian form: $\dot{F}=\{F, H\}$, where $H=\langle k, \omega\rangle/2$ is the kinetic energy of the body, and the bracket $\{\ ,\ \}$ is defined by the equalities $\{k_1, k_2\}=-k_3$, $\{k_2, k_3\}=-k_1$, $\{k_3, k_1\}=-k_2$. This bracket is degenerate: indeed, function $F=\langle k, k\rangle$ commutes with all functions defined on $R^3=\{k\}$. However, we get a nondegenerate Poisson bracket by restricting $\{\ ,\ \}$ to the level surface $F=|c|^2$, diffeomorphic to the two-dimensional sphere S^2. The symplectic manifold S^2 carries the sought-for Hamiltonian system; its Hamiltonian is the total energy $\langle k, \omega\rangle/2$, restricted to S^2.

Let us describe the classical method of reducing Euler's problem to a Hamiltonian system with one degree of freedom, which uses special canonical variables. Let $oXYZ$ be a stationary trihedron with the origin at the fixed point, and let $oxyz$ be a moving coordinate system (whose axes are the principal axes of the body). The position of the rigid body in fixed space is specified by the three *Euler angles*: ϑ (the angle of nutation), made by the oZ and oz axes, φ (the angle of pure rotation), made by the ox axis and the line of intersection of the planes oxy and oXY (called the nodal line), and ψ (the angle of precession), made by the oX axis and the nodal line. On $SO(3)$ the angles ϑ, φ, ψ form a coordinate system similar to geographic coordinates on the sphere: it is singular at poles (where $\vartheta=0, \pi$) and multi-valued along a meridian. Let $p_\vartheta, p_\varphi, p_\psi$ be the momenta canonically conjugate to the coordinates ϑ, φ, ψ. If the rigid body rotates in an axially-symmetric force field with symmetry axis oZ, then the Hamiltonian does not depend on ψ. In this case reduction can be regarded as "elimination of the node", that is, elimination of the cyclic variable ψ that specifies the position of the nodal line in fixed space.

Now let us introduce "special canonical variables" L, G, H, l, g, h. To this end, we denote by Σ the plane passing through the point o and orthogonal to the angular momentum vector of the body. Then L is the projection of

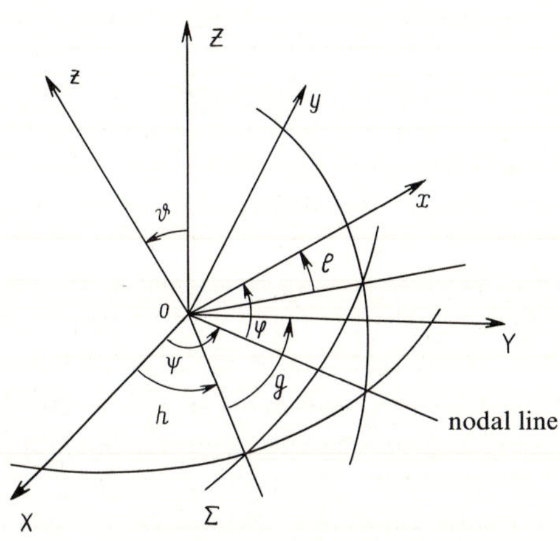

Fig. 20. Special canonical variables

the momentum on the oz axis, G the magnitude of the momentum, H the projection of the momentum on the oZ axis, l the angle made by the oz axis and the line of intersection of Σ with the plane oxy, g the angle made by the lines of intersection of Σ with the planes oXY and oxy, and h the angle between the oX axis and the line of intersection of Σ with the plane oXY.

Proposition 4. *The coordinate change $\vartheta, \varphi, \psi, p_\vartheta, p_\varphi, p_\psi \to l, g, h, L, G, H$ is a "homogeneous" canonical transformation:*

$$p_\vartheta\, d\vartheta + p_\varphi\, d\varphi + p_\psi\, d\psi = L\, dl + G\, dg + H\, dh.$$

This result is due to H. Andoyer; noncanonical variables analogous to the elements L, G, H, l, g, h were used by Poisson in his analysis of rotational motion of celestial bodies [48].

It readily follows from the definition of the special canonical variables that $A_1\omega_1 = \sqrt{G^2 - L^2}\sin l$, $A_2\omega_2 = \sqrt{G^2 - L^2}\cos l$, $A_3\omega_3 = L$. Consequently, in Euler's problem the Hamiltonian takes the form

$$\frac{1}{2}(A_1\omega_1^2 + A_2\omega_2^2 + A_2\omega_3^2) = \frac{1}{2}\left(\frac{\sin^2 l}{A_1} + \frac{\cos^2 l}{A_2}\right)(G^2 - L^2) + \frac{L^2}{2A_3}.$$

For a fixed value of the angular momentum G_0, the variables L and l vary in the annulus $|L| \leqq G_0$, $l \bmod 2\pi$. The level lines of the Hamiltonian are shown in Fig. 21. The curves $L = \pm G_0$ correspond to the singular points of the Euler equations, representing constant rotations of the body around the principal axis oz. The variables L, l may be naturally regarded as geographic symplectic coordinates on the reduced phase space S^2.

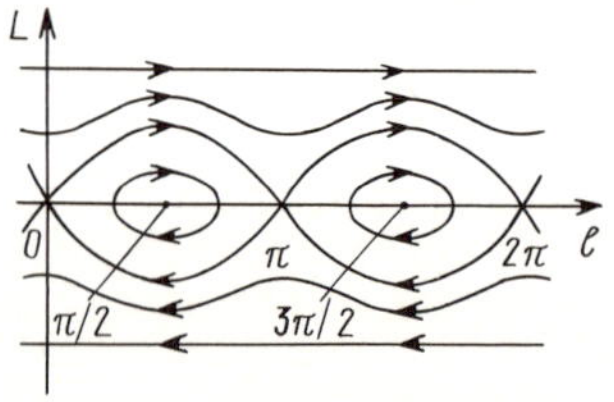

Fig. 21

We now consider from the viewpoint of reduction the three-body-problem, which has (in the spatial case) nine degrees of freedom. We show that using the six first integrals of momentum and angular momentum the equations governing the motion of three bodies interacting gravitationally can be reduced to a system with four degrees of freedom. Using also the energy integral we are led to concluding that the three-body problem reduces to the investigation of a dynamical system on a certain seven-dimensional manifold. In the case where the three bodies lie permanently in a fixed plane, the dimension of this manifold equals five. These results go back to Lagrange and Jacobi.

We pass to barycentric coordinates and use first the three-dimensional commutative group of translations to reduce the dimension of the Hamiltonian equations of motion from 18 to 12. The reduced system inherits the symmetry group $SO(3)$ from the original system. Fixing the value of the angular momentum, we obtain equations of motion on a nine-dimensional invariant manifold. On taking its quotient by the isotropy subgroup of rotations around the constant angular momentum vector, we get the required Hamiltonian system with eight-dimensional phase space. Now the problem is basically to effect the indicated reduction explicitly.

First let us eliminate the motion of the center of mass. Let r_k be the position vectors of the point masses m_k in a barycentric coordinate system, so that $\sum m_k r_k = 0$. In order to use this relation for lowering the order of the differential equations of motion

$$m_k \ddot{r}_k = V'_{r_k} \quad (1 \leqq k \leqq 3), \qquad V = \sum_{i<j} \frac{m_i m_j}{|r_{ij}|}, \tag{9}$$

we introduce the relative position vectors $\xi = r_2 - r_1$, $\eta = r_3 - \zeta$, where $\zeta = (m_1 r_1 + m_2 r_2)/(m_1 + m_2)$ is the center of mass of the points m_1 and m_2. Let $\mu = m_1 m_2/(m_1 + m_2)$ and $\nu = (m_1 + m_2) m_3/\sum m_k$.

Proposition 5. *If $r_k(t)$ is a motion of three point masses which interact gravitationally, then the functions $\xi(t)$ and $\eta(t)$ satisfy the equations*

$$\mu \ddot{\xi} = W'_{\xi}, \qquad \nu \ddot{\eta} = W'_{\eta}, \qquad W(\xi, \eta) = V|_{\xi, \eta}. \tag{10}$$

These equations have the first integral

$$\mu(\xi \times \dot{\xi}) + \nu(\eta \times \dot{\eta}) = \sum m_k (r_k \times \dot{r}_k) = c.$$

Equations (10) describe the motion of "fictitious" material points with masses μ and ν. Proposition 5 is readily generalized to the case of arbitrary $n > 3$. Equations (10) with 6 degrees of freedom are, needless to say, Hamiltonian.

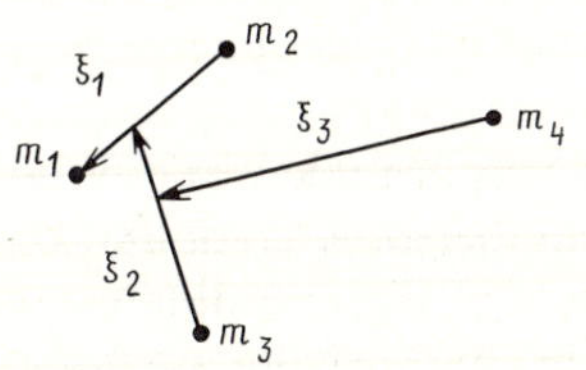

Fig. 22. Elimination of the center of mass in the n-body problem

Elimination of the angular momentum ("elimination of the node") can be carried out for equations (10). However, it is simpler to formulate the final result independently, in a form symmetric with respect to the masses

m_1, m_2, m_3. Let Σ be Laplace's "invariable plane": it contains the center of mass and is orthogonal to the constant angular momentum vector c. Let Π denote the plane passing through the points m_1, m_2, m_3. We let ϑ_i denote the exterior angle of triangle $m_1 m_2 m_3$ at the vertex m_i, and use Δ for the area of this triangle. Then

$$\sin\vartheta_i = \frac{2\Delta}{\rho_j\rho_k}, \qquad \cos\vartheta_i = \frac{\rho_i^2-\rho_j^2-\rho_k^2}{2\rho_j\rho_k}$$
$$\Delta = \frac{\prod(\rho_j+\rho_k-\rho_i)^{1/2}}{4\cdot\left(\sum\rho_i\right)^{-1/2}}, \tag{11}$$

where (i, j, k) runs through the cyclic permutations of $(1, 2, 3)$, and ρ_i denotes the length of the side opposite m_i. Let γ be the angle between the planes Π and Σ; for planar motions $\gamma\equiv 0$.

Proposition 6. *For a fixed value of the angular momentum:* $\sum m_k(r_k\times\dot r_k)=c$ *equations* (9) *reduce, in a barycentric coordinate system, to the following Hamiltonian system with four degrees of freedom:*

$$\dot\Gamma=-H'_\gamma, \quad \dot\gamma=H'_\Gamma, \quad \dot P_k=-H'_{\rho_k}, \quad \dot\rho_k=H'_{P_k} \quad (1\leqq k\leqq 3), \tag{12}$$

with

$$H(\Gamma, P_1, P_2, P_3, \gamma, \rho_1, \rho_2, \rho_3)=\frac{|c|^2\sin\gamma}{4\Delta}\sum\frac{\rho_i^2}{m_i}\sin^2\left(\frac{\Gamma}{|c|\sin\gamma}+\frac{\vartheta_j-\vartheta_k}{3}\right)+$$
$$+\sum\frac{P_j^2+P_k^2-2P_jP_k\cos\vartheta_i}{2m_i}+|c|\cos\gamma\sum\left(\frac{P_j}{\rho_k}-\frac{P_k}{\rho_j}\right)\frac{\sin\vartheta_i}{3m_i}+$$
$$+|c|^2\cos^2\gamma\sum\frac{\rho_j^2+\rho_k^2-\rho_i^2/2}{36m_i\rho_j^2\rho_k^2}-\sum\frac{m_jm_k}{\rho_i};$$

here the quantities Δ, ϑ_1, ϑ_2, *and* ϑ_3 *are given by formulas* (11) *as functions of* ρ_1, ρ_2, ρ_3, *and the symbol* $\sum f_{ijk}$ *designates the sum* $f_{123}+f_{231}+f_{312}$.

This result is due to van Kampen and Wintner [98]. The proof rests on elementary, though cumbersome computations. The expressions for the momenta Γ and P_s in terms of the coordinates and velocities of the moving masses are very complicated and usually are not employed.

In the case of planar motion, the first two equations of system (12) reduce to the equalities $\Gamma=\gamma=0$, and we obtain a Hamiltonian system with three degrees of freedom.

If $c=0$ equations (12) form a natural Hamiltonian system with three degrees of freedom (cf. Theorem 13).

§3. Relative Equilibria and Bifurcations of Invariant Manifolds

3.1. Relative Equilibria and the Amended Potential. We return to the investigation of a Hamiltonian system (M, ω^2, H) with symmetry group G, and assume that the action of G on the phase space M is Poisson. Let $(\tilde{M}_c, \tilde{\omega}^2, \tilde{H})$ be the reduced Hamiltonian system in the sense of §2.2.

Definition. The phase curves of the given Hamiltonian system on M with momentum constant $P_G = c$, which are taken by the projection $M \to \tilde{M}_c$ into equilibrium positions of the reduced Hamiltonian system, are called *relative equilibria* or *stationary motions* (of the original system).

Example 15. Consider rotations of a rigid body in an axially-symmetric force field. Let c be a fixed value of the angular momentum of the body relative to the symmetry axis of the field. The equations of motion of the reduced system can be written in the form

$$A\dot{\omega} = A\omega \times \omega - e \times V', \quad \dot{e} = e \times \omega; \quad \langle A\omega, e\rangle = c, \quad \langle e, e\rangle = 1, \tag{13}$$

where $V(e)$ is the force function. Obviously, at the equilibrium positions of the reduced system, $e = \text{const}$, and hence $\omega = \lambda e$. The multiplier λ is uniquely determined from the equation $\langle A\omega, e\rangle = c$: it equals $c/\langle Ae, e\rangle$. Since $e = \text{const}$, the angular velocity ω is also constant. From the first of equations (13) we obtain the equation that gives the relative equilibria with momentum c:

$$c^2(Ae \times e) + (V' \times e)\langle Ae, e\rangle^2 = 0, \qquad \langle e, e\rangle = 1.$$

This fact was first noted by O. Staude in 1894. In a stationary motion (relative equilibrium) the rigid body rotates uniformly with angular velocity $|\omega| = |c|/\langle Ae, e\rangle$ around the symmetry axis of the force field. △

Proposition 7. *The phase curve $x(t)$ of the Hamiltonian system (M, ω^2, H) with symmetry group G is a relative equilibrium if and only if $x(t) = g^t(x(0))$, where g^t is a one-parameter subgroup of G.*

◁ If $x(t) = g^t(x_0)$ and g^t is a subgroup of G, then the projection $M \to \tilde{M}_c$ takes the solution $x(t)$ into an equilibrium position of the reduced system. Conversely, let $x(t) = h^t(x_0)$ be a relative equilibrium of the Hamiltonian system with Hamiltonian H and initial condition $x(0) = x_0$. We claim that h^t is a subgroup of G. In fact, let g^t be a one-parameter subgroup of G such that

$$\left.\frac{d}{dt}\right|_{t=0} g^t(x_0) = \dot{x}(0) \left(= \left.\frac{d}{dt}\right|_{t=0} h^t(x_0)\right). \tag{14}$$

Since G is a symmetry group, the groups h^t and g^t commute, and hence, by (14), $x(t) = g^t(x_0)$. ▷

In Example 15 above the trajectories of stationary motions are orbits of the group $SO(2)$ of rotations around the symmetry axis of the force field.

For natural mechanical systems with symmetry one can indicate a more effective criterion for stationarity of motions. Let $(M, \langle\, , \rangle, V)$ be a natural mechanical system with symmetry group G (in the sense of §2.1): M is the total space of a smooth principal bundle with base N and structure group G.

Proposition 8. *If the symmetry group G is commutative, then $y \in N$ is a relative equilibrium position (i.e., the projection of a relative equilibrium on the base N) with momentum constant $c \in \mathscr{G}^*$ if and only if y is a critical point of the amended potential $\tilde{U}_c$: $N \to R$.*

This assertion is a consequence of Theorem 14 and of the definition of relative equilibria. Since, for instance, every smooth function on the sphere has at least two critical points, Proposition 8 has the following corollary:

Corollary. *A rigid body with a fixed point rotating in any axially-symmetric conservative force field has at least two distinct stationary rotations for every value of the angular momentum.*

In the general case the number of distinct stationary motions may be estimated using, for example, the Morse inequalities. However, in concrete problems one can usually obtain more exact information (see §§3.3–3.4).

3.2. Invariant Manifolds, Regions of Possible Motions, and Bifurcation Sets. Let (M, ω^2, H, G) be a Hamiltonian system with Poisson symmetry group G. Since the Hamiltonian H is a first integral, it is natural to combine it with the momentum integral P: $M \to \mathscr{G}^*$ and consider the smooth *"energy-momentum" mapping* $H \times P$: $M \to R \times \mathscr{G}^*$.

Definition. The *bifurcation set* Σ of the Hamiltonian system with symmetry (M, ω^2, H, G) is the set of points $(h, c) \in R \times \mathscr{G}^*$ such that the mapping $H \times P$ is not a trivial fibre bundle over any of the neighborhoods of (h, c).

In particular, Σ contains the set Σ' of critical values of the energy-momentum mapping. However, Σ is in general larger than Σ': an example is provided by the bifurcation set of Kepler's problem, exhibited in Ch. 2, §1.

Proposition 9. *The critical points of the mapping $H \times P$: $M \to R \times \mathscr{G}^*$ on a regular momentum level coincide with the relative equilibria.*

This simple result proves useful in studying the structure of bifurcation sets.

Definition. For fixed values of the energy $h \in R$ and of the momentum $c \in \mathscr{G}^*$, we call the set $I_{h,c} = (H \times P)^{-1}(h, c)$ an *invariant* (or *integral*) *manifold* of the Hamiltonian system (M, ω^2, H, G).

It is clear that the only values (h, c) for which the invariant levels $I_{h,c}$ are not necessarily smooth manifold are those in Σ. Since the action of G leaves the function H invariant, the isotropy subgroup G_c acts on $I_{h,c}$ and thus the quotient manifold $\tilde{I}_{h,c}=I_{h,c}/G_c$ is meaningful. If c is a regular value of the momentum mapping, then $\tilde{I}_{h,c}$ coincides with the energy level $\tilde{H}^{-1}(h)$ of the reduced Hamiltonian system $(\tilde{M}_c, \tilde{\omega}^2, \tilde{H})$. It is therefore natural to call the set $\tilde{I}_{h,c}$ a *reduced invariant* (or *integral*) *manifold*. For example, in the spatial three-body problem the typical manifolds $\tilde{I}_{h,c}$ are seven-dimensional, whereas in the planar problem they are five-dimensional. Since the map $H \times P$ is a fibration over each component of the set $R \times \mathcal{G}^* \setminus \Sigma$, the topological type of the invariant manifolds $\tilde{I}_{h,c}$ may change only when the point (h, c) crosses the bifurcation set Σ.

Thus, the study of the original Hamiltonian system with symmetry reduces to investigating the map $H \times P$ and the structure of the phase flows on the reduced invariant manifolds $\tilde{I}_{h,c}$.

Let us examine in more detail the structure of the "energy-momentum" mapping for a natural mechanical system $(M, \langle\,,\rangle, V)$ with symmetry group G; we shall not assume that the action of G is free. Let Λ denote the set of all points $x \in M$ for which the isotropy subgroup G_x (i.e., the set of all $g \in G$ such that $g(x)=x$) has positive dimension. The set Λ is closed in M. For example, in the spatial three-body problem Λ consists of the triplets of points lying on a line. In the planar problem Λ reduces to the point $r_1=r_2=r_3=0$ (as usual, we assume that the barycenter is the origin of the coordinate system).

Let $\mathcal{I}: \dot{x} \mapsto \langle \dot{x}, v_X \rangle$ be the momentum mapping. By Lemma 2, for every point $x \in M \setminus \Lambda$ and every $c \in \mathcal{G}^*$ there is a unique vector $w_c(x)$ such that $\mathcal{I}(w_c)=c$ and $\langle w_c, v_X \rangle = 0$ for all $X \in \mathcal{G}$. Recall that the amended potential $U_c: M \to R$ was defined in §2.1 by the formula $-V+\langle w_c, w_c \rangle/2$.

Proposition 10. *The amended potential has the following properties:*

1) $U_c(x)=\min\limits_{v \in \mathcal{I}_x^{-1}(c)} H(v)$, *where* $H(v)=\dfrac{\langle v, v\rangle}{2}-V(x)$ *is the total energy of the system and* $\mathcal{I}_x=\mathcal{I}|_{T_xM}$;

2) *on* $M\setminus\Lambda$ *the set of critical points of the function* $H|_{\mathcal{I}^{-1}(c)}: \mathcal{I}^{-1}(c) \to R$ *coincides with* $w_c(\Gamma)$, *where* Γ *denotes the set of critical points of the amended potential* $U_c: M\setminus\Lambda \to R$;

3) $\Sigma' = \{(h, c): h \in U_c(\Gamma)\}$;

and

4) $\pi(I_{h,c})=U_c^{-1}(-\infty, h]$, *where* $\pi: TM \to M$ *is the canonical projection.*

This proposition was formulated by S. Smale; in concrete situations it was used earlier by various authors. Assertion 2) sharpens Proposition 9.

Definition. The set $\pi(I_{h,c}) \subset M$ is called the *region of possible motions* for fixed values of energy h and momentum c.

If the group G is commutative, assertion 4) of Proposition 10 can be replaced by

4′) *$\pi'(\tilde{I}_{h,c}) \subset \tilde{U}_c^{-1}(-\infty, h]$, where $\pi': TN \to N$ is the canonical projection, $N = M/G$ is the reduced configuration space, and $\tilde{U}_c: N \to R$.*

If M is compact, then $\Sigma = \Sigma'$ and inclusion 4′) becomes an equality. In the noncompact case this is no longer true: a counterexample is provided by the n-body problem in three-dimensional space. It is interesting to note that in the planar n-body problem the region of possible motions is described by the inequality $U_c \leqq h$ (Proposition 7 in Ch. 1, § 14).

3.3. The Bifurcation Set in the Planar Three-Body Problem

Proposition 11. *For arbitrary values of the masses in the planar three-body problem:*

(1) *the set of critical values Σ' of the mapping $H \times \mathcal{I}: TM \to R^2\{h, c\}$ is the union of four cubic curves given by equations $hc^2 = \alpha_i < 0$ $(1 \leqq i \leqq 4)$;*

(2) *the bifurcation set Σ is the union of Σ' and the coordinate axes $h = 0$ and $c = 0$.*

◁ If U is the potential energy in the three-body problem, then the amended potential U_c is obviously equal to $U + c^2/2I$, where I is the moment of inertia of the mass points with respect to the barycenter (cf. Ch. 1, § 1). At a relative equilibrium, dU_c is proportional to dI, hence the three points from a central configuration (Ch. 2, § 3.1). For a fixed value $|c| \neq 0$ there are exactly five such configurations: three collinear and two triangular. In the case of the latter, the triangle is necessarily equilateral, and these two equilateral triangle configurations differ only in the order of the point masses at the vertices. Let ω be the constant angular velocity of rotation of a central configuration. Then obviously $|c| = I|\omega|$, $T = I\omega^2/2$, and

$$h = T + U = \frac{c^2}{2I} + U.$$

Since all the configurations of a given type are similar, we may assume that $I = \alpha^2 I_0$ and $U = \alpha^{-1} U_0$. The similarity ratio α can be found from the equality $2T = U$, which in turn is a consequence of Lagrange's identity $\ddot{I} = 2T - U$. This yields $\alpha = c^2/I_0 U_0$, so that at a relative equilibrium $hc^2 = \alpha_k < 0$. By assertion 2), Proposition 10, the bifurcation set Σ contains the curves with equations $hc^2 = \alpha_k$ $(1 \leqq k \leqq 5)$. Among the five numbers $\alpha_1, \ldots, \alpha_5$ at least two must coincide (namely, those corresponding to Lagrange's equilateral solutions). It is clear that the bifurcation set Σ also contains the lines $h = 0$ and $c = 0$ (as in Kepler's problem). As Smale showed, Σ does not contain any other points (see [37]). ▷

Smale's work [37] offers information on the topological structure of the invariant manifolds for (h, c) belonging to various components of the set $R^2 \setminus \Sigma$.

3.4. Bifurcation Sets and Invariant Manifolds in the Motion of a Heavy Rigid Body with a Fixed Point. Let $A_1 \geqq A_2 \geqq A_3$ be the principal moments of inertia of the rigid body and let x_1, x_2, x_3 be the coordinates of the center of mass relative to the principal axes. If ω is the angular velocity of the body and e the vertical unit vector (both given in body coordinates), then $H = \langle A\omega, \omega\rangle/2 + \varepsilon\langle x, e\rangle$ and $\mathscr{I} = \langle A\omega, e\rangle$, where $A = \operatorname{diag}(A_1, A_2, A_3)$. Our task is to describe the bifurcation diagram Σ in the plane $R^2 = \{h, c\}$, and the topological structure of the reduced invariant manifolds $\tilde{I}_{h,c}$. It is useful to examine first the particular degenerate case where $\varepsilon = 0$ (Euler's problem). The relative equilibrium positions are the critical points of the reduced potential $\tilde{U}_c = c^2/(2\langle Ae, e\rangle)$ on the unit sphere $\langle e, e\rangle = 1$. If the body is asymmetric $(A_1 > A_2 > A_3)$, there are exactly six such points: $(\pm 1, 0, 0)$, $(0, \pm 1, 0)$, $(0, 0, \pm 1)$. Corresponding to these are uniform rotations of the rigid body around its principal axes. Since at a relative equilibrium $\omega = ce/\langle Ae, e\rangle$ (see Example 15), the energy h and the momentum c are related by one of the equalities $h = c^2/2A_j$ $(1 \leqq j \leqq 3)$. Since the configuration space of the rigid body (the group $SO(3)$) is compact, the bifurcation set is the union of the three indicated parabolas. In the case of dynamic symmetry the number of parabolas is smaller; if $A_1 = A_2 = A_3 = A$, then Σ coincides with the parabola $h = c^2/2A$. Let $B_{h,c} = \{\tilde{U}_c < h\}$ be the region of possible motions on the Poisson sphere. The classification of the regions $B_{h,c}$ and of the reduced invariant manifolds $\tilde{I}_{h,c}$ in Euler's problem is given by the following proposition:

Proposition 12. *Let $A_1 > A_2 > A_3$. Then:*
(1) *if $h < c^2/2A_1$, then $B_{h,c} = \emptyset$ and $\tilde{I}_{h,c} = \emptyset$;*
(2) *if $c^2/2A_1 < h < c^2/2A_2$, then $B_{h,c} = D^2 \cup D^2$ and $\tilde{I}_{h,c} = 2S^3$;*
(3) *if $c^2/2A_2 < h < c^2/2A_3$, then $B_{h,c} = D^1 \times S^1$ and $\tilde{I}_{h,c} = S^2 \times S^1$;*
(4) *if $c^2/2A_3 < h$, then $B_{h,c} = S^2$ and $\tilde{I}_{h,c} = SO(3)$.*

The description of the topological structure of reduced invariant manifolds is based on the following remark: $\tilde{I}_{h,c}$ is diffeomorphic to the fibre bundle with base $B_{h,c}$ and fiber S^1, in which the whole fibre over each point x of the boundary $\partial B_{h,c}$ is identified with x.

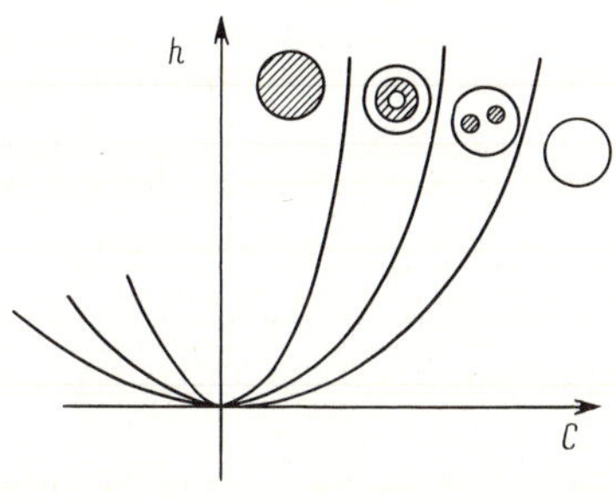

Fig. 23. Bifurcation diagram in Euler's problem

In the general case, where the center of mass does not coincide with the fixed point, the complete description of bifurcation sets and invariant manifolds becomes a considerably more difficult task. This problem is studied in detail in works of Ya.V. Tatarinov [185]. To illustrate we give a number of pictures, taken from [185], which demonstrate the mechanism of the metamorphoses that the bifurcation diagram undergoes when the center of mass shifts from a general position in the plane $x_3=0$ to the axis $x_1=x_2=0$. The numbers in these pictures indicate the "multi-valued genus" of the regions of possible motions on the Poisson sphere. More precisely, we say that the connected region $B_{h,c}$ has genus l if $B_{h,c}$ is diffeomorphic to the sphere S^2 with l disjoint discs removed. If the region of possible motions is disconnected, we assign to it the multi-valued genus $l_1, l_2, \ldots$, where each l_k is the genus of one of its components (since in the case under consideration $l_k \leqq 3$ for all k, there is no danger of confusion). The topology of the invariant manifolds in uniquely determined by the structure of the region of possible motions (i.e., by its genus).

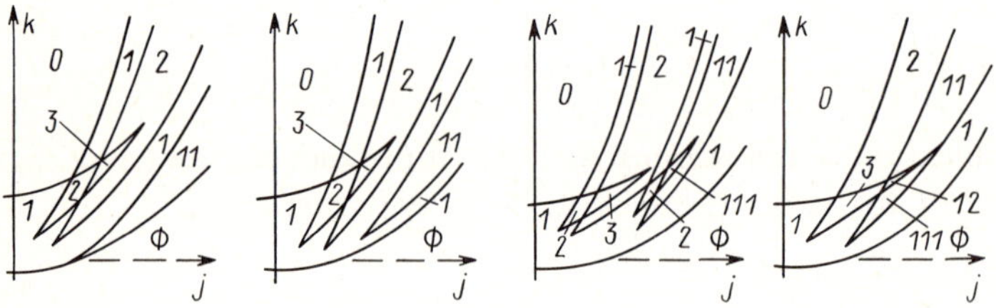

Fig. 24

Chapter 4
Integrable Systems and Integration Methods

§1. Brief Survey of Various Approaches to the Integrability of Hamiltonian Systems

Differential equations, including Hamiltonian equations, are customarily classified into integrable and nonintegrable. "When, however, one attempts to formulate a precise definition of integrability, many possibilities appear, each with a certain intrinsic theoretic interest" (D. Birkhoff, Dynamical Systems). In this section we give a brief survey of various approaches to integrability of Hamiltonian systems, "not forgetting the dictum of Poincaré, that a system of differential equations is only more or less integrable" (ibid.).

1.1. Quadratures. Integration by quadratures of a system of differential equations in R^n is the search for its solutions by a finite number of "algebraic" operations (including inversion of functions) and "quadratures", i.e., calculation of integrals of known functions. The following result connects the integration by quadratures of Hamiltonian systems with the existence of a sufficiently rich set of first integrals.

Theorem 1 (V.V. Kozlov and N.N. Kolesnikov [21]). *Regard R^{2n} as a symplectic manifold with the standard symplectic structure. Suppose that the system with Hamiltonian $H: R^{2n} \times R\{t\} \to R$ has n first integrals $F_1, \ldots, F_n: R^{2n} \times R\{t\} \to R$ (i.e., $F'_{it} + \{F_i, H\} = 0$), such that $\{F_i, F_j\} = \sum c^k_{ij} F_k$, $c^k_{ij} = \text{const}$. If:*

1) *on the set $M_f = \{(x, t) \in R^{2n} \times R: F_i(x, t) = f_i,\ 1 \leqq i \leqq n\}$ the functions $F_1, \ldots, F_n$ are independent;*

2) *$\sum c^k_{ij} f_k = 0$ for all $i, j = 1, \ldots, n$;*

3) *the Lie algebra $\mathscr{A}$ of linear combinations $\sum \lambda_i F_i$, $\lambda_i \in R$, is solvable, then the solutions of the Hamiltonian system $\dot{x} = \mathfrak{I}\, dH$ that lie on M_f can be found by quadratures.*

Corollary. *If a Hamiltonian system with n degrees of freedom has n independent first integrals in involution (i.e., the Lie algebra $\mathscr{A}$ is commutative) then it can be integrated by quadratures.*

This assertion was first proved by E. Bour for autonomous canonical systems and subsequently generalized by J. Liouville to the nonautonomous case.

Suppose that functions H and F_i are independent of time. Then H is also a first integral, say $H=F_1$. The theorem of integration by quadratures is, of course, valid in this case too, and the condition $\{F_1, F_i\}=0$ can be replaced by the weaker condition $\{F_1, F_i\}=c^1_{1i}F_1$, $1\leqq i\leqq n$.

The proof of Theorem 1 is based on the following result of S. Lie:

Theorem 2. *Suppose that the vector fields $X_1, \ldots, X_n$ are linearly independent in a small domain $U\subset R^n\{x\}$ and generate a Lie algebra under commutation, and also that $[X_1, X_i]=\lambda_i X_1$. Then the differential equation $\dot{x}=X_1(x)$ can be integrated by quadratures in U.*

◁ We prove this assertion in the simplest case $n=2$. The proof in the general case is analogous (see [74]). The equation $\dot{x}=X_1(x)$, $x\in U$, can be integrated if we find a first integral $F(x)$ for which $dF(x)\neq 0$ in the domain U. Since $X_1(x)\neq 0$, $x\in U$, such a function obviously exists (at least locally). If $X_1(F)=0$, then $X_2(F)$ is again a first integral, because $X_1(X_2(F))=X_2(X_1(F))+\lambda_2 X_1(F)=0$. Obviously, $X_2(F)=f(F)$ locally, where $f(\cdot)$ is a smooth function of one variable, $f\neq 0$. Set

$$G(F)=\int_{F_0}^{F}\frac{dz}{f(z)}.$$

Since $X_1(G)=0$ and $X_2(F)=dG(X_2(F))=X_2(F)/f(F)=1$, the system of equations

$$X_1(F)=a_{11}\frac{\partial F}{\partial x_1}+a_{12}\frac{\partial F}{\partial x_2}=0, \qquad X_2(F)=a_{21}\frac{\partial F}{\partial x_1}+a_{22}\frac{\partial F}{\partial x_2}=1$$

has a solution. Solving this system with respect to F'_{x_1} and F'_{x_2} we obtain F through an additional integration. Since $X_2(F)=1$, it follows that $dF\neq 0$. ▷

To prove Theorem 1 (in the autonomous case) we consider the n Hamiltonian flows $\mathfrak{J}\,dF_i$. In view of conditions 1 and 2, they are tangent to the manifold M_f and linearly independent everywhere on M_f. Since $\{F_i, F_j\}=\sum c^k_{ij}F_k$, it follows that $[\mathfrak{J}\,dF_i, \mathfrak{J}\,dF_j]=\sum c^k_{ij}\mathfrak{J}\,dF_k$. Hence, the tangent fields $\mathfrak{J}\,dF_i$ form a solvable algebra; moreover, $[\mathfrak{J}\,dH, \mathfrak{J}\,dF_i]=\lambda_i\mathfrak{J}\,dH$ (with $\lambda_i=c^1_{1i}=\text{const}$). Theorem 1 now follows from Lie's Theorem 2.

Remark. Theorem 2 in turn can be deduced from Theorem 1. To this end, consider the n functions $F_i(x, y)=y\cdot X_i(x)$, $i=1, \ldots, n$, defined on $U\times R^n$. If $[X_i, X_j]=\sum c^k_{ij}X_k$, then obviously $\{F_i, F_j\}=\sum c^k_{ij}F_k$. The manifold $M_0=\{(x, y)\colon F_1=\ldots=F_n=0\}=\{(x, y)\colon y=0\}$ is invariant under the Hamiltonian system with Hamiltonian F_1. Theorem 2 now follows upon applying the autonomous variant of Theorem 1 on M_0 and identifying M_0 with U. △

The nonautonomous theorem can be deduced from the autonomous version by the following device: Hamilton's equations

$$\dot{x}=H'_y, \quad \dot{y}=-H'_x; \qquad H=H(x,y,t)$$

can be written in the extended phase space of variables x, y, h, t as a canonical system with the Hamiltonian $K(x,y,h,t)=H(x,y,t)-h$:

$$\dot{x}=K'_y, \quad \dot{y}=-K'_x, \quad \dot{h}=K'_t, \quad \dot{t}=-K'_h.$$

Denoting the Poisson bracket in the extended symplectic space $R^{2n}\{x,y\}\times R^2\{h,t\}$ by $\{\ ,\ \}_*$, we have

$$\{F_i(x,y,t), F_j(x,y,t)\}_*=\{F_i,F_j\}=\sum c_{ij}^k F_k,$$

and

$$\{F_i(x,y,t), K(x,y,h,t)\}_*=\{F_i, H-h\}=\frac{\partial F_i}{\partial t}+\{F_i,H\}=0.$$

It remains to observe that the functions $F_1, \ldots, F_n$ and K are independent.

Example 1. Consider the motion on a line of three point masses which attract one another with a force inverse proportional to the cube of the mutual distances. Let m_i be the masses, x_i the coordinates, and $p_i=m_i\dot{x}_i$ the momenta of the points. The potential energy of interaction equals

$$U=\sum_{i<j}\frac{a_{ij}}{(x_i-x_j)^2}, \qquad a_{ij}=\text{const}.$$

Functions $F_1=\sum p_i^2/2m_i+U$, $F_2=\sum p_i x_i$, and $F_3=\sum p_i$ are independent and $\{F_1,F_3\}=0$, $\{F_2,F_3\}=F_3$, $\{F_1,F_2\}=2F_1$. Since the corresponding Lie algebra $\mathscr{A}$ is solvable, the motions lying on the zero levels of the total energy and momentum can be found by quadratures. This is readily achieved directly. If the points have equal masses and the coefficients a_{ij} $(i<j)$ are all equal, then the equations of motion can be integrated by quadratures globally (see Example 8, Ch. 4). We note that in the example considered here the potential U can be replaced by an arbitrary homogeneous function of degree -2. $\triangle$

1.2. Complete Integrability. Let M be a symplectic manifold and $F_1, \ldots, F_n$ independent functions on M which generate a finite-dimensional subalgebra of the Lie algebra $C^\infty(M)$ (so that $\{F_i,F_j\}=\sum c_{ij}^k F_k$ with $c_{ij}^k=\text{const}$). At each point $x\in M$ where the functions $F_1, \ldots, F_n$ are independent, the vectors $\mathfrak{I}\,dF_i$ span an n-dimensional linear subspace $\Pi(x)$ of T_xM. The distribution of planes $\Pi(x)$ is involutive (i.e., if $X(x), Y(x)\in\Pi(x)$, then $[X(x),Y(x)]\in\Pi(x)$). Therefore, by Frobenius' theorem, through every point $x\in M$ there passes a maximal integral manifold N_x of distribution Π. These manifolds are sometimes immersed in M in a rather complicated way; for example, they are not necessarily closed. If $n=\dim M/2$, then among the integral manifolds of Π there are the closed surfaces $M_f=\{x\in M\colon F_i(x)=f_i, \sum c_{ij}^k f_k=0\}$. If $x\in M_f$, then N_x coincides with one of the components of M_f. In the particular case where the functions $F_1, \ldots, F_n$ pairwise commute, almost the entire manifold M is "foliated" into closed invariant manifolds M_f.

Theorem 3. *Suppose the smooth functions $F_1, \ldots, F_n: M \to R$ are (pairwise) in involution and* $\dim M = 2n$. *If*:

1) *$F_1, \ldots, F_n$ are independent on M_f,*

and

2) *the Hamiltonian vector fields $\mathfrak{I}\, dF_i$ $(1 \leqq i \leqq n)$ are complete on M_f (i.e., every integral curve of $\mathfrak{I}\, dF_i$ is defined on the entire real axis), then:*

a) *every component of M_f is diffeomorphic to $T^k \times R^{n-k}$;*

b) *there are coordinates $\varphi_1, \ldots, \varphi_k \bmod 2\pi$, $y_1, \ldots, y_{n-k}$ on $T^k \times R^{n-k}$ in which Hamilton's equation $\dot{x} = \mathfrak{I}\, dF_i$ on M_f takes the form*

$$\dot{\varphi}_m = \omega_{mi}, \quad \dot{y}_s = c_{si} \qquad (\omega, c = \text{const}).$$

Each of the Hamiltonian systems with Hamiltonians $F_1, \ldots, F_n$ is said to be *completely integrable.*

◁ We will merely sketch the proof of this theorem. Consider the n one-parameter groups $g_i^{t_i}$, $t_i \in R$, which are the phase flows of the n Hamiltonian vector fields $\mathfrak{I}\, dF_i$. By condition 2), the value $g_i^{t_i}(x)$, $x \in M_f$, is defined for all t_i. The groups g_i and g_j commute, since $\mathfrak{I}\, dF_i$ and $\mathfrak{I}\, dF_j$ commute on M_f. It follows that an action of the Abelian group $R^n = \{t_1, \ldots, t_n\}$ is defined on M_f by

$$g^t(x) = g_1^{t_1} \ldots g_n^{t_n}(x), \qquad t = (t_1, \ldots, t_n).$$

From condition 1) and the connectedness of M_f one can deduce that the action of R^n on M_f is free and transitive. Consequently, M_f is diffeomorphic to the quotient manifold R^n/Γ, where Γ is the isotropy subgroup of R^n at some point $x \in M_f$ (i.e., $\Gamma = \{t \in R^n,\ g^t x = x\}$). Since the tangent vector fields $\mathfrak{I}\, dF_i$ are linearly independent on M_f, Γ is a discrete subgroup of R^n, isomorphic to Z^k $(0 \leqq k \leqq n)$. Therefore, $M_f \simeq R^n/Z^k = T^k \times R^{n-k}$. The uniformly varying "global" coordinates $\varphi \bmod 2\pi$ can be expressed as linear functions of $t_1, \ldots, t_n$. The details of the proof can be found in [6]. ▷

Remark. If the algebra $\mathscr{A}$ of first integrals is not commutative, then the closed invariant levels M_f are diffeomorphic to the quotient of the simply-connected Lie group with algebra $\mathscr{A}$ by some discrete subgroup. However, the implementation of this general remark rests on the yet unsolved problem of classifying Lie groups and algebras. △

In the theory and practice of completely integrable systems the most interesting case is when the level set M_f is compact. Then $k = n$ and $M_f \simeq T^n$. The uniform motion on the *torus* $T^n = \{\varphi_1, \ldots, \varphi_n \bmod 2\pi\}$ according to the law $\varphi_i = \varphi_i^0 + \omega_i t$ $(1 \leqq i \leqq n)$ is called *conditionally-periodic* (or *quasi-periodic*), and the number $\omega_1, \ldots, \omega_n$ are its *frequencies.* A torus carrying a motion with frequencies $\omega_1, \ldots, \omega_n$ is called *nonresonant* if the frequencies are *indepen-*

dent (over the field of rationals), i.e., if from the equality $\sum k_i\omega_i=0$ with k_i integers it follows that all $k_i=0$. On nonresonant tori the phase trajectories are everywhere dense. This is a simple consequence of the following general result of H. Weyl:

Theorem 4. *Suppose $f\colon T^n \to R$ is a Riemann-integrable function and the numbers $\omega_1, \ldots, \omega_n$ are independent over the rationals. Then for every point $\varphi^0 \in T^n$ the limit*

$$\lim_{s\to\infty} \frac{1}{s} \int_0^s f(\omega t+\varphi_0)\, dt$$

exists and equals

$$\frac{1}{(2\pi)^n} \int_{T^n} f(\varphi)\, d\varphi_1 \wedge \ldots \wedge d\varphi_n.$$

Suppose, in particular, that f is the indicator function of a Jordan-measurable domain $D \subset T^n$. Applying Theorem 4 to f, we obtain the following assertion: *the mean sojourn time of a phase trajectory in the domain D is proportional to the measure of D*. This indicates the uniform distribution of trajectories on nonresonant tori. If the torus is *resonant*, then the phase trajectories fill out densely tori of smaller dimension.

1.3. Normal Forms. Consider the Hamiltonian system

$$\dot{z}=\mathfrak{I}\, dH(z), \qquad z=(p, q)\in R^{2n},$$

in the neighborhood of the point $z=0$. Suppose that the real-analytic function H is represented by a convergent power series in z which starts with terms of order two: $H=\sum_{k\geqq 2} H_k$. Then the point $z=0$ is obviously an equilibrium position.

The eigenvalues of the linearized system $\dot{z}=\mathfrak{I}\, dH_2$ can be of the following four types: real pairs $(a, -a)$, $a\neq 0$; purely imaginary pairs $(ib, -ib)$, $b\neq 0$; quadruplets $\pm a \pm ib$, $a\neq 0$, $b\neq 0$; and zero as a multiple eigenvalue (see Ch. 6, §2.3). If eigenvalues of the first or third kind are present, the equilibrium position $z=0$ is evidently unstable. We shall assume that the eigenvalues of the linearized system are purely imaginary and distinct. One can show that in this case there is a linear canonical coordinate transformation $p, q \mapsto x, y$ which reduces the quadratic term H_2 to the form

$$\tfrac{1}{2}\sum \alpha_s(x_s^2+y_s^2).$$

The eigenvalues are precisely $\pm i\alpha_1, \ldots, \pm i\alpha_n$.

Theorem 5 (G.D. Birkhoff). *If $\alpha_1, \ldots, \alpha_n$ are independent over the field of rationals, then there is a formal canonical transformation $x, y \mapsto \xi, \eta$, given by formal power series*

$$x=u(\xi, \eta)=\xi+\ldots, \qquad y=v(\xi, \eta)=\eta+\ldots, \tag{1}$$

taking $H(x, y)$ into a Hamiltonian $K(\rho)$ which is a formal power series in the variables $\rho_s=\xi_s^2+\eta_s^2$.

If series (1) converge, then the equations with Hamiltonian H are readily integrated. In fact, the functions $\rho_1, \ldots, \rho_n$, which are convergent power series in the variables x, y, form a complete set of independent first integrals in involution. It follows from the canonical equations

$$\dot{\xi}_s=\Omega_s\eta_s, \quad \dot{\eta}_s=-\Omega_s\xi_s, \quad \Omega_s=2K'_{\rho_s},$$

that $\xi_s(t)$ and $\eta_s(t)$ are linear combinations of $\sin\Omega_s t$ and $\cos\Omega_s t$. Therefore, the original coordinates x and y are *conditionally-periodic functions* of time with frequencies $\Omega_1, \ldots, \Omega_n$.[23] In particular, the equilibrium position $z=0$ is stable.

Remark. The assumption of Birkhoff's theorem is insufficient to guarantee the Lyapunov stability of the equilibrium position of a Hamiltonian system. In the infinitely differentiable case a counter-example is given in [86]. No example has yet been found for analytic Hamiltonians.

Theorem 6 (see [172]). *If the system with Hamiltonian $H=\sum_{k\geq 2} H_k$ has n analytic first integrals in involution*

$$G_m=\tfrac{1}{2}\sum \kappa_{ms}(x_s^2+y_s^2)+\sum_{k\geq 2} G_{mk}$$

and $\det(\kappa_{ms})\neq 0$, *then Birkhoff's transformation* (1) *converges.*

From this result it is clear why (following Birkhoff) we call integrable those Hamiltonian systems for which the Birkhoff transformation converges. Postponing the discussion of convergence until Chapter 7, we mention here that, as a rule, Birkhoff's series diverge.

Theorem 5 can be generalized to the case where the numbers $\alpha=(\alpha_1, \ldots, \alpha_n)$ are rationally dependent. To this end, we consider all vectors $j=(j_1, \ldots, j_n)$ with integer components such that $\langle j, \alpha\rangle=\sum j_s\alpha_s=0$. They form a free Abelian group Γ whose rank we denote by r. If the numbers α_s are independent then obviously $r=0$.

Let us make a formal canonical change of variables $x, y\mapsto\xi, \eta$ of type (1). In the new variables ξ, η the Hamiltonian $H(x, y)$ is represented by a formal power series $K(\xi, \eta)$. Passing to the complex variables $\zeta_s=\xi_s+i\eta_s$, $\bar{\zeta}_s=\xi_s-i\eta_s$, we expand K in a series in the products

$$\zeta^k\bar{\zeta}^l=\prod_{s=1}^{n}\zeta_s^{k_s}\bar{\zeta}_s^{l_s}.$$

Definition. The formal series $K(\xi, \eta)$ is in *normal form* if its expansion contains only terms $\zeta^k\bar{\zeta}^l$ with $k-l\in\Gamma$.

[23] A function $g\colon R\{t\}\to R$ is said to be *conditionally-periodic* (or *quasi-periodic*) with frequencies $\omega_1, \ldots, \omega_n$ if $g(t)=f(\omega_1 t, \ldots, \omega_n t)$, where $f\colon T^n\to R$.

In particular, if the numbers $\alpha_1, \ldots, \alpha_n$ are independent, then the normal form of the Hamiltonian contains only terms of the type

$$\zeta^k \bar{\zeta}^k = (\xi_1^2 + \eta_1^2)^{k_1} \ldots (\xi_n^2 + \eta_n^2)^{k_n}.$$

Proposition 1. *The series $K(\xi, \eta)$ is in normal form if and only if $\mathscr{D}(K) = 0$, where*

$$\mathscr{D} = \sum_{s=1}^{n} \alpha_s \left(\xi_s \frac{\partial}{\partial \eta_s} - \eta_s \frac{\partial}{\partial \xi_s} \right).$$

The proof is an immediate consequence of the formula

$$\mathscr{D}(\zeta^k \bar{\zeta}^l) = i \langle \alpha, k - l \rangle (\zeta^k \bar{\zeta}^l).$$

Theorem 7. *There exists a formal canonical transformation of type* (1) *which reduces the given Hamiltonian $H(x, y)$ to normal form, i.e., $\mathscr{D}(K) = 0$.*

The proof of this result can be found in [27]. For $\Gamma = \{0\}$ it reduces to Birkhoff's theorem.

We show that in the under case consideration one can find $n - r$ independent commuting formal first integrals of the form

$$G = \tfrac{1}{2} \sum \beta_s (\xi_s^2 + \eta_s^2),$$

where the vector $\beta = (\beta_1, \ldots, \beta_n)$ is orthogonal to all the vectors in the group Γ. In fact,

$$\frac{d}{dt} G = \sum \beta_s \left(\xi_s \frac{\partial K}{\partial \eta_s} - \eta_s \frac{\partial K}{\partial \xi_s} \right) = 0$$

whenever $\beta \perp \Gamma$. Since rank $\Gamma = r$, one can find exactly $n - r$ linearly independent such vectors β.

Example 2. Let us apply these considerations to the Hamiltonian system with Hamiltonian

$$H = \tfrac{1}{2}(y_1^2 + y_2^2 + x_1^2 + x_2^2 + 2x_1^2 x_2 - \tfrac{2}{3} x_2^3).$$

This system was studied in detail by M. Hénon and C. Heiles using numerical methods. Here $n = 2$ and $\alpha_1 = \alpha_2 = 1$. Hence, the group Γ is defined by the equality $j_1 + j_2 = 0$, and rank $\Gamma = 1$. To produce a first integral independent of H we may take $\beta_1 = \beta_2 = 1$ to get $G = (\xi_1^2 + \eta_1^2 + \xi_2^2 + \eta_2^2)/2$. If, according to Theorem 7, H is transformed to normal form, then

$$K = \tfrac{1}{2}(\xi_1^2 + \eta_1^2 + \xi_2^2 + \eta_2^2) + \ldots$$

has the same initial terms as G. One can check by a straightforward computation (which takes into account the terms of order $\geqq 2$) that the functions K and G are indeed independent. For a discussion of the numerical results of Hénon and Heiles in connection with the construction of a formal integral, the reader is referred to the works of Gustavson [96] and Moser [27].

The transformation to normal form can be performed not only in the neighborhood of equilibrium points, but also, for example, in the neighborhood of periodic trajectories. The foregoing discussion is valid in this case as well, with necessary modifications. In the next chapter we shall consider several variants of perturbation theory, in which the Hamiltonian is also transformed to a certain "normal" form. As in the case of Birkhoff's normal forms, the series of perturbation theory are in general divergent.

§2. Completely Integrable Systems

In this section we continue the study of Hamiltonian systems having a complete set of independent first integrals in involution.

2.1. Action-Angle Variables

Theorem 8. *Suppose that the conditions of Theorem* 3 *are fulfilled and the invariant manifold* M_f *is compact. Then:*

1) *a small neighborhood of* M_f *in the ambient symplectic manifold* M *is diffeomorphic to the direct product* $D \times T^n$, *where* D *is a small domain in* R^n;

2) *in* $D \times T^n$ *there exist symplectic coordinates* $I, \varphi \bmod 2\pi$ $(I \in D,\ \varphi \in T^n)$ *in which functions* $F_1, \ldots, F_n$ *depend only on* I, *and the symplectic structure has the form* $dI \wedge d\varphi$.

In particular, in variables $I, \varphi \bmod 2\pi$ the Hamiltonian H of the given completely integrable system with invariant tori has the form $H = H(I)$. Consequently, Hamilton's equations become

$$\dot{I} = -H'_\varphi = 0, \qquad \dot{\varphi} = H'_I = \omega(I). \tag{2}$$

Hence, $I(t) = I_0$ and $\omega(I, t) = \omega(I_0)$. The variables I which "label" the invariant tori in $D \times T^n$ are called *"action"* variables, and the uniformly varying coordinates φ are called *"angle"* variables.

Example 3. Consider a Hamiltonian system with one degree of freedom having the Hamiltonian $H: R^2\{p, q\} \to R$. Let h_0 be a regular value of H and suppose that the level curve $H = h_0$ is bounded. Then for h sufficiently close to h_0 the levels $M_h = \{H = h\}$ are diffeomorphic to one-dimensional tori (circles). Obviously, on each curve M_f there is an angular coordinate $\varphi \bmod 2\pi$ that changes uniformly in time. In this problem the canonically conjugate action variable is $\Pi(h)/2$, where $\Pi(h)$ denotes the area of the domain in R^2 bounded by M_h. This follows from the readily verifiable formula

$$dp \wedge dq = \frac{1}{2\pi}\, d\Pi \wedge d\varphi.$$

By Green's formula,

$$I(h)=\frac{1}{2\pi}\iint_{H\leq h} dp\wedge dq=\frac{1}{2\pi}\oint_{H=h} p\,dq.$$

Thus, the variable I has the dimension of action (in the sense of Hamilton), which explains the terminology.

If $H=(a^2p^2+b^2q^2)/2$, then M_f is an ellipse which bounds an area equal to $\Pi(h)=2\pi h/ab=2\pi h/\omega$ $(\omega=ab)$. Consequently, for the harmonic oscillator the action variable is the ratio of energy to the frequency of oscillations. The corresponding angle variable is, of course, the phase of the harmonic oscillations. △

◁ We now prove Theorem 8. In the neighborhood of the torus $M_f\simeq T^n$ one can take as coordinates the functions $I_i=F_i$ and the angles $\varphi_i \bmod 2\pi$, whose existence is guaranteed by Theorem 3. Since the differentials dF_i are linearly independent, it follows that functions I_i, φ_i $(1\leqq i\leqq n)$ define a diffeomorphism of a neighborhood of M_f onto the direct product $D\times T^n$ (where D is a domain in $R^n=\{I\}$). Consider the nondegenerate matrix of Poisson brackets

$$\left[\begin{array}{c|c}\{I_i, I_j\} & \{I_i, \varphi_j\}\\ \hline \{\varphi_i, I_j\} & \{\varphi_i, \varphi_j\}\end{array}\right]=\left[\begin{array}{c|c}0 & a_{ij}\\ \hline -a_{ji} & b_{ij}\end{array}\right].$$

According to Theorem 3, the brackets $\{I_i, \varphi_j\}$ are constant on M_f, and hence $a_{ij}=a_{ij}(I)$. Let us show that functions b_{ij} also depend only on I. In fact, by the Jacobi identity

$$\{F_m, \{\varphi_i, \varphi_j\}\}+\{\varphi_i, \{\varphi_j, F_m\}\}+\{\varphi_j, \{F_m, \varphi_i\}\}=0.$$

The brackets $\{F_m, b_{ij}\}=\alpha_{ij}^m$ are independent of φ. On the other hand,

$$\alpha_{ij}^m=\sum_s \frac{\partial b_{ij}}{\partial\varphi_s}\{F_m, \varphi_s\}=\sum_s \frac{\partial b_{ij}}{\partial\varphi_s} a_{ms}.$$

Since $\det(a_{ms})\neq 0$, from here we obtain $\partial b_{ij}/\partial\varphi_s$ as functions of I only. Therefore, $b_{ij}=\{\varphi_i, \varphi_j\}=\sum_s f_{ij}^s(I)\,\varphi_s+g_{ij}(I)$. Since $d\varphi_i$ are single-valued 1-forms in the neighborhood of M_f, it follows that $f_{ij}^s\equiv 0$.

Next we make the change of variables $I_s=I_s(J_1, \ldots, J_n)$ to obtain $\{J_i, \varphi_j\}=\delta_{ij}$. To this end we must solve the system of equations

$$a_{ij}(I)=\{I_i, \varphi_j\}=\sum \frac{\partial I_i}{\partial J_s}\delta_{sj}=\frac{\partial I_i}{\partial J_j}.$$

The solvability condition for this system,

$$\frac{\partial a_{ij}}{\partial J_s}=\frac{\partial a_{is}}{\partial J_j}\Leftrightarrow\sum_k \frac{\partial a_{ij}}{\partial I_k}a_{ks}=\sum_k \frac{\partial a_{is}}{\partial I_k}a_{kj},$$

follows from the Jacobi identity, applied to the functions I_i, φ_j, and φ_k.

If the φ_i do not commute, we must pass to new angle coordinates $\psi_i \bmod 2\pi$ via $\varphi_i=\psi_i+f_i(J)$. The functions f_i are determined from the following system of equations

$$b_{ij}=\frac{\partial f_i}{\partial J_j}-\frac{\partial f_j}{\partial J_i} \qquad (1\leqq i,j\leqq n).$$

The condition that this system be locally solvable is that the 2-form $\sum b_{ij}\,dI_j\wedge dI_j$ be closed. It is fulfilled because the original symplectic 2-form is closed. This completes the proof of the existence of action-angle coordinates $J, \psi \bmod 2\pi$. ▷

Remark. Let p, q be symplectic coordinates in R^{2n} and let $(\gamma_1, \ldots, \gamma_n)$ be fundamental cycles on M_f depending continuously on the constants $f=(f_1, \ldots, f_n)$. Since the form $p\,dq-I\,d\varphi$ is closed, the difference

$$\oint_{\gamma_s} p\,dq-\oint_{\gamma_s} I\,d\varphi=\oint_{\gamma_s} p\,dq-2\pi I_s$$

is constant. Consequently,

$$I_s=\frac{1}{2\pi}\oint_{\gamma_s} p\,dq \qquad (1\leqq s\leqq n), \tag{3}$$

since the action variables themselves are defined to within an additive constant. Formulas (3) are most effective in the analysis of systems with separable variables (see §3). △

A Hamiltonian system with Hamiltonian function $H(I)$ is said to be *nondegenerate* (in the domain $D\times T^n$) if the Jacobian

$$\det\left(\frac{\partial\omega}{\partial I}\right)=\det\left(\frac{\partial^2 H}{\partial I^2}\right)\neq 0$$

in D. In this case almost all (with respect to the Lebesgue measure) invariant tori are nonresonant, whereas the resonant tori are dense in $D\times T^n$.

The system is said to be *properly degenerate* if $\det\left(\frac{\partial\omega}{\partial I}\right)\equiv 0$.

One reason for degeneracy may be that the number of first integrals defined on the entire phase space is larger than n (of course, in this case not all the integrals are in involution). For example, in Euler's problem on the rotation of a rigid body under its own inertia, which has three degrees of freedom, there exist four independent integrals. Their common levels foliate the three-dimensional invariant tori into two-dimensional tori. This situation is described by a generalization of Theorem 8. Thus, suppose that $F_1, \ldots, F_{n+k}$ are independent first integrals of the Hamiltonian system with Hamiltonian H, and as above, let $M_f=\{x\in M\colon F_i(x)=f_i,\ 1\leqq i\leqq n+k\}$. We assume that M_f is connected and compact.

Theorem 9 (N.N. Nekhoroshev [158]). *Suppose that the first $n-k$ functions F_i are in involution with all the F_i. Then:*

1) *the levels M_f are diffeomorphic to $(n-k)$-dimensional tori;*

2) *in the neighborhood of each level M_f there are symplectic coordinates I, p, $\varphi \bmod 2\pi$, q, such that*

$$I_s = I_s(F_1, \ldots, F_{n-k}), \qquad 1 \leqq s \leqq n-k,$$

while p and q depend on all the F_i.

The symplectic coordinates whose existence is guaranteed by Theorem 9 may be referred to as *generalized action-angle variables.*

Example 4. Consider Kepler's problem in three-dimensional space. This Hamiltonian system with three degrees of freedom has four $(n+k=3+1)$ first integrals: H (total energy), M^2 (squared magnitude of the angular momentum), M_y and M_z (the projections of the angular momentum on the y- and z-axes). The two functions H and M^2 (here $n-k=3-1$) commute with all the integrals, so that Theorem 9 is applicable. The generalized action-angle coordinates in Kepler's problem are usually denoted by L, G, Θ, l, g, ϑ, and are called *Delaunay elements.* If a, e, and i designate respectively the major semiaxis, the eccentricity, and the inclination of elliptic orbits, then

$$L = \sqrt{\gamma a}, \qquad G = \sqrt{\gamma a(1-e^2)}, \qquad \Theta = G \cos i.$$

Further, ϑ is the longitude of the ascending node, $g+\vartheta$ the longitude of perihelion, and l the mean anomaly. In these variables $H = -\gamma^2/2L^2$, $M^2 = G^2$ and $M_z = \Theta$. The expression of M_y in terms of Delaunay elements is more cumbersome. For details we refer the reader to [15]. In our problem the role of the variables I, φ and p, q is played by the elements L, G, l, g, and respectively Θ, ϑ. △

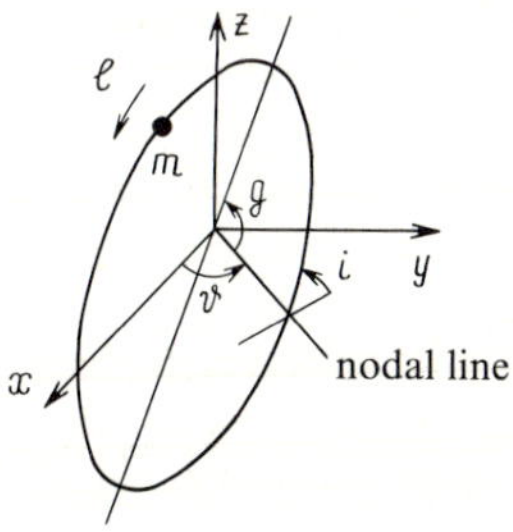

Fig. 25. Delaunay elements

In conclusion, we address a problem connected with time averaging in integrable Hamiltonian systems. Let $f: D \times T^n \to R$ be a continuous function. We examine the behavior of f along the solutions of the Hamiltonian system

(2). To this end we take its average over the torus

$$\lambda(I)=(2\pi)^{-n}\int_0^{2\pi}\dots\int_0^{2\pi} f(I,\varphi)\,d\varphi_1\wedge\dots\wedge d\varphi_n$$

and its time average

$$g(I,\varphi)=\lim_{s\to\infty}\frac{1}{s}\int_0^s f(I,\omega t+\varphi)\,dt.$$

By Theorem 4, the last limit always exists, but if the frequencies $(\omega_1(I),\dots,\omega_n(I))=\omega(I)$ are dependent, then its value depends on the initial phase φ. Moreover,

$$(2\pi)^{-n}\int_{T^n} g(I,\varphi)\,d\varphi=\lambda(I)$$

for all $I\in D$. On nonresonant tori, $g(I,\varphi)=\lambda(I)$ for all $\varphi\in T^n$. It follows that the time average is generally a discontinuous function in the domain $D\times T^n$.

Proposition 2. *If the torus $I=I_0$ is nonresonant, then*

$$\lim_{I\to I_0} g(I,\varphi)=\lambda(I_0)$$

uniformly in $\varphi\in T^n$.

Therefore, $g(I,\varphi)$ resembles the classical example, given by Riemann, of a function which is continuous at the rational points and discontinuous at the irrational points. The proof of Proposition 2 can be found in [20].

2.2. Noncommutative Sets of First Integrals. Suppose M is a $2n$-dimensional symplectic manifold and $F_1,\dots,F_k\colon M\to R$ are smooth independent functions. Hence, the linear span $\mathscr{G}$ of $F_1,\dots,F_k$ over the field R has dimension k. Let us assume that $\mathscr{G}$ is closed under the Poisson bracket: $\{F_i,F_j\}=\sum c_{ij}^s F_s$. This means that $\mathscr{G}$ has a structure of a finite-dimensional real Lie algebra. We define the rank of algebra $\mathscr{G}$, denoted rank $\mathscr{G}$, as the maximum of the rank of the matrix $(a_{ij}(F_1,\dots,F_k))$, where $a_{ij}=\{F_i,F_j\}$, on M (cf. Ch. 3, §2.2).

Theorem 10. *Suppose that on the level set $M_f=\{x\in M\colon F_i(x)=f_i,\ 1\leqq i\leqq k\}$ the differentials dF_i are linearly independent, and that the algebra $\mathscr{G}$ is such that*

$$\dim\mathscr{G}+\operatorname{rank}\mathscr{G}=\dim M. \tag{4}$$

If M_f is connected and compact, then it is diffeomorphic to the r-dimensional torus T^r, with $r=\operatorname{rank}\mathscr{G}$. Furthermore, if $F_1,\dots,F_k$ are first integrals of the Hamiltonian system with Hamiltonian H, then one can choose angle coordinates $\varphi_1,\dots,\varphi_r \bmod 2\pi$ on M_f in which Hamilton's equations $\dot{x}=\mathfrak{I}\,dH(x)$ on T^r take the form: $\dot{\varphi}_s=\omega_s=\mathrm{const}$.

This result, kindred to Theorem 9, was obtained by Mishchenko and Fomenko [141]. It can be deduced from the Lie-Cartan theorem (Ch. 6,

§2.2, Theorem 16), which, under the assumptions of Theorem 10, guarantees the existence, in the neighborhood of the point $F_i = f_i$, of k functions $\Phi_1, \ldots, \Phi_k$ of the first integrals $F_1, \ldots, F_k$ such that

$$\{\Phi_i, \Phi_{r+i}\} = 1, \quad 1 \leqq i \leqq r,$$

whereas all the other brackets $\{\Phi_i, \Phi_j\} = 0$. It follows that the set $N_c = \{x \in M: \Phi_m(x) = c_m, 1 \leqq m \leqq 2r\}$ is a symplectic submanifold of M, and that the restrictions of $\Phi_{2r+1}, \ldots, \Phi_k$ to N_c form a complete set of commuting integrals. The conclusion of Theorem 10 is now seen to be a consequence of Theorem 3.

Example 5. Consider the motion of a point mass in a central force field. The corresponding Hamiltonian system has four independent integrals: H, M_x, M_y, and M_z. The functions M_x, M_y, M_z generate a Lie algebra isomorphic to $so(3)$. H commutes with M_x, M_y, and M_z. If the constant values of the "area integrals" are not all zero, the rank of the matrix of Poisson brackets (a_{ij}) is obviously two. In this case equality (4) is satisfied, and hence Theorem 10 applies. △

In Example 5, as in all known examples described by Theorem 10, one can find a complete set of first integrals in involution. This fact is not accidental, as the next theorem shows.

Theorem 11. *If M is compact, then under the assumptions of Theorem* 10 *one can find $n = \dim M/2$ independent first integrals $\Phi_1, \ldots, \Phi_n$ in involution; they are polynomials in $F_1, \ldots, F_k$.*

In the work [142] of Mishchenko and Fomenko, where this result is proved, it is also conjectured that the compactness constraint on M can be dropped. We note that the complete integrability in the neighborhood of the level set M_f follows, of course, from the Lie-Cartan theorem, but in comparison with Theorem 11 this is a weaker result.

2.3. Examples of Completely Integrable Systems

a) The Euler-Poisson equations

$$A\dot{\omega} = A\omega \times \omega + e \times r, \quad \dot{e} = e \times \omega,$$

describing the rotations of a heavy rigid body fixed at a point, contain 6 parameters: three eigenvalues A_1, A_2, A_3 of the inertia operator, and three coordinates r_1, r_2, r_3 of the center of mass. As shown in Chapter 3 (Example 14), the Euler-Poisson equations are Hamiltonian on the four-dimensional invariant manifolds

$$M_c = \{(\omega, e) \in R^6: \langle A\omega, e \rangle = c, \ \langle e, e \rangle = 1\}.$$

These equations on M_c always have one first integral, the energy integral. Hence, for complete integrability it suffices to have one more independent integral. We list the known integrability cases:

1) Euler's case (1750): $r_1=r_2=r_3=0$. The new integral, $\langle A\omega, A\omega\rangle$, is the square of the length of the angular momentum vector.

2) Lagrange's case (1788): $A_1=A_2$, $r_1=r_2=0$. The new integral, ω_3, is the projection of the angular momentum on the axis of dynamic symmetry.

3) Kovalevskaya's case (1889): $A_1=A_2=2A_3$, $r_3=0$. The first integral discovered by Kovalevskaya is

$$(\omega_1^2-\omega_2^2-\nu e_1)^2+(2\omega_1\omega_2-\nu e_2)^2; \quad \nu=r/A_3,\ r^2=r_1^2+r_2^2.$$

4) Goryachev-Chaplygin's case (1900): $A_1=A_2=4A_3$, $r_3=0$, and the constant is $c=\langle A\omega, e\rangle=0$. In contrast to cases 1)–3), here an integrable Hamiltonian system is obtained only on the invariant manifold M_0.

We note that all these integrable cases form manifolds of one and the same codimension 3 in the six-dimensional parameter space A_i, r_j.

In the first two cases the equations of motions were studied in detail from various points of view in classical works of Euler, Poinsot, Lagrange, Poisson, and Jacobi. The Kovalevskaya case is nontrivial in many respects. It was found by imposing the condition that the solutions of the Euler-Poisson equations be meromorphic in the complex time plane. The Goryachev-Chaplygin case is considerably simpler: it can be integrated by separation of variables. To see this, we write the Hamiltonian in the special canonical variables L, G, l, g (see Ch. 3, §2.3):

$$H=\frac{G^2+3L^2}{8A_3}+r\left(\frac{L}{G}\sin l\cos g+\cos l\sin g\right).$$

Consider the canonical transformation

$$L=p_1+p_2,\quad G=p_1-p_2,\quad q_1=l+g,\quad q_2=l-g.$$

In the new symplectic coordinates p, q

$$H=\frac{p_1^3-p_2^3}{2A_3(p_1-p_2)}+r\left(\frac{p_1}{p_1-p_2}\sin q_1+\frac{p_2}{p_1-p_2}\sin q_2\right).$$

Equating this expression to h and multiplying by p_1-p_2 we see that it separates:

$$p_1^3/2A_3+rp_1\sin q_1-hp_1=p_2^3/2A_3-rp_2\sin q_2-hp_2.$$

We put

$$p_1^3/2A_3+rp_1\sin q_1-Hp_1=\Gamma,\quad p_2^3/2A_3-rp_2\sin q_2-Hp_2=\Gamma. \tag{5}$$

Function Γ is a first integral of the equations of motion. In the traditional variables ω, e it has the form

$$\Gamma=2A_3^2 f,\quad \text{where } f=\omega_3(\omega_1^2+\omega_2^2)-\nu\omega_1 e_3 \quad (\nu=r/A_3).$$

We write down a closed system of equations for the variables p_1, p_2:

$$\dot{p}_1 = -\frac{\partial H}{\partial q_1} = \frac{r p_1}{p_1 - p_2} \cos q_1, \qquad \dot{p}_2 = -\frac{\partial H}{\partial q_2} = \frac{r p_2}{p_1 - p_2} \cos q_2,$$

or, taking account of (5),

$$\dot{p}_1 = \frac{\pm\sqrt{\Phi(p_1)}}{p_1 - p_2}, \qquad \dot{p}_2 = \frac{\pm\sqrt{\Phi(p_2)}}{p_1 - p_2}, \tag{6}$$

where $\Phi(z) = r^2 z^2 - (\Gamma + hz - z^3/2A_3)^2$ is a polynomial of degree 6. The solutions of these equations can be expressed through hyperelliptic functions of time. The variables p_1 and p_2 vary in the disjoint intervals $[a_1, b_1]$ and $[a_2, b_2]$, where a_i, b_i are the adjacent roots of the polynomial $\Phi(z)$, between which it takes positive values. If a_i (b_i) is a multiple root, then the solution $p_i(t)$ tends to a_i (respectively, to b_i) as $t \to +\infty$ or $t \to -\infty$. Multiple roots correspond to the case where the integrals H and Γ are not independent. Below we consider the typical case in which the polynomial $\Phi(z)$ has only simple roots.

We introduce angular variables $\varphi_1, \varphi_2 \bmod 2\pi$ by the formulas

$$\varphi_i = \frac{\pi}{\tau_i} \int_{a_i}^{p_i} \frac{dz}{\pm\sqrt{\Phi(z)}}, \qquad \tau_i = \int_{a_i}^{b_i} \frac{dz}{\sqrt{\Phi(z)}}. \tag{7}$$

In the new variables equations (6) take the form

$$\dot{\varphi}_i = \frac{\pi}{2\tau_i(p_1(\varphi_1) - p_2(\varphi_2))} \qquad (i = 1, 2), \tag{8}$$

where $p_i(z)$ are the real-valued hyperelliptic functions of period 2π defined by relations (7). Since the trajectories of (8) on $T^2 = \{\varphi_1, \varphi_2 \bmod 2\pi\}$ are straight lines, the ratio of frequencies of the corresponding conditionally-periodic motions equals the ratio τ_1/τ_2 of the real periods of the hyperelliptic integral

$$\int_{z_0}^{z} \frac{dz}{\sqrt{\Phi(z)}}.$$

This remarkable fact holds also for the equations of Kovalevskaya's problem (see [20]).

b) Since equations of form (8) are frequently encountered in the study of integrable problems of classical mechanics, they deserve a closer examination. To this end we consider differential equations on $T^n = \{x_1, \ldots, x_n \bmod 2\pi\}$ of the somewhat more general form

$$\dot{x}_i = \frac{\omega_i}{f(x)}, \qquad 1 \leqq i \leqq n, \tag{9}$$

where $\omega_i=\text{const}\neq 0$, and f is a smooth (or analytic) positive function on T^n. Equations (9) admit the invariant measure

$$\operatorname{mes}(D)=\int_D f(x)\,dx_1\wedge\ldots\wedge dx_n.$$

We let Λ denote the total measure of T^n. Next, we apply the averaging operator

$$\frac{1}{\Lambda}\int_{T^n}(\cdot)\,dx_1\wedge\ldots\wedge dx_n$$

to the right-hand side of (9). This yields the equations

$$\dot{\varphi}_i=\Omega_i=\frac{\omega_i}{\Lambda}=\text{const}\qquad(1\leqq i\leqq n). \tag{10}$$

Proposition 3. *Suppose that the partial differential equation*

$$\left\langle\frac{\partial R}{\partial x},\omega\right\rangle=f-\Lambda \tag{11}$$

has a smooth (analytic) solution $R(x_1,\ldots,x_n)\colon T^n\to\mathbb{R}$. *Then there exists a smooth (respectively, analytic) change of variables* $x\mapsto\varphi$ *reducing system* (9) *to form* (10).

◁ The required change of variables is given by the formula

$$\varphi_i=x_i+\frac{\omega_i}{\Lambda}R(x_1,\ldots,x_n).\ \triangleright$$

Let

$$f(x)=\frac{\Lambda}{(2\pi)^n}+\sum_{k\neq 0}f_k e^{i\langle k,x\rangle}.$$

If $\omega_1,\ldots,\omega_n$ are incommensurable (over the rationals) we can write the formal equality

$$R(x)=\sum_{k\neq 0}\frac{f_k}{i\langle k,\omega\rangle}e^{i\langle k,x\rangle}. \tag{12}$$

This series defines a smooth function $R(x)$ if, say, $f(x)$ is a trigonometric polynomial.

Proposition 4. *If* f *is a smooth function, then equation* (11) *has a smooth solution for almost every collection of numbers* $\omega_1,\ldots,\omega_n$.

◁ In fact, the Fourier coefficients f_k decrease with the growth of k faster than any power, and the following estimate holds for almost every ω:

$$|\langle k,\omega\rangle|\geqq c/|k|^\gamma,\qquad c,\gamma=\text{const}.$$

Therefore, in this case series (12) converges absolutely and its sum is a smooth periodic function. ▷

If $f(x)=\sum_s f_s(x_s)$ (as in the case of system (8)), then equation (11) separates, and hence, whatever the arithmetic properties of the frequencies $\omega_1, \ldots, \omega_n$, it always has the solution:

$$R=\sum_{s=1}^{n} \frac{F_s(x_s)-I_s x_s}{\omega_s}, \qquad F_s(x)=\int_0^x f_s(t)\,dt, \qquad I_s=\frac{1}{2\pi} F_s(2\pi).$$

In particular, equations (8) can always be reduced to the form (10), whose existence is guaranteed by Theorem 3.

c) The problem of the motion of a rigid body in an ideal fluid is far richer in integrable cases. The equations of motions can be written in Kirchhoff's form (Ch. 1, §2.4):

$$\dot{k}=k\times\omega+e\times u, \qquad \dot{e}=e\times\omega, \tag{13}$$

where $\omega=H'_k$, $u=H'_e$, $H=\langle Ak, k\rangle/2+\langle Bk, e\rangle+\langle Ce, e\rangle/2$. Since the matrices A, B, and C are symmetric, A can be diagonalized: $A=\operatorname{diag}(a_1, a_2, a_3)$. Thus, in the general case Kirchhoff's problem contains 15 parameters. Equations (13) have always three independent integrals: $F_1=H$, $F_2=\langle k, e\rangle$ and $F_3=\langle e, e\rangle$. As in the case of the heavy top, the integration of these equations reduces to seeking a fourth independent first integral. We indicate two integrability cases, discovered by A. Clebsch (1871) and V.A. Steklov (1893). In Clebsch's case it is assumed that $B=0$, $C=\operatorname{diag}(c_1, c_2, c_3)$, and

$$a_1^{-1}(c_2-c_3)+a_2^{-1}(c_3-c_1)+a_3^{-1}(c_1-c_2)=0.$$

The supplementary integral of Kirchhoff's equations is

$$k_1^2+k_2^2+k_3^2-a_1 e_1^2-a_2 e_2^2-a_3 e_3^2.$$

In Steklov's case $B=\operatorname{diag}(b_1, b_2, b_3)$, $C=\operatorname{diag}(c_1, c_2, c_3)$ and, in addition

$$b_j=\mu(a_1 a_2 a_3)a_j^{-1}+\nu, \qquad c_1=\mu^2 a_1(a_2-a_3)^2+\nu', \ldots \quad (\mu, \nu, \nu'=\text{const}).$$

The fourth integral is

$$\sum_j (k_j^2-2\mu(a_j+\nu)k_j e_j)+\mu^2((a_2-a_3)^2+\nu'')e_1^2+\text{circular permutations}.$$

The parameters ν, ν', and ν'' are not essential: their appearance is connected with the presence of the classical integrals F_2 and F_3.

d) The motion of n point vortices in the plane (Ch. 1, §3.4) is described by a Hamiltonian system with n degrees of freedom. Hamilton's equations possess the four first integrals H,

$$P_x=\sum \Gamma_s x_s, \quad P_y=\sum \Gamma_s y_s \quad \text{and} \quad M=\sum \Gamma_s(x_s^2+y_s^2)/2;$$

here Γ_s is the intensity of the s-th vortex. The Poisson brackets of these integrals are readily calculated: $\{P_x, P_y\}=-\sum \Gamma_s$, $\{P_x, M\}=-P_y$, $\{P_y, M\}=P_x$. Therefore, the problem of n vortices is completely integrable for $n\leqq 3$. The case $n=1$ is trivial, for $n=2$ one can take H and M as independent commuting integrals, and for $n=3$, the integrals H, M, and $P_x^2+P_y^2$. In the problem of four vortices there are as many first integrals as degrees of freedom, but not all of them commute. One can show, however, that if the sum of the intensities Γ_s of the vortices is zero, then those solutions of the equations of motion, on which the integral P_x and P_y have null constant values, can be found by quadratures.

§3. Some Methods of Integrating Hamiltonian Systems

The common feature of the various approaches to the integration of Hamiltonian systems discussed in §1 is the existence of a complete set of independent commuting first integrals. In this section we indicate several general methods of searching for first integrals ("conservation laws"). We begin with the simplest and most effective of them.

3.1. Method of Separation of Variables. Consider the system of canonical equations

$$\dot{p}=-H'_q, \quad \dot{q}=H'_p, \quad (p, q)\in R^{2n} \tag{14}$$

with Hamiltonian $H(p, q)$. Suppose we can find a canonical transformation $g\colon p, q\mapsto x, y$ such that in the new coordinates x, y the function $H(p, q)=K(x)$ does not depend on y. Then the canonical equations are readily integrated:

$$p=p(x, y), \quad q=q(x, y); \quad x=x_0, \; y=y_0+\omega(x_0)t, \; \omega(x)=K'_x. \tag{15}$$

If $\det(\partial p/\partial x)\neq 0$, the canonical transformation g can be given locally by a generating function $S(x, q)$:

$$p=S'_q, \quad y=S'_x.$$

Thus, the task of integrating equation (14) becomes that of finding a generating function S which, in view of the equality $p=S'_q$, should satisfy the nonlinear partial differential equation

$$H(S'_q, q)=K(x). \tag{16}$$

This equation is obtained from the *Hamilton-Jacobi equation*

$$V'_t+H(V'_q, q)=0$$

through the substitution $V(q, t)=-Kt+S(q)$. We emphasize that in equation (16) function K is considered to be undetermined and that to specify it uniquely one must impose supplementary conditions (see, for example, the utilization of equation (16) in perturbation theory, Ch. 5). It is customary to set $K(x_1, \ldots, x_n)=x_1$; the trajectories of the system with this Hamiltonian are straight lines in $R^{2n}=\{x, y\}$. A solution S of equation (16) such that

$$\det(\partial^2 S/\partial q\, \partial x) \neq 0,$$

is called a *complete integral* of this equation.

Theorem 12. *If a complete integral $S(x, q)$ of equation* (16) *is known, then the canonical equations $\dot{p}=-H'_q$, $\dot{q}=H'_p$ can be integrated by quadratures. In this case the n functions $x_1(p, q), \ldots, x_n(p, q)$, determined from the equation $p=\partial S(x, q)/\partial q$, form a complete set of independent first integrals in involution.*

The proof of this theorem of Jacobi follows from Proposition 9 of Ch. 1 and formulas (15). The method of integrating Hamilton's equation based on Theorem 12 was proposed by Jacobi in 1837. Jacobi used earlier results of Hamilton. The Hamilton-Jacobi method can be traced back to investigations of Pfaff and Cauchy in the theory of characteristics of partial differential equations.

Definition. If equations (16) have a complete integral of the form $S(x, q) = \sum_k S_k(q_k, x_1, \ldots, x_n)$, then we say that the variables $q_1, \ldots, q_n$ are *separable*.

Proposition 5. *Suppose that in some symplectic coordinates $(p, q)=(p_1, \ldots, p_n, q_1, \ldots, q_n)$ the Hamiltonian $H(p, q)$ has one of the following forms:*

1) $H=f_n(f_{n-1}(\ldots f_2(f_1(p_1, q_1), p_2, q_2), \ldots, p_{n-1}, q_{n-1}), p_n, q_n)$,

or

1′) $H=\sum f_s(p_s, q_s)/\sum g_s(p_s, q_s)$.

Then the functions

2) $F_1=f_1(p_1, q_1)$, $F_2=f_2(f_1(p_1, q_1), p_2, q_2), \ldots, F_n=H$,

and, respectively,

2′) $F_0=H$, $F_s=f_s(p_s, q_s)-H g(p_s, q_s)$, $1 \leqq s \leqq n$,

form a complete set of first integral in involution of the Hamiltonian system with Hamiltonian H.

Consider, for example, case 1′). We set $K=x_0$ and write equation (16):

$$\sum_k x_0 g_k\left(\frac{\partial S}{\partial q_k}, q_k\right)-f_k\left(\frac{\partial S}{\partial q_k}, q_k\right)=0.$$

Its complete integral can be found as a sum

$$\sum_k S_k(q_k, x_0, x_k),$$

where S_k, as a function of q_k, satisfies the ordinary differential equation

$$x_0 g_k\left(\frac{d S_k}{d q_k}, q_k\right)-f_k\left(\frac{d S_k}{d q_k}, q_k\right)=x_k, \qquad \sum_{k \geqq 1} x_k=0.$$

The variables $x_0, x_1, \ldots, x_n$ are first integrals in involution (Theorem 12). Generally, any n of them are independent.

Proposition 5 describes the simplest and most common types of separation of variables. Practically, we have already used separation of symplectic coordinates of type 1′) in solving the Goryachev-Chaplygin problem (§2.3). We note that cases 1) and 1′) of Proposition 5 may occur simultaneously, and also that other, more complicated, kinds of separation of variables are possible.

Example 6 (P. Stäckel, 1895). Let Φ and Φ_{ij} denote the determinant of the matrix $(\varphi_{ij}(q_j))$ $(1 \leqq i, j \leqq n)$ and respectively the cofactor of the entry φ_{ij}. Suppose that in symplectic coordinates $p_1, \ldots, p_n, q_1, \ldots, q_n$ the Hamiltonian H has the form

$$H(p, q) = \sum_{s=1}^{n} \Phi_{1s}(q) f_s(p_s, q_s)/\Phi(p, q); \tag{17}$$

then Hamilton's equations can be integrated. Setting $K(x) = x_1$, we write equation (16) in the form:

$$\sum_m \Phi_{1m}\left[\sum_k x_k \varphi_{km}(q_m) - f_m\left(\frac{\partial S}{\partial q_m}, q_m\right)\right] = 0.$$

Its complete integral can be found as a sum

$$S(x, q) = \sum_m S_m(q_m, x_1, \ldots, x_n),$$

where S_m, as a function of q_m, satisfies the equation

$$f_m\left(\frac{dS_m}{dq_m}, q_m\right) = \sum_k x_k \varphi_{km}(q_m).$$

One can show that the n functions

$$F_k(p, q) = \sum_m \frac{\Phi_{km} f_m}{\Phi}$$

form a complete set of first integrals in involution of the Hamiltonian system with Hamiltonian (17). △

There is, of course, no general rule for searching for separable variables. "Therefore, we must go in the opposite direction and, after finding some remarkable substitution, look for problems to which it can be successfully applied" (C.G.J. Jacobi, Vorlesungen über Dynamic). We indicate here such a "remarkable substitution", related to *elliptic coordinates* in R^n.

Let $0 < a_1 < a_2 < \ldots < a_n$ be distinct positive numbers. For each $x = (x_1, \ldots, x_n) \in R^n$ the equation

$$f(\lambda) = \sum_s \frac{x_s^2}{a_s - \lambda} = 1 \tag{18}$$

defines n real numbers $\lambda_1, \ldots, \lambda_n$ (its roots), which separate $a_1, \ldots, a_n$ (Fig. 26).

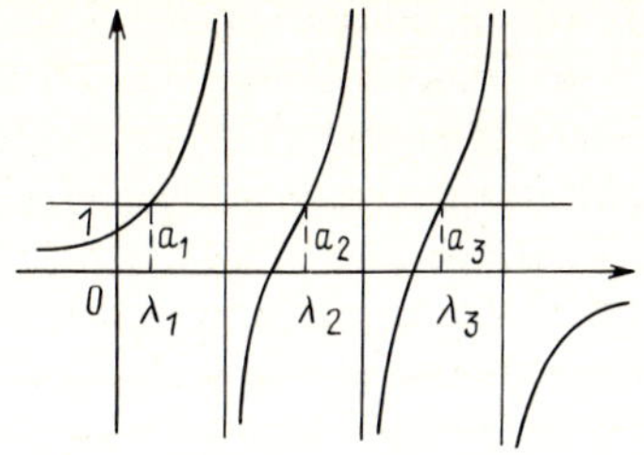

Fig. 26

The numbers $\lambda_1, \ldots, \lambda_n$ serve as curvilinear coordinates in R^n and are called Jacobi's *elliptic coordinates*. One can show that

$$x_i^2 = \prod_{s=1}^{n} (a_i - \lambda_s) \left\{ \prod_{\substack{s=1 \\ s \neq i}}^{n} (a_i - a_s) \right\}^{-1}. \tag{19}$$

Using this formula one can readily derive the relation $4\sum_s' \dot{x}_s^2 = \sum_s' M_s \dot{\lambda}_s^2$, with

$$M_s = \prod_{i \neq s} (\lambda_i - \lambda_s) \{ \prod_i (a_i - \lambda_s) \}^{-1}. \tag{19'}$$

Notice the interesting duality of formulas (19) and (19′).

Now let us pass to the symplectic coordinates λ_s, $\mu_s = M_s \dot{\lambda}_s / 4$. The energy of free motion of a point mass in R^n thus takes the form

$$H = \tfrac{1}{2} \sum \dot{x}_s^2 = 2 \sum \mu_s^2 / M_s(\lambda). \tag{20}$$

Here it is not immediately clear how the variables λ, μ could be separated. However, we can resort to the following formula of Jacobi: the sum

$$\sum_{s=1}^{n} \frac{\lambda_s^m}{\prod_{i \neq s} (\lambda_s - \lambda_i)} = \begin{cases} 0, & \text{if } m < n-1, \\ 1, & \text{if } m = n-1. \end{cases}$$

This formula permits us to transform expression (20) into

$$\sum_s \frac{\sum_{m=0}^{n-1} F_m \lambda_s^m}{\prod_{i \neq s} (\lambda_s - \lambda_i)} = 2 \sum_s \frac{\mu_s^2 \prod_j (\lambda_s - a_j)}{\prod_{i \neq s} (\lambda_s - \lambda_i)}.$$

Here $F_{n-1} = H$, while $F_0, F_1, \ldots, F_{n-2}$ are still arbitrary. Now the variables λ, μ separate: we can put

$$\sum_{m=0}^{n-1} F_m \lambda_s^m = 2 \mu_s^2 \prod_{j=1}^{n} (\lambda_s - a_j).$$

From this system of equations we find $F_0, F_1, \ldots, F_{n-2}$ as functions of λ, μ; they provide a complete set of independent first integrals in involution.

At first glance this result may seem trivial: the complete integrability of the problem of free motion of a point in R^n is obvious from the outset. However, the formulas obtained above yield a result of Jacobi which is far from obvious: the problem of the motion of a point on the surface of a multidimensional ellipsoid in the absence of external forces is completely integrable (according to Maupertuis' principle, the trajectories of the moving point coincide with the geodesics of the ellipsoid). In fact, let the value of the variable λ_1 be fixed, say, $\lambda_1=0$. Then $\lambda_2, \ldots, \lambda_n$ are curvilinear orthogonal coordinates on the surface of the $(n-1)$-dimensional ellipsoid $\sum x_s^2/a_s=1$. The Hamiltonian of the problem of geodesics is given by formula (20), in which we must substitute $\lambda_1=0$ and $\mu_1=0$. The separation of variables $\lambda_2, \mu_2, \ldots, \lambda_n, \mu_n$ is effected according to the scheme indicated above. We remark that for the two-dimensional ellipsoid the Hamiltonian has form 1′) given in Proposition 5. Upon fixing the value of one of the variables $\lambda_2, \ldots, \lambda_n$ we may, by the same method, establish the complete integrability of the problem of geodesics on multidimensional hyperboloids of arbitrary types. The results of the qualitative analysis of the behavior of geodesics on the surface of a two-dimensional ellipsoid, based on Jacobi's formulas, can be found in [8]. Jacobi showed that the problem of free motion on an ellipsoid remains integrable if the point is subjected to the action of an elastic force whose direction passes through the center of the ellipsoid.

As yet another example of application of elliptic coordinates we consider the planar motion of a point mass in the attraction field of *two fixed centers*. This problem was integrated by Euler in 1760. Let x_1, x_2 be Cartesian coordinates in the plane of motion and let $(0, c)$ and $(0, -c)$ be the coordinates of the two centers of attraction $(c>0)$. We pass to elliptic coordinates in $R^2=\{x_1, x_2\}$, assuming that $a_2-a_1=2c$. This means, in particular, that for each fixed value of λ, the equation

$$\frac{x_1^2}{a_1-\lambda}+\frac{x_2^2}{a_2-\lambda}=1$$

defines a conic with foci at the fixed centers. In symplectic coordinates λ, μ the Hamiltonian of this problem is

$$H=2\frac{(a_1-\lambda_1)(a_2-\lambda_1)}{\lambda_2-\lambda_1}\mu_1^2+2\frac{(a_1-\lambda_2)(a_2-\lambda_2)}{\lambda_1-\lambda_2}\mu_2^2+U(\lambda_1, \lambda_2),$$

where U is the interaction potential. Let r_1, r_2 be the distances from the moving mass to the two centers. Using formula (19) for $n=2$ it is a simple matter to check that

$$r_1^2=(x_2+c)^2+x_1^2=(\sqrt{a_2+\lambda_1}+\sqrt{a_2+\lambda_2})^2,$$
$$r_2^2=(x_2-c)^2+x_1^2=(\sqrt{a_2+\lambda_1}-\sqrt{a_2+\lambda_2})^2.$$

Consequently,

$$U=\frac{\gamma_1}{r_1}+\frac{\gamma_2}{r_2}=\frac{\gamma_1 r_2+\gamma_2 r_1}{r_1 r_2}=\frac{(\gamma_1+\gamma_2)\sqrt{a_2+\lambda_1}-(\gamma_1-\gamma_2)\sqrt{a_2+\lambda_2}}{\lambda_1-\lambda_2}.$$

In the end we see that the variables λ_1, μ_1 and λ_2, μ_2 separate and hence, according to Proposition 5, the problem of two fixed centers is integrable. Lagrange showed that the integrability is preserved if a supplementary elastic force, directed to the midpoint of the segment joining the two fixed centers of attraction, is allowed to act on the point. A qualitative investigation of the problems of two centers can be found in Charlier's book [15].

To conclude this subsection we mention two important problems which can be solved by separation of variables:

1) Kepler's problem in a homogeneous force field; that is, the motion of a point mass under the action of the gravitational attraction of a fixed center and one other force of constant magnitude and direction. Here separation of variables is achieved by introducing *"parabolic" coordinates:* these are obtained as the limit of elliptic coordinates when one of the foci approaches infinity. This problem was solved by Lagrange in 1766.

2) The problem of the motion of a point mass on the sphere $\langle x, x\rangle=1$, $x\in R^n$, in a force field with quadratic potential $U(x)=\langle Ax, x\rangle$. Here separation of variables is achieved using elliptic coordinates. For $n=3$ this problem was first considered by C. Neumann in 1859. The case of arbitrary n is discussed in detail in [29].

3.2. Method of $L-A$ (Lax) Pairs. This method works when the differential equation $\dot{x}=f(x)$, $x=(x_1, \ldots, x_n)$, can be represented as a matrix equation of the form:

$$\dot{L}=[A, L]=AL-LA; \tag{21}$$

here the entries of the square (and possibly complex) matrices A and L are functions of $x_1, \ldots, x_n$. Equation (21) arises naturally in many problems of mechanics and physics.

Example 7. Euler's equation $\dot{M}=M\times\omega$ can be put in form (21) upon setting

$$L=\begin{bmatrix} 0 & M_1 & -M_2 \\ -M_1 & 0 & M_3 \\ M_2 & -M_3 & 0 \end{bmatrix}, \quad A=\begin{bmatrix} 0 & \omega_1 & -\omega_2 \\ -\omega_1 & 0 & \omega_3 \\ \omega_2 & -\omega_3 & 0 \end{bmatrix},$$

where $(M_1, M_2, M_3)=M$ and $(\omega_1, \omega_2, \omega_3)=\omega$. The eigenvalues of L are 0 and $\pm i\langle M, M\rangle$; they are constant on the trajectories of Euler's equation.

The next result shows that this last fact is not accidental.

Proposition 6. *The eigenvalues of the operator L are first integrals of equation* (21).

◁ We have $L(t+s)=e^{sA}L(t)e^{-sA}+o(s)$ as $s\to 0$. Hence, $\det(L(t+s)-\lambda E)=\det(L(t)-\lambda E)+o(s)$. It follows that $\det(L(t)-\lambda E)$ does not depend on t. ▷

In practice it is, of course, more convenient to deal with the symmetric polynomials in λ given by the coefficients of the characteristic equation $\det(L-\lambda E)=0$, rather than with the eigenvalues λ themselves. To determine, in each concrete case, whether the first integrals found by this method are independent and form a complete set requires a separate investigation.

Example 8 (the Toda lattice or chain). Consider n particles on the line with coordinates $x_1, \ldots, x_n$, whose motion is governed by the system of equations

$$\ddot{x}_i=-U'_{x_i}, \quad U=\sum_{k=1}^{n}\exp(x_k-x_{k+1}), \quad x_{n+1}\equiv x_1. \tag{22}$$

It was shown by M. Henon, H. Flaschka, and S.V. Manakov (1974) that this system admits an $L-A$ pair representation. For example, one can take

$$L=\begin{bmatrix} b_1 & a_1 & 0 & \ldots & 0 & a_n \\ a_1 & b_2 & a_2 & \ldots & 0 & 0 \\ 0 & a_2 & b_3 & \ldots & 0 & 0 \\ \cdot & \cdot & \cdot & \ldots & \cdot & \cdot \\ 0 & 0 & 0 & \ldots & b_{n-1} & a_{n-1} \\ a_n & 0 & 0 & \ldots & a_{n-1} & b_n \end{bmatrix}, \quad A=\begin{bmatrix} 0 & a_1 & 0 & \ldots & 0 & a_n \\ -a_1 & 0 & a_2 & \ldots & 0 & 0 \\ 0 & -a_2 & 0 & \ldots & 0 & 0 \\ \cdot & \cdot & \cdot & \ldots & \cdot & \cdot \\ 0 & 0 & 0 & \ldots & 0 & a_{n-1} \\ a_n & 0 & 0 & \ldots & -a_{n-1} & 0 \end{bmatrix},$$

where $2a_k=\exp((x_k-x_{k+1})/2)$ and $2b_k=-\dot{x}_k$. It follows from Proposition 6 that the eigenvalues of L are first integrals of equations (22); also, one can show that they are independent and commute. In [148] Moser used the same method to establish the integrability of the system of points of equal masses on the line with the pair-interaction potential

$$\sum_{k<l} u(x_k-x_l),$$

where $u(z)$ is one of the functions z^{-2}, $(\sin z)^{-2}$, $(\sinh z)^{-2}$. F. Calogero has generalized this result by establishing the integrability of the equation of motion of point masses with potential $\wp(z)$ (see also [165]). The potentials considered by Moser are degenerate cases of Weierstrass' $\wp$-function.

The motion of certain systems of point masses on the line may be regarded as the motion of a single point in the Weyl chamber of the Lie algebra A_k under the influence of a potential which grows at the walls of the camera. Similar problems for other simple Lie algebras (with the same potentials) are also integrable (as proved by A.M. Perelomov and M.A. Ol'shanetskii). As it turned out, the problem of the free motion of a rigid body in R^n is also integrable (S.V. Manakov).

We remark that an $L-A$ representation was found for almost all problems of classical mechanics previously integrated by other methods. Also, various algebro-geometric constructions were found that explained the existence of "hidden" conservation laws. The existence of an $L-A$ representation is useful not only in finding first integrals, but also in explicitly integrating the equations of motion. For a discussion of various aspects of the contemporary theory of integration of Hamiltonian systems we refer the reader to [29, 87, 45] (translator's note: see also the forthcoming book, *A Systematic Approach Towards Solving Integrable Systems* by M. Adler and P. van Moerbeke, Academic Press).

§4. Nonholonomic Integrable Systems

4.1. Differential Equations with Invariant Measure. Consider the differential equation

$$\dot{x}=f(x), \qquad x\in R^n \tag{23}$$

and let g^t be its phase flow. We shall assume that equation (23) has an integral invariant with smooth density $M(x)$, i.e.,

$$\int_{g^t(D)} M(x)\,dx=\int_D M(x)\,dx \tag{24}$$

for every measurable domain $D\subset R^n$ and all t. We recall the well-known result of Liouville on the existence of an integral invariant.

Proposition 7. *The smooth function $M\colon R^n\to R$ is the density of an invariant*

$$\int M(x)\,dx$$

of equation (23) *if and only if*

$$\operatorname{div}(Mf)=0. \tag{25}$$

If $M(x)>0$ for all x, formula (24) defines a measure on R^n which is *invariant* under the flow g^t. The existence of an invariant measure facilitates the integration of the given differential equation; for instance, if $n=2$ the equation is integrable by quadratures. In fact, equality (25) implies the local solvability of the system of equations

$$F'_{x_1}=Mf_1, \qquad F'_{x_2}=-Mf_2.$$

From this system the function $F(x_1, x_2)$ is found by simple quadratures. It remains to remark that F is a first integral of system (23). Following Euler, the function M is also called an *integrating factor.*

Theorem 13. *Suppose that the system of equations* (23) *with invariant measure* (24) *has $n-2$ first integrals $F_1, \ldots, F_{n-2}$. Suppose also that $F_1, \ldots, F_{n-2}$ are independent on the invariant set $M_c=\{x\in R^n\colon F_s(x)=c_s,\ 1\leqq s\leqq n-2\}$. Then:*

1) *the solutions of equations* (23) *lying on* M_c *can be found by quadratures. Moreover, if* L_c *is a compact connected component of the level set* M_c *and* $f(x)\neq 0$ *on* L_c, *then*

2) L_c *is a smooth manifold diffeomorphic to a two-torus,*

3) *one can find angular coordinates* x, y mod 2π *on* L_c *in which equations* (23) (*on* L_c) *take the form*

$$\dot{x}=\frac{\lambda}{\Phi(x, y)}, \qquad \dot{y}=\frac{\mu}{\Phi(x, y)}, \tag{26}$$

where λ, μ *are constant and* Φ *is a smooth positive* 2π-*periodic function in* x *and* y.

We indicate the key steps of the proof. Since the vector field f is tangent to M_c, the differential equations (23) can be restricted to M_c. The resulting equations on M_c have an invariant measure (see Ch. 1, §3.6, where the explicit formula for the invariant measure is also given). The integrability by quadratures on M_c now follows from Euler's remark. Conclusion 1) of Theorem 13 (first noted by Jacobi) is thus proved. Conclusion 2) is a well-known topological fact: every compact, connected, orientable, two-dimensional manifold on which there is a tangent field with no singular points is diffeomorphic to the two-torus. Conclusion 3) is actually *Kolmogorov's theorem* [107] (1953) on the reduction of differential equations with smooth invariant measure on the torus.

Example 9. Consider Chaplygin's problem on the rolling of a balanced, but dynamically asymmetric ball on a rough horizontal plane (Ch. 3, §1.2, Example 5). The motion of the ball is described by the following system of equations in $R^6=R^3\{\omega\}\times R^3\{\gamma\}$:

$$\dot{k}+\omega\times k=0, \qquad \dot{\gamma}+\omega\times\gamma=0, \qquad k=I\omega+ma^2\gamma\times(\omega\times\gamma). \tag{27}$$

Here I, m, and a are respectively the inertia tensor, the mass, and the radius of the ball. These equations have an invariant measure with the density

$$M=\{(ma^2)^{-1}-\langle\gamma, (I+ma^2 E)^{-1}\gamma\rangle\}^{-1/2}, \qquad E=(\delta_{ij}).$$

Thanks to the existence of the four independent first integral $F_1=\langle k, k\rangle$, $F_2=\langle k, \gamma\rangle$, $F_3=\langle\gamma, \gamma\rangle=1$, $F_4=\langle k, \omega\rangle$, equations (27) can be integrated by quadratures. We remark that system (27) has no equilibrium positions on the regular (noncritical) level sets M_c. In fact, if $\dot{\gamma}=0$ then the vectors ω and γ are linearly dependent, which in turn implies the linear dependence of the differentials dF_2 and dF_4.

Equations (27) are most easily integrated in the case where the constant value of the "area" integral F_2 is zero. In elliptic coordinates u, v on the Poisson sphere $\langle\gamma, \gamma\rangle=1$, the equations of motion on M_c can be reduced to the form

$$\dot{u}=\frac{\sqrt{P_5(u)}}{u(u^{-1}-v^{-1})\,\Phi(u,v)}, \qquad \dot{v}=\frac{\sqrt{P_5(v)}}{v(u^{-1}-v^{-1})\,\Phi(u,v)},$$

$$\Phi=\sqrt{(\alpha-u)(\alpha-v)}.$$

The coefficients of the 5-th degree polynomial P_5 and the constant α depend on the parameters of the problem and the constant values of the first integrals (for details see [14]). The variables u, v assume values in those distinct closed intervals on which the polynomial P_5 is nonnegative. The uniformizing change of coordinates (7) (§2.3), where function $\Phi(z)$ must be replaced by $P_5(z)\,z^{-2}$, introduces angle coordinates on M_c in which the equations of motion take form (26). Moreover, the ratio λ/μ (the rotation number of the tangent vector field) equals the ratio of the real periods of the Abelian integral

$$\int \frac{z\,dz}{\sqrt{P_5(z)}}.$$

In contrast to equations (8), in Chaplygin's problem the variables x, y are not separable. △

According to Propositions 3 and 4, for almost every choice of a pair (λ, μ) the differential equations (26) can be reduced in some angle coordinates x', y' to the form

$$\dot{x}'=\lambda', \quad \dot{y}'=\mu' \qquad (\lambda', \mu'=\text{const}). \tag{28}$$

Proposition 8. *Let* λ *and* μ *be incommensurable over the rationals, and let* $\Phi(x, y)=\sum \Phi_{mn} \exp i(mx+ny)$. *If by a smooth change of angle variables* $x, y \mapsto x', y'$ *system* (26) *can be transformed into system* (28), *then*

$$\sum_{|m|+|n| \neq 0} \left|\frac{\Phi_{mn}}{m\lambda+n\mu}\right|^2 < \infty. \tag{29}$$

In the general case (when the Fourier expansion of function Φ contains all the harmonics) the set of those points $(\lambda, \mu)\in R^2$ for which series (29) diverges is dense in R^2. A discussion of problems concerning the reducibility of equations (26) can be found in Kolmogorov's work [107].

In conclusion, we indicate a number of qualitative properties of the solutions of system (26) which are independent of the arithmetic nature of the numbers λ and μ. Let x_0, y_0 denote the initial values of x, y. It follows from Theorem 4 (under the assumption that λ and μ are incommensurable) that

$$x=x_0+\frac{\lambda}{\langle\Phi\rangle}t+X(t, x_0, y_0), \qquad y=y_0+\frac{\mu}{\langle\Phi\rangle}t+Y(t, x_0, y_0),$$

where

$$\langle\Phi\rangle=\frac{1}{(2\pi)^2}\int_0^{2\pi}\int_0^{2\pi}\Phi\,dx\,dy, \qquad \lim_{t\to\infty}\frac{X}{t}=\lim_{t\to\infty}\frac{Y}{t}=0.$$

The distance d between the points of $T^2=\{x, y \bmod 2\pi\}$ is measured in the metric dx^2+dy^2.

Theorem 14 ([20]). *Suppose that λ and μ are incommensurable. Then:*

a) *for every $\varepsilon>0$ and $T>0$ there is a $\tau>T$ such that*

$$|X(\tau, x_0, y_0)|<\varepsilon, \qquad |Y(\tau, x_0, y_0)|<\varepsilon, \qquad d\{(x(\tau), y(\tau)), (x_0, y_0)\}<\varepsilon$$

for all $(x_0, y_0)\in T^2$;

b) *if $\Phi(x_0, y_0)\neq\langle\Phi\rangle$, then the function $X^2(t, x_0, y_0)+Y^2(t, x_0, y_0)$ vanishes infinitely many times as $t\to\infty$;*

c) *there exist points $(x_0, y_0)\in T^2$, with $\Phi(x_0, y_0)=\langle\Phi\rangle$, such that simultaneously $X(t, x_0, y_0)\geqq 0$ $(\leqq 0)$, $Y(t, x_0, y_0)\geqq 0$ (respectively, $\leqq 0$) for all t.*

From assertion a) it follows, in particular, that (for incommensurable λ and μ) the ergodic phase flow of system (26) is not mixing.[24] We remark that if the rotation number λ/μ is abnormally fast approximated by rational numbers, then system (26) may be weakly mixing (or, equivalently, have continuous spectrum) (see [107]).

4.2. Some Solved Problems of Nonholonomic Mechanics

a) Let us show that the problem of a balanced, dynamically asymmetric ball rolling on a rough surface remains integrable when it is assumed that the particles of the ball are attracted by this plane with a force proportional to distance. Since the center of mass of the ball coincides with its geometric center, the potential can be written as

$$U(\gamma)=\frac{\varepsilon}{2}\int\langle r, \gamma\rangle^2\, dm=\frac{\varepsilon}{2}\langle J\gamma, \gamma\rangle, \tag{30}$$

where γ is the vertical unit vector, r the position vector of the particles of the ball, and J the inertia tensor of the ball relative to its center. The forces of attraction produce the torque

$$\int r\times(\varepsilon\langle r, \gamma\rangle\gamma)\, dm=\varepsilon\int\langle r, \gamma\rangle(r\times\gamma)\, dm=U'_\gamma\times\gamma=\varepsilon(J\gamma\times\gamma).$$

To obtain the moment of the forces relative to the point of contact, we must add the moment of the total force,

$$\varepsilon\int\langle r, \gamma\rangle\gamma\, dm=\varepsilon\langle\int r\, dm, \gamma\rangle\gamma,$$

which vanishes because the center of mass and the geometric center coincide. It follows that the equations describing the rolling of the ball can be written in the form:

$$\dot{k}+\omega\times k=-\varepsilon(J\gamma\times\gamma), \qquad \dot{\gamma}+\omega\times\gamma=0 \tag{31}$$

(cf. equations (5) in Ch. 3, §1.2). They have four independent first integrals: $F_1=\langle k, \omega\rangle+\varepsilon\langle J\gamma, \gamma\rangle$, $F_2=\langle k, \gamma\rangle$, $F_3=\langle\gamma, \gamma\rangle=1$, and $F_4=\langle k, k\rangle-\langle A\gamma, \gamma\rangle$,

[24] See Vol. 2 of this Encyclopaedia, Ch. 4, §1.

where $A = \operatorname{diag}(A_1, A_2, A_3)$, $A_1 = \varepsilon(J_2 + ma^2)(J_3 + ma^2)$, ... (circular permutations). Since equations (31) admit the invariant measure with density (27), it follows, by Theorem 13, that they are integrable.

We remark that the problem of the rotation of a rigid body fixed at a point in the axially-symmetric force field with potential (30) is also integrable: in addition to the classical first integrals F_1, F_2, F_3 it also has the integral F_4, in which one must put $a = 0$. The integral F_4 was first found by F. Tisserand in 1872, in connection with his investigation of the rotation of celestial bodies. The fact is that the potential of a rigid body in a central Newtonian force field coincides with (30) to within $O(\rho^4/R^4)$, where ρ is the typical dimension of the body and R the distance from the body to the center of attraction. It was first noticed by V.A. Steklov (1902) that the Euler-Poisson equations with potential (30) are identical in form with Kirchhoff's equations describing the motion of a rigid body in an ideal fluid in Clebsch's case (1871). Moreover, the integral F_4 is identical with the integral found by Clebsch.

b) As another example consider a homogeneous disc rolling without sliding on a rough horizontal plane subject to the force of gravity. This problem was solved by P. Appell and D. Korteweg in 1899. Consider a moving coordinate system $Oxyz$, with origin O at the center of the disc, such that the Oz axis is orthogonal to the plane of the disc, the Ox axis remains always horizontal, and the Oy axis contains the contact point. Let u, v, w and p, q, r be the projections on these axes of the barycenter's velocity and of the angular velocity of rotation of the disc, respectively. If a denotes the radius of the rolling disc, then

$$u = -ar, \qquad v = 0, \qquad w = ap.$$

Let m and A, C denote the mass of the disc and respectively its moments of inertia relative to the axes Ox and Oz. The following equations of motion hold (see [5]):

$$(C + ma^2)\dot{r} = ma^2 pq, \qquad A\dot{q} = (Cr - Aq \cot \vartheta)p,$$

where ϑ is the angle between the Oz axis and the vertical. Since $p = \dot{\vartheta}$, this system yields the following linear differential equations from which the integrals $r = r(\vartheta)$ and $q = q(\vartheta)$ of the equations of motion are determined:

$$(C + ma^2)\frac{dr}{d\vartheta} = ma^2 q, \qquad \frac{dq}{d\vartheta} = Cr - Aq \cot \vartheta. \tag{32}$$

Appell and Korteweg observed that $q(\vartheta)$ and $r(\vartheta)$ can be expressed through Gauss' hypergeometric series. The angle ϑ, as a function of t, can be found by a simple quadrature from the energy integral

$$m(u^2 + v^2 + w^2) + A(p^2 + q^2) + Cr^2 = -2mga \sin \vartheta + h, \tag{33}$$

where g is the gravitational acceleration and h an arbitrary constant. Using these formulas one can show that the inequality $0 < \vartheta(t) < \pi$ holds for almost all initial conditions (more precisely, for all points except those lying on a certain hypersurface in the phase space). This result explains the astonishing ability of the disc to roll on the plane without falling. In fact, let $q_1(\vartheta)$, $r_1(\vartheta)$ be a solution of the linear system (32) which is bounded for $\vartheta\to 0$ $(\vartheta\to\pi)$. If $q_2(\vartheta)$, $r_2(\vartheta)$ is another solution, then, by Liouville's formula,

$$q_1(\vartheta)\,r_2(\vartheta)-q_2(\vartheta)\,r_1(\vartheta)=c\exp\left(\int_{\pi/2}^{\vartheta}\cot x\,dx\right)=c/\sin\vartheta,\qquad c=\text{const}.$$

Since the right-hand side of this equality tends to $+\infty$ as $\vartheta\to 0$ (or π), it follows that every solution which is bounded for $\vartheta\to 0$ (or $\vartheta\to\pi$) is linearly dependent of the solution $q=q_1(\vartheta)$, $r=r_1(\vartheta)$. Thus, almost certainly $q^2+r^2\to\infty$ at the endpoints of the interval $(0,\pi)$. Since the right-hand member of the energy integral (33) is bounded, it is clear that $\varepsilon<\vartheta(t)<\pi-\varepsilon$ (where $\varepsilon>0$) for all t.

The Appell-Korteweg method can be used to solve the related problem of the motion of a sharp-edged homogeneous circular disc on a smooth horizontal ice surface. The nonholonomic constraint is that the velocity of the disc's contact point be parallel to its horizontal diameter. In contrast to the Appell-Korteweg problem, here the first integrals are expressible in terms of elementary functions and have a simple mechanical meaning: the projections of the angular momentum of the disc (calculated relative to its center) on the vertical and on the Oz-axis are conserved. For almost all initial conditions the disc never falls on the ice and the trajectory of its contact point is bounded. More precisely, the contact point describes a closed curve which, in turn, rotates as a rigid body around a point with constant angular velocity (for details see [111]).

c) Following G.K. Suslov, we also consider the rotation of a rigid body fixed at a point and subject to the nonintegrable constraint $\langle a,\omega\rangle=0$, where a is a constant vector. Suppose the body rotates in an axially-symmetric force field with potential $U(\gamma)$. Applying the method of Lagrange multipliers (Ch. 1, §2.5), we write the equations of motion in the form

$$A\dot\omega+\omega\times A\omega=\gamma\times U'_\gamma+\lambda a,\qquad \dot\gamma+\omega\times\gamma=0. \tag{34}$$

Now using the constraint equation $\langle a,\omega\rangle=0$ we find the Lagrange multiplier λ as a function of ω and γ:

$$\lambda=-\langle a, A^{-1}(A\omega\times\omega)+A^{-1}(\gamma\times U'_\gamma)\rangle/\langle a, A^{-1}a\rangle.$$

Equations (34) always have three independent first integrals:

$$F_1=\langle A\omega,\omega\rangle/2+U(\gamma),\qquad F_2=\langle\gamma,\gamma\rangle,\quad\text{and}\quad F_3=\langle a,\omega\rangle.$$

For real motions $F_1=1$ and $F_3=0$. Thus, the task of integrating equations (34) reduces to that of finding an invariant measure (the existence of which is by no means obvious) and a fourth independent integral. Let us consider the particular case where a is an eigenvector of the operator A. Under this assumption the phase flow of system (34) preserves the "standard" measure in $R^6 = R^3\{\omega\} \times R^3\{\gamma\}$. Suppose that the body rotates in a homogeneous force field: $U(\gamma) = \langle b, \gamma \rangle$. If $\langle a, b \rangle = 0$, then equations (34) admits the fourth integral $F_4 = \langle A\omega, b \rangle$, and hence are integrable by quadratures. This case was noted by E.I. Kharlamova in 1957. We mention another integrable case: if the force function is given by formula (30), then the equations of motion admit a fourth first integral, namely, the one found by Tisserand (see a)).

To conclude we note that in nonholonomic mechanics only relatively few exactly solvable problems are known (practically, all available information can be found in [14, 5, 30]). However, even in the simplest of these problems the behavior of the system can be quite unexpected. Examples (which were already discussed in Ch. 1, §2.5) are provided by the skate on an inclined plane and the homogeneous ball inside a vertical cylinder.

Chapter 5
Perturbation Theory for Integrable Systems

In nature one often encounters systems that differ from integrable ones by small perturbations. Thus, for example, the problem of the motion of planets around the Sun can be regarded as a perturbation of the integrable problem of the motion of noninteracting point masses around a fixed center of attraction. Methods for studying such systems have been developed, which form what is known as *perturbation theory*. These methods are usually simple and effective. They often enable us to describe the perturbed motion almost as completely as the unperturbed motion. Some of them were already proposed and applied by Lagrange and Laplace in their investigations of secular perturbations of planets. The justification of various methods in perturbation theory is rather difficult. Justification questions have been approached only relatively recently, and some of them are still far from being completely solved.

In the present chapter we consider methods of perturbation theory which are grouped around the averaging principle and the idea of dividing the motion into a slow evolution (or drift) and rapid oscillations.

§1. Averaging of Perturbations

1.1. The Averaging Principle. If an integrable conservative system is subjected to a small perturbation then the quantities that are first integrals of the unperturbed system start "evolving" slowly. Over time intervals of order 1 this evolution is small. Over time intervals of order $1/\varepsilon$, where ε is the magnitude of the perturbation, the evolution may become considerable (of order 1). The fundamental principle formulated below, named the *averaging principle*, permits us to write in closed form equations governing the evolution which contain only slowly changing variables.

We shall assume that the joint level surfaces of the single-valued first integrals of the unperturbed integrable system are tori carrying conditionally periodic motions (as in the case of complete integrability), and that the equations of the unperturbed motion can be written in the form

$$\dot{I}=0, \qquad I\in B\subset R^n,$$
$$\dot{\varphi}=\omega(I), \qquad \varphi\in T^m. \tag{1}$$

Here $I=(I_1, \ldots, I_n)$ are first integrals of the problem, the equation $I=\text{const}$ singles out an invariant m-dimensional torus, $\varphi=(\varphi_1, \ldots, \varphi_m)$ are coordinates (phases) on this torus, $\omega(I)=(\omega_1(I), \ldots, \omega_m(I))$ are the frequencies of the unperturbed motion, and B is the domain of R^n in which the variables I are defined.

Example 1. Suppose the unperturbed system describes the motion of the noninteracting planets around the Sun on Keplerian ellipses. Then the first integrals I_j are the major semiaxes, eccentricities, inclinations, and longitudes of the nodes and pericenters, while the phases φ_j are the mean anomalies of the planets. △

A small perturbation of the given system results in the addition of small terms in the equations of motion. In variables I, φ the equations describing the perturbed motion take the form

$$\begin{aligned} \dot{I} &= \varepsilon f(I, \varphi, \varepsilon), \\ \dot{\varphi} &= \omega(I) + \varepsilon g(I, \varphi, \varepsilon). \end{aligned} \tag{2}$$

The functions f and g are 2π-periodic in φ. In equations (2) the variables I and the phases φ are called *slow* and *fast* variables, respectively.

In applications one is usually interested mainly in the slow variables. The *averaging principle* consists of replacing the system of perturbed equations (2) by the *averaged system*

$$\dot{J} = \varepsilon F(J), \qquad F(J) = (2\pi)^{-m} \oint_{T^m} f(J, \varphi, 0)\, d\varphi, \tag{3}$$

for the purpose of providing an approximate description of the evolution of the slow variables over time intervals order $1/\varepsilon$.

In this way one obtains a closed system describing the slow motion, which is considerably simpler than the original system (2); for example, for its numerical integration the step can be chosen $1/\varepsilon$ times larger than for (2). For this reason the averaging principle is extremely productive and widely used in practice.

Let $I(t)$ and $J(t)$, with $I(0)=J(0)$, be the slow motion in the original and averaged system, respectively. According to the averaging principle, we must replace $I(t)$ by $J(t)$. To justify this recipe (which does not always lead to the correct answer) we must find conditions ensuring that for all $0 \leqq t \leqq 1/\varepsilon$, $|I(t)-J(t)| \to 0$ as $\varepsilon \to 0$. If the last relation holds, then it would be desirable to have upper estimates for $|I(t)-J(t)|$ on the interval $0 \leqq t \leqq 1/\varepsilon$. Sometimes one even succeeds in finding such estimates on considerably larger time intervals. These problems are still far from complete solution; they are discussed in the following paragraphs of this section.

Example 2. Consider the system of equations

$$\dot{I} = \varepsilon(a + b\cos\varphi), \qquad \dot{\varphi} = \omega,$$

and the corresponding averaged system

$$\dot{J}=\varepsilon a.$$

Here

$$I(t)=I_0+\varepsilon a t+\varepsilon b[\sin(\omega t+\varphi_0)-\sin\varphi_0]/\omega$$

and

$$J(t)=I_0+\varepsilon a t.$$

The solutions of the exact system oscillate about those of the averaged system with frequency ω and an amplitude of order ε (Fig. 27). *Averaging* simply means ignoring the purely periodic term in the right-hand side of the equation for I. This term has the same order as the remaining one; however, it oscillates and leads only to small oscillations of the solution. The remaining term leads to a drift which over time $1/(\varepsilon a)$ modifies I by a quantity of order 1.

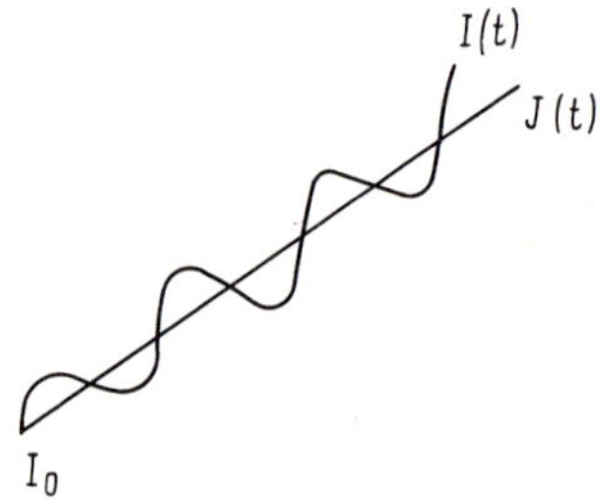

Fig. 27

The averaging principle is based on the idea that in general the oscillating terms ignored through the averaging procedure lead only to small oscillations which are superimposed on the drift described by the averaged system. △

The averaging principle formulated above was used by Lagrange and Laplace in the theory of secular perturbations of planetary orbits. Following their works this principle has become a standard tool in celestial mechanics. It was subsequently rediscovered by van der Pol and used by him to solve problems in the theory of nonlinear oscillations. The wide-scale application of the averaging principle was stimulated by the investigations of L.I. Mandel'shtam, N.D. Papaleksi, N.M. Krylov, N.N. Bogolyubov, and Yu.A. Mitropol'skii. The history of the principle is rather intricate; for an exposition we refer the reader to the introductory sections of monograph [139]. Nowadays the averaging principle is used in many variants (and sometimes under different names) in various applied fields.

Suppose now that in the equations of unperturbed motion the frequencies are identically commensurable, i.e., that one or several relations hold of the form $(k,\omega)=0$, where $k=(k_1,\ldots,k_m)\in Z^m\setminus\{0\}$, $(k,\omega)=k_1\omega_1+\ldots+k_m\omega_m$. Then the trajectories of the unperturbed motion on the torus T^m fill up tori of dimension smaller than m and averaging over the entire T^m cannot, generally speaking, yield a correct description of the motion. Another argument is that the Fourier series expansion of the right-hand side of the perturbed system contains non-oscillatory harmonics, and there are no grounds for discarding them. If over a sufficiently long time interval the frequencies are almost commensurable, then again we may find that averaging over the entire torus is not legitimate. In such cases, named resonant, one resorts to the following procedure, called *partial averaging*. Suppose given one or several *resonance relations* (or simply, *resonances*), i.e., equalities of the form $(k, \omega)=0$ with irreducible integer coefficient vectors k. We let K denote the subgroup of the integer lattice Z^m generated by these vectors, and let $r=\operatorname{rank} K$. The harmonic $\exp(i(k, \omega))$ is said to be *resonant* if $k \in K$. The *partially-averaged system* (taking into account the given resonances) is, by definition,

$$\begin{aligned} \dot{J} &= \varepsilon F_K(J, \varphi), \\ \dot{\varphi} &= \omega(J) + \varepsilon G_K(J, \varphi), \end{aligned} \tag{4}$$

where F_K and G_K denote the sums of the resonant harmonics in the Fourier series of the functions $f(I, \varphi, 0)$ and $g(I, \varphi, 0)$, respectively.

To justify the term "partial averaging" we subject the equations of perturbed motion to the change of phase $\vartheta = R\varphi$, where R is a unimodular matrix with integer entries, the first r rows of which belong to K (such a matrix exists according to [73]). Let $\gamma=(\gamma_1, \ldots, \gamma_r)$ denote the first r components of ϑ, and use $\chi=(\chi_1, \ldots, \chi_{m-r})$ for the remaining components. Averaging the resulting system over the variables χ we obtain equations equivalent to (4).

This argument also shows that in order to describe the variation of J in (4) it suffices to consider the system of $n+r$ equations for J and γ. The J's are slow variables, whereas the γ's are semifast: they vary slowly in the vicinity of the phase points where the resonance relations hold, and fastly far away from these points.

Example 3. Consider the system of equations

$$\begin{aligned} \dot{I}_1 &= -\varepsilon \sin(\varphi_1 - \varphi_2), & \dot{I}_2 &= \varepsilon[\cos(\varphi_1 - \varphi_2) + \sin \varphi_2], \\ \dot{\varphi}_1 &= 1 + I_1, & \dot{\varphi}_2 &= 1. \end{aligned}$$

Take $\gamma = \varphi_1 - \varphi_2$. The phase portrait of this system in the plane I_1, γ is shown in Fig. 28. The region of oscillations is of width $4\sqrt{\varepsilon}$. Independent

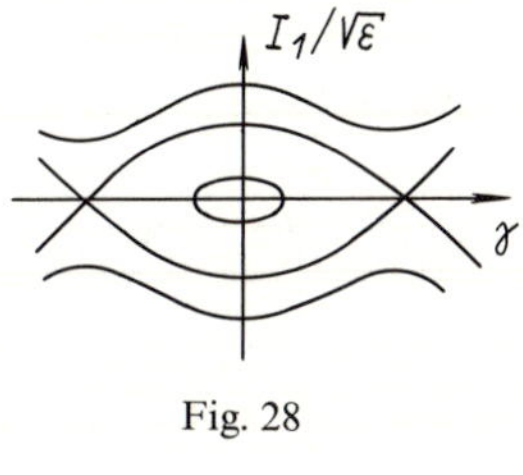

Fig. 28

averaging over the variables φ_1, φ_2 for $I_1(0)=0$ and $\gamma(0)=0$ leads over time $1/\varepsilon$ to an error of order 1. Partial averaging with the resonance $\omega_1-\omega_2=0$ accounted for gives the system

$$\dot{J}_1=-\varepsilon \sin\gamma, \quad \dot{J}_2=\varepsilon\cos\gamma, \quad \dot{\gamma}=J_1,$$

which describes the motion over time $1/\varepsilon$ with accuracy of order ε. △

1.2. Procedure for Eliminating Fast Variables in the Absence of Resonances. The basic role in all the problems connected with the averaging principle is played by changes of coordinates which permit us to eliminate, with a given degree of accuracy, the fast phases from the equations of perturbed motion, and in this way separate the slow from the rapid motion. Such changes of coordinates were constructed by Lindstedt, Bolin, Delaunay, Newcomb, Poincaré, Von Zeipel, Krylov and Bogolyubov. We shall presently describe them in the nonconservative case; the conservative case is treated in §2.

The desired change of coordinates $I, \varphi \mapsto J, \psi$ is sought as a formal series

$$\begin{aligned} I&=J+\varepsilon u_1(J,\psi)+\varepsilon^2 u_2(J,\psi)+\dots \\ \varphi&=\psi+\varepsilon v_1(J,\psi)+\varepsilon^2 v_2(J,\psi)+\dots, \end{aligned} \tag{5}$$

where the functions u_j and v_j are 2π-periodic in ψ. We try to choose u_j and v_j so that in the new coordinates the right-hand sides of the given equations will not contain fast phases, i.e., so that they will take the form

$$\begin{aligned} \dot{J}&=\varepsilon F_0(J)+\varepsilon^2 F_1(J)+\dots \\ \dot{\psi}&=\omega(J)+\varepsilon G_0(J)+\varepsilon^2 G_1(J)+\dots. \end{aligned} \tag{6}$$

Let us assume that the right-hand sides of the perturbed equations (2) are analytic in all the arguments. Subjecting them to the coordinate change (5), replacing $\dot{J}$ and $\dot{\psi}$ in the resulting relations by expressions (6), and equating the terms of the same order in ε, we obtain the following system of relations:

$$\begin{aligned} F_0(J)&=f(J,\psi,0)-\frac{\partial u_1}{\partial\psi}\,\omega, \\ G_0(J)&=g(J,\psi,0)+\frac{\partial\omega}{\partial J}\,u_1-\frac{\partial v_1}{\partial\psi}\,\omega, \\ F_i(J)&=X_i(J,\psi)-\frac{\partial u_{i+1}}{\partial\psi}\,\omega, \quad i\geqq 1, \\ G_i(J)&=Y_i(J,\psi)+\frac{\partial\omega}{\partial J}\,u_{i+1}-\frac{\partial v_{i+1}}{\partial\psi}\,\omega. \end{aligned} \tag{7}$$

The functions X_i, Y_i are uniquely determined by the terms $u_1, v_1, \dots, u_i, v_i$ in expansion (5).

To write the solutions of system (7) constructed above, we introduce new notations. Suppose that the function $h(J,\psi)$ is 2π-periodic in ψ, and expand it in a Fourier series:

$$h(J, \psi) = h_0(J) + \sum_{k \neq 0} h_k(J) \exp(i(k, \psi)).$$

We denote

$$\langle h \rangle^{\psi} = h_0(J), \qquad \{h\}^{\psi} = \sum_{k \neq 0} \frac{h_k}{i(k, \omega)} \exp(i(k, \psi)). \tag{8}$$

Obviously,

$$\frac{\partial}{\partial \psi} \{h\}^{\psi} \omega = h - \langle h \rangle^{\psi}.$$

We call the operator $\{\cdot\}^{\psi}$ the *integration operator*. It is not defined when the denominators in (8) vanish or are very small in comparison with the numerators. The difficulties connected with the presence of these *small denominators* (or *small divisors*) are of fundamental importance in perturbation theory. We shall temporarily ignore them and assume that the integration operator can be applied to all the functions encountered below. Then the solution of system (7) is given by the formulas

$$\begin{aligned}
F_0(J) &= \langle f(J, \psi, 0) \rangle^{\psi}, \\
u_1(J, \psi) &= \{f(J, \psi, 0)\}^{\psi} + u_1^0(J), \\
G_0(J) &= \left\langle g(J, \psi, 0) + \frac{\partial \omega}{\partial J} u_1(J, \psi) \right\rangle^{\psi}, \\
v_1(J, \psi) &= \left\{ g(J, \psi, 0) + \frac{\partial \omega}{\partial J} u_1(J, \psi) \right\}^{\psi} + v_1^0(J), \\
F_i &= \langle X_i \rangle^{\psi}, \qquad u_{i+1} = \{X_i\}^{\psi} + u_{i+1}^0, \quad i \geqq 1, \\
G_i &= \left\langle Y_i + \frac{\partial \omega}{\partial J} u_{i+1} \right\rangle^{\psi}, \qquad v_{i+1} = \left\{ Y_i + \frac{\partial \omega}{\partial J} u_{i+1} \right\}^{\psi} + v_{i+1}^0.
\end{aligned} \tag{9}$$

Here $u_i^0(J)$, $v_i^0(J)$ are arbitrary functions; the customary choice is $u_i^0 = v_i^0 = 0$.

On truncating the series for the coordinate change (5) at the terms of order $r \geqq 1$, we obtain a coordinate change which reduces the equations of perturbed motion to the form

$$\dot{J} = \varepsilon F_{\Sigma}(J, \varepsilon) + \varepsilon^{r+1} \alpha(J, \psi, \varepsilon),$$
$$\dot{\psi} = \omega(J) + \varepsilon G_{\Sigma}(J, \varepsilon) + \varepsilon^{r+1} \beta(J, \psi, \varepsilon).$$

Thus, it turns out that the phases appear only in the terms of order ε^{r+1}. If these terms are ignored, the system of equations for J detaches itself from the rest of the equations. If its solutions can be found, then the evolution of the phase ψ is determined by means of a quadrature. Returning to the original variables, we see that the variation of I is in fact a slow drift (described by the equation for J) on which small rapid oscillations (described by means of the change of coordinates) are superimposed in exactly the same manner as in Example 1 and Fig. 27. The evolution of φ can be represented as a rotation with slowly varying frequency, on which oscillations are again superimposed. In the first approximation this procedure leads to the averaged system (to which the equations that describe approximately the variation of the phases are adjoined).

In the foregoing discussion it was assumed that in the formulas for the change of coordinates the denominators $(k, \omega(J))$ do not vanish in the domain considered. This assumption is fulfilled in single-frequency systems with non-vanishing frequency, in systems with constant incommensurable frequencies, in systems with finitely many harmonics in the perturbation term, and also in some other cases (see §§1.4–1.6 below). However, for general multi-frequency systems this condition fails. Here we encounter two difficulties.

Firstly, in the general case the denominators $(k, \omega(J))$, $k \in Z^m \setminus \{0\}$, vanish on an everywhere dense set, so that formulas (8), (9) do not even allow us to determine the functions u_i, v_i. This obstacle is circumvented by using the following modified change of coordinates. The perturbing terms εf, εg can be expressed as

$$\varepsilon f = f^{(1)} + f^{(2)} + \dots, \qquad \varepsilon g = g^{(1)} + g^{(2)} + \dots,$$

where $f^{(m)}$, $g^{(m)}$ are trigonometric polynomials in φ whose absolute values are bounded from above by a quantity of order ε^m. The elimination of fast phases described above is now carried out as if $f^{(m)}$ and $g^{(m)}$ were terms of order ε^m in the expansion of the perturbation in powers of ε (one ignores the fact that $f^{(m)}/\varepsilon^m$ and $g^{(m)}/\varepsilon^m$ themselves depend on ε). Then in each step of the procedure there arise only finitely many small denominators. Accordingly, there is only a finite number (depending on ε and i) of resonant surfaces $(k, \omega(J)) = 0$, on which the function u_i and v_i are not defined. In each finite order of approximation the map $J, \psi \mapsto I, \varphi$ is, as it turns out, not defined on a finite number of surfaces (the number depends on ε and on the order of approximation). Outside some neighborhood of these surfaces (usually of width $\sqrt{\varepsilon}$) the differences $|I - J|$, $|\varphi - \psi|$ are sufficiently small (usually of order $\sqrt{\varepsilon}$, too), and the map $J, \psi \mapsto I, \varphi$ defines a genuine change of coordinates. Subjecting the equations of motion to this coordinate change we can "push" the phases to terms of higher order in ε.

The second difficulty alluded to above is that during evolution the frequencies $\omega(I)$ themselves vary slowly. As a result, over time $1/\varepsilon$ the phase point may repeatedly visit the neighborhood of the resonant surfaces. Consequently, even the first-approximation coordinate change is not defined, in general, along the entire trajectory over time $1/\varepsilon$. The indicated coordinate change nevertheless remains the main tool for analysing the motion between resonances. The phenomena occurring when a resonant surface is crossed are considered below, in §§1.7 and 1.8. The situation can be roughly described as follows: the total measure of the resonant domains is actually small; consequently, for most initial conditions the motion in these domains cannot have a strong effect on the evolution, and the averaging principle enables us to describe most of the trajectories.

Remark 1. There exist several variants of the procedure for constructing a coordinate change which separates fast from slow motions. In particular, such a coordinate change may be sought not in the form of a series, as above, but as a composition of successive coordinate changes (substitutions). The first coordinate change is given by the formulas

$$I = J + \varepsilon u_1(J, \psi), \qquad \varphi = \psi + \varepsilon v_1(J, \psi),$$

where u_1, v_1 are the first terms of series (5), defined above. In the new coordinates, the phases appear in the perturbed equations in terms of order ε^2. Then an analogous coordinate change "pushes" the phases to the terms of order ε^3, and so on. This method yields the same formal coordinate change (5), but it is often more convenient technically. For Hamiltonian perturbations the method of successive changes of coordinates enjoys the remarkable property of quadratic convergence: the second coordinate change eliminates the phases from the terms of order less than ε^4, the third from the terms of order less than ε^8, and so on (see §2.2 B below).

2. Yet another variant of the procedure for separating slow from fast motions is based on the following considerations: a transformation $x = (I, \varphi) \mapsto (J, \psi) = y$ of the form (5) is the shift by "time" ε along the solutions of a formal system of differential equations

$$\frac{dy}{d\varepsilon} = W(y, \varepsilon), \qquad W = W_1(y) + \varepsilon W_2(y) + \dots .$$

Conversely, every such system of equations generates a shift transformation of the form (5). Therefore, instead of the functions u_i, v_i appearing in (5) we may seek the functions W_i that give the corresponding system of differential equations. The resulting procedure enjoys the desirable feature that it provides sufficiently simple general formulas for the higher-order approximations. For details see [94]. △

1.3. Procedure for Eliminating Fast Variables in the Presence of Resonances. Suppose given a subgroup $K \subset Z^m$ of vectors with integer components which specifies the possible resonances. Let us try to construct a change of variables $I, \varphi \mapsto J, \psi$ so that in the new variables the phases will appear in the right-hand sides of the perturbed equations only in combinations (k, ψ) with $k \in K$. To this end we use series (5).

The equations for J, ψ must have the form

$$\begin{aligned} \dot{J} &= \varepsilon F_0(J, \gamma) + \varepsilon^2 F_1(J, \gamma) + \dots \\ \dot{\psi} &= \omega(J) + \varepsilon G_0(J, \gamma) + \dots, \end{aligned} \tag{10}$$

where $\gamma = (\gamma_1, \dots, \gamma_r)$ are semifast variables (independent combinations of phases ψ with coefficients from K), and $r = \operatorname{rank} K$. Consider also the fast variables $\chi = (\chi_1, \dots, \chi_{m-r})$, such that $(\gamma_1, \dots, \gamma_r, \chi_1, \dots, \chi_{m-r}) = R\psi$, where R is the unimodular matrix introduced in the end of §1.1. Performing the indicated change of variables in the perturbed equations, using expressions (10) for $\dot{J}$ and $\dot{\psi}$, and equating the terms of the same order in ε, we obtain a system of relations of form (7), but now the functions F_i, G_i appearing in their right-hand sides depend on both J and γ. A particular[25] solution of this system is given by the formulas

$$F_0(J,\gamma)=\langle f(J,\psi,0)\rangle^{\chi},$$

$$u_1(J,\psi)=\{f(J,\psi,0)-F_0(J,\gamma)\}^{\psi}+u_1^0(J),$$

$$G_0(J,\gamma)=\left\langle g(J,\psi,0)+\frac{\partial\omega}{\partial J}u_1(J,\psi)\right\rangle^{\chi},$$

$$v_1(J,\psi)=\left\{g(J,\psi,0)+\frac{\partial\omega}{\partial J}u_1(J,\psi)-G_0(J,\gamma)\right\}^{\psi}+v_1^0(J),$$

$$F_i=\langle X_i\rangle^{\chi}, \qquad u_{i+1}=\{X_i-F_i\}^{\psi}+u_{i+1}^0(J),$$

$$G_i=\left\langle Y_i+\frac{\partial\omega}{\partial J}u_{i+1}\right\rangle^{\chi}, \qquad v_{i+1}=\left\{Y_i+\frac{\partial\omega}{\partial J}u_{i+1}-G_i\right\}^{\psi}+v_{i+1}^0(J).$$

Here the symbol $\langle\cdot\rangle^{\chi}$ designates averaging over the variables χ (it is first necessary to express ψ in terms of γ and χ); $\{\cdot\}^{\psi}$ designates the integration operator (8), as in §1.2; finally, u_i^0, v_i^0 are arbitrary functions of J. The formulas contain only denominators (k,ω) with $k\notin K$.

On truncating the series giving the coordinate change at the terms of order r, we obtain a system of equations in which the fast phases appear only in terms of order ε^{r+1}. Omitting these terms we obtain a truncated system which can be used to describe the motion near the resonant surfaces $(k,\omega(J))=0$, $k\in K$. In the first approximation this system is identical to the system obtained by partial averaging of system (4) with the given resonances account for (provided that we take $u_1^0=0$).

1.4. Averaging in Single-Frequency Systems. We consider a system of perturbed equations (2) which contains only one phase. We shall assume that the frequency $\omega(I)$ of variation of this phase does not vanish: $\omega(I)>c^{-1}>0$, where $c=\text{const}$. In this case the system is also called a *system with fastly rotating phase*. Such systems are among the main objects of perturbation theory. Here are some examples:

– oscillatory systems with one degree of freedom subjected to small non-conservative perturbations (for example, a pendulum with a small perturbing torque);

– oscillatory systems with one degree of freedom whose parameters change slowly (for example, a pendulum with slowly varying length [188]);

– systems acted upon by small rapidly oscillating excitations which are time-periodic;

[25] In the general solution, u_i^0 and v_i^0 depend on J, γ, and the functions F_i, G_i change.

– motion in the two-body problem in the presence of a small perturbation (a small "pull" [90], resistance of medium);

– rotation of a rigid body around its center of mass in the presence of a small perturbing torque that depends on the position of the body in space (pull produced by jet engines mounted on the body, resistance of the medium [47, 154]); here the unperturbed motion is described by the Euler-Poisson equations and there are two fast phases, but one of them, namely, the one characterizing the body's precession about the angular momentum vector, does not appear in the equations;

– motion of a charged particle in a magnetic field which changes only slightly over the length of the Larmor radius [60, 161]; in the unperturbed system the magnetic field is constant and the particle moves on a Larmor circle which in turn drifts along the lines of force of the field, and the role of the fast phase is played by the angular coordinate of the point on this circle.

Reference [188] contains numerous examples of analysis of motion in single-frequency systems using averaging.

In the single-frequency case the justification of the averaging principle is practically complete. Below we give results on the accuracy of averaging over time intervals of order $1/\varepsilon$, on properties of the higher-order approximations in the procedure for eliminating the fast variable, and on the connections between the invariant manifolds (stationary points, cycles, invariant tori) of the exact and averaged systems.

Remark. Hereafter, in the statement of theorems we omit the natural conditions ensuring continuability of solutions (absence of blow-up): we shall always assume that for $0 \leqq t \leqq 1/\varepsilon$ the solution $J(t)$ of the averaged system does not approach too closely the boundary of the domain in which the system is defined. △

Suppose that in the perturbed equations (2) the frequency ω and the perturbations f, g are smooth functions in their domain of definition $B \times S^1 \times [0, \varepsilon_0]$, and that in this domain they are bounded, together with their first-order derivatives, by a constant C.

Theorem 1. *For sufficiently small* ε *the difference between the slow motion* $I(t)$ *in the exact system and the motion* $J(t)$ *in the averaged system remains small over time* $1/\varepsilon$: *if* $I(0)=J(0)$, *then*

$$|I(t)-J(t)|<c_1\varepsilon, \qquad 0 \leqq t \leqq 1/\varepsilon,$$

where c_1 *is a constant that depends on the constants* c *and* C *introduced above.*

◁ The first-approximation coordinate change given in §1.2 differs from the identity by a quantity of order ε. It takes the exact system into the averaged system to which a small perturbation (of order ε^2) is added. Over time $1/\varepsilon$, the deviation of the value of the slow variable from its value in the averaged system caused by this perturbation is of order at most ε. Returning to the original variables we obtain the desired result. ▷

This theorem was proved by P. Fatou and, by a different method, by L.I. Mandel'shtam and N.D. Papaleksi (see [139]). The proof given here, based on the elimination of the fast phase through a change of variables, is due to N.N. Bogolyubov.[26]

If the perturbed system is analytic, the procedure described in § 1.2 permits us to eliminate the phase from its right-hand side to any finite order in ε. However, the series obtained in this way diverge in general, so that complete elimination of the phase and separation of the fast and slow motions cannot be achieved. It turns out that the phase can be eliminated up to an exponentially small error. In the general case this error is an essentially irremovable presence in any variant of perturbation theory. Let us formulate a more exact result concerning elimination of phase. We shall assume that the right-hand side of the perturbed system admit an analytic continuation to a complex δ-neighborhood of the domain in which it is defined, and that this continuation remains bounded in modulus by the constant C. Suppose that in this neighborhood $|\omega|>c^{-1}>0$.

Theorem 2 ([157]). *There exists an analytic change of coordinates*

$$I=J+\varepsilon u(J, \psi, \varepsilon), \qquad \varphi=\psi+\varepsilon v(J, \psi, \varepsilon), \qquad |u|+|v|<c_1,$$

which reduces the perturbed system to the form

$$\dot{J}=\varepsilon(\Phi(J, \varepsilon)+\alpha(J, \psi, \varepsilon)),$$
$$\dot{\psi}=\Omega(J, \varepsilon)+\varepsilon \beta(J, \psi, \varepsilon),$$

where

$$|\alpha|+|\beta|<c_2 \exp(-c_3^{-1}/\varepsilon),$$

and

$$|\Phi-\langle f\rangle|+|\Omega-\omega|<c_4 \varepsilon;$$

here $c_1, \ldots, c_4$ are positive constants that depend on c, C, δ, and ε_0. (The notation for the constants is chosen so that the estimates are preserved when c and C are increased.)

The exponential additional term cannot be removed, since when any curve which is initially close to a circle $I=\text{const}$ evolves under the phase flow, its projection on the space of slow variables grows, generally speaking, at least as $\varepsilon t c_5 \exp(-c_6^{-1}/\varepsilon)$.

By investigating the averaged system we can often establish the existence of limit cycles and invariant tori of the original system and also calculate them approximately.

[26] In point of fact, Bogolyubov proved by this method a similar theorem for more general systems (which he called *systems in standard form*), of which particular cases are the single-frequency systems and systems in which the perturbations are quasi-periodic in time [11, 12].

Theorem 3 ([12]). *Suppose the averaged system has a nondegenerate*[27] *equilibrium. Then the exact system has a limit cycle along which the slow variables vary within a neighborhood of order* ε *of the equilibrium. If all the eigenvalues of the averaged system linearized about this equilibrium have negative real parts, then the limit cycle is asymptotically stable. If the real part of at least one of the eigenvalues is positive, then the cycle is unstable.*

◁ We take the phase as a new independent variable (the new time). Consider two maps of R^n into itself: T_0, the shift along the solutions of the averaged system by 2π in the new time, and T_1, the same transformation for the system obtained from the exact system through the first-approximation change of coordinates described in §1.2. Under the maps T_0 and T_1 each point is displaced by a quantity of order ε. Also, T_0 and T_1 differ by a quantity of order ε^2. The map T_0 has a nondegenerate fixed point J_*. By the implicit function theorem, for sufficiently small ε, T_1 has a fixed point $J = J_* + O(\varepsilon)$. The latter clearly serves as an initial condition for the required limit cycle. ▷

Example 4. The *van der Pol equation:*

$$\ddot{x} = -x + \varepsilon(1 - x^2)\dot{x},$$

describes oscillations with a small nonlinear "friction", which is positive for large amplitudes and negative for small ones. The unperturbed equation $\ddot{x} = -x$ can be written in the standard form $\dot{I} = 0$, $\dot{\varphi} = -1$, with $2I = x^2 + \dot{x}^2$ and $\varphi = \arg(x + i\dot{x})$. The perturbed equation for I is

$$\dot{I} = \varepsilon(1 - x^2)\dot{x}^2 = 2\varepsilon I(1 - 2I\cos^2\varphi)\sin^2\varphi.$$

Finally, the averaged equation is

$$\dot{J} = \varepsilon(J - J^2/2).$$

It has a repelling equilibrium $J = 0$ and an attracting one $J = 2$. Corresponding to the former is the equilibrium $x = 0$ of the original system. To the equilibrium $J = 2$ there corresponds, according to the Theorem 3, a stable limit cycle of the original equation, close to the circle $x^2 + \dot{x}^2 = 4$ (Fig. 29). △

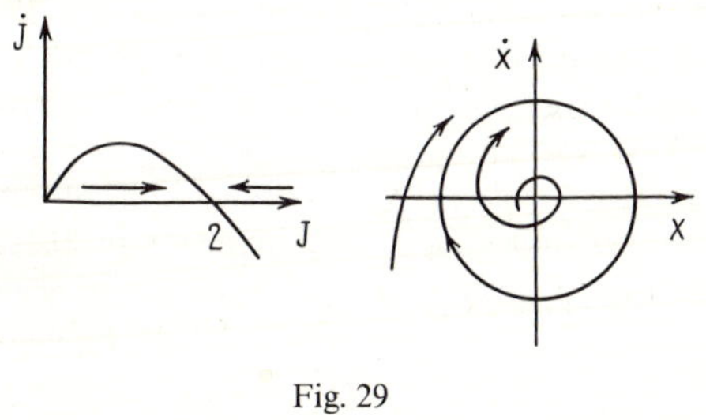

Fig. 29

[27] An equilibrium is said to be *nondegenerate* if the system linearized about it has no zero eigenvalues.

Suppose that the averaged system has a limit cycle and that the characteristic exponents of the system linearized about this cycle have nonzero real parts (except for one, which corresponds to the shift along the cycle and is equal to zero). Then the exact system has a two-dimensional *invariant torus*, stable or unstable together with the cycle; along this torus the slow variables change within a neighborhood of order ε of the limit cycle [12, 120]. The procedure for eliminating the fast variable permits us to obtain a formal series expansion of such an invariant torus in powers of ε. To this end it suffices to insert the expansion of the limit cycle of system (6), which describes the slow motion, in the expansion of the coordinate change (5). As a rule, the series thus obtained diverge (see Proposition 1 and Example 5 below). However, they have an asymptotic meaning: upon truncating these series at the terms of order ε^r we get an approximation of the invariant torus with error $O(\varepsilon^{r+1})$.

The motion on an invariant two-torus created from a limit cycle of the averaged system is characterized by the *rotation number* $\mu(\varepsilon)=\lim_{t\to+\infty}\frac{\vartheta(t)}{\varphi(t)}$, where ϑ, φ (mod 2π) are coordinates on the torus [9]; the rotation number was introduced by Poincaré. If $\mu(\varepsilon)$ is irrational, then the motion is conditionally-periodic and every trajectory is dense on the torus. If the rotation number is rational then there are cycles on the torus; if these cycles are nondegenerate, then they are even in number (half stable and half unstable), and the remaining trajectories are attracted by them for $t\to\pm\infty$. For generic systems, the rotation number $\mu(\varepsilon)$ is a continuous piecewise-constant function of ε on an open everywhere dense set (a kind of *Cantor stairs*, only here the total relative measure of the intervals of constancy on $[0,\varepsilon_0]$ tends to zero as $\varepsilon_0\to 0$). The existence of constancy intervals is connected with the presence of nondegenerate cycles on the torus: for a small change in ε these cycles do not vanish, and consequently the rotation number does not change. In a generic system, as $\varepsilon\to 0$ an infinite sequence of bifurcations consisting of birth and death of cycles occurs on the torus. These various phenomena are not detected through the formal procedure of perturbation theory.

We remark that in an analytic system a limit cycle born out of an equilibrium position of the averaged system is analytic and depends analytically on ε (this follows from the proof of Theorem 3). When a torus is born out of a limit cycle of the averaged system the situation is completely different.

Proposition 1 ([120]). *An invariant torus of the perturbed system is, as a rule (in the general position case), not analytic in the parameter ε. For a dense open set of values of ε the torus has only a finite number of derivatives with respect to the phase variables (this number, however, increases when ε decreases).*

◁ Suppose that for some $\varepsilon=\varepsilon_*$ there exists an asymptotically stable invariant torus such that the rotation number $\mu(\varepsilon_*)$ is rational and the correponding cycles are nondegenerate. For simplicity we assume that the phase space is three-dimensional. The section of the torus by the plane $\varphi = \text{const}$ is shown in Fig. 30, borrowed from [147]. Linearizing the perturbed system about a limit cycle and converting it to a system with constant coefficients (according to the Floquet-Lyapunov theory), we transform the cycle into a saddle or a node, depending on whether it is unstable or stable. The invariant torus consists of the unstable (outgoing) whiskers of the saddle, joined at the nodes. It is clear that a small variation of ε does not change this picture. However, at a node all the curves shown in Fig. 30, except for four that we shall refer to as principal, have, as a rule, only a finite number of derivatives. This is seen even in the linear theory: relative to the principal axes, the phase curves have the form $y=|cx|^{\lambda_1/\lambda_2}$, where λ_1 and λ_2 are the eigenvalues of the reduced linearized systems; λ_1 and λ_2 characterize the rate at which the torus attracts trajectories from the ambient space, and respectively the speed with which the trajectories on the torus approach each other. Since the sections of the unstable whiskers of the saddles do not necessarily coincide with principal curves of the nodes, the torus has a finite number of derivatives (finite smoothness). Nevertheless, this number is very large and grows rapidly as ε decreases, since $\lambda_1/\lambda_2 > c_1^{-1}\exp(c_2^{-1}/\varepsilon)$ (this follows from Theorem 2: if an exponentially small term is added to the system, then all the trajectories on the torus become cycles).

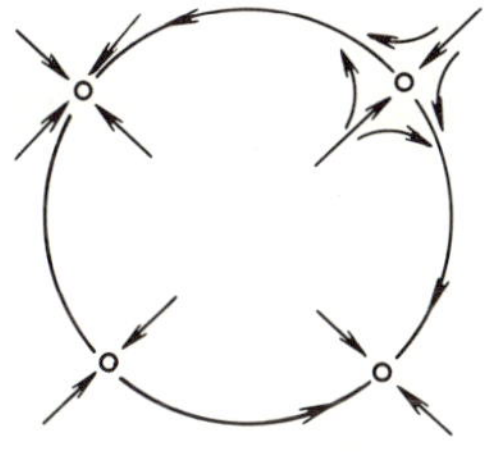

Fig. 30

Further, if there are arbitrarily small values of ε for which the torus is not analytic in the phase variables, then it is also not analytic in ε. In fact, as we observed above, the procedure for eliminating the fast variable permits us to write a formal expansion of the torus in powers of ε. The coefficients of this expansion are all analytic functions in the phase variables. If the torus were analytic in ε in a neighborhood of zero, then this formal expansion would necessarily coincide with the true one (as the asymptotic expansion of an analytic function), and would converge; hence, its sum would be an analytic function of the phase variables. Thus, in the case under consideration the torus is not analytic in ε. ▷

The lack of analyticity of the torus in ε can be demonstrated by very simple examples, in which we nevertheless have analyticity in the phase variables.

Example 5 ([61]). Let ρ and $\vartheta \bmod 2\pi$ be polar coordinates in the plane of slow variables. Consider the perturbed equations

$$\dot{\rho} = -\varepsilon[\rho - 1 + P(\vartheta, \varphi)], \qquad \dot{\vartheta} = \varepsilon, \ \dot{\varphi} = 1,$$

where

$$P = \sum a_k \exp(i(k_1\vartheta + k_2\varphi)), \qquad a_k = \exp(-|k|), \ k_2 \neq 0,$$

in the annulus $1/2 < \rho < 2$.

Averaging over φ we get

$$\dot{\rho} = -\varepsilon(\rho - 1), \qquad \dot{\vartheta} = \varepsilon.$$

The circle $\rho = 1$ is a limit cycle of the averaged system. It is readily verified that the equation

$$\rho = 1 - \varepsilon \sum \frac{a_k \exp(i(k_1\vartheta + k_2\varphi))}{i(k_1\varepsilon + k_2) + \varepsilon}, \qquad k \neq 0$$

defines an invariant torus of the exact system. In an arbitrarily small complex disc $|\varepsilon| < \varepsilon_0$ one can find a value of ε for which one of the denominators in the last expression vanishes. It follows that the indicated invariant torus is not analytic in ε. △

If the averaged system has a degenerate equilibrium position (cycle) then the problem of the existence and stability of a periodic solution (respectively, a torus) of the exact system can, as a rule[28], be settled using the higher-order approximations of the procedure for eliminating the fast variables.

Example 6 (Stability of the topmost equilibrium of a *pendulum with oscillating point of suspension* [59, 100]). The equation of motion of a pendulum whose point of suspension performs vertical sinusoidal oscillations has the following form in the presence of viscous friction:

$$\ddot{\vartheta} + \nu\dot{\vartheta} + (g - a\omega^2 \sin \omega t)\, l^{-1} \sin \vartheta = 0.$$

Here ϑ is the angular deviation of the pendulum from the vertical, a and ω are the amplitude and the frequency of the oscillations of the point of suspension, l the length of the pendulum, g the gravitational acceleration, and ν the damping coefficient. We put $\tau = \omega t$, $K^2 = gl/(\omega^2 a^2)$, $\delta = \nu K \sqrt{l/g}$, and $\varepsilon = \sqrt{a/l}$. We shall assume that K and δ are quantities of order 1 and that ε is small. The equation of motion linearized about the topmost equilibrium position is written:

$$x'' + \varepsilon^2 \delta x' + \varepsilon^2(\sin \tau - K^2 \varepsilon^2) x = 0$$

(the prime denotes differentiation with respect to τ). Introducing the variable $y = x'/\varepsilon$, we write this equation in the system form appropriate for averaging:

$$x' = \varepsilon y,$$
$$y' = -\varepsilon^2 \delta y + \varepsilon(-\sin \tau + K^2 \varepsilon^2) x.$$

[28] For nonconservative perturbations.

In the averaged system the equilibrium position $x=y=0$ is degenerate. The first three steps of the procedure for eliminating fast phases reduce the system to the form:

$$x'=\varepsilon y+O(\varepsilon^4)$$
$$y'=-\varepsilon^2\delta y+\varepsilon^3(K^2-1/2)x+O(\varepsilon^4).$$

Ignoring the terms of order ε^4 we obtain a truncated system in which the equilibrium position $x=y=0$ is stable for $K<1/\sqrt{2}$ and unstable for $K>1/\sqrt{2}$ (for sufficiently small values of ε). It is not hard to show that for damping $\delta\neq 0$ and sufficiently small ε thc topmost equilibrium position of the original pendulum is stable (respectively, unstable) for the indicated values of K.[29]

The phenomenon of stabilization of the topmost equilibrium of the pendulum under vibrations of the point of suspension was discovered by Bogolyubov [59] and Kapitsa [100]. Subsequently, the possibility of stabilization through vibrations was investigated in various works (a list of which is given in [139]), the best known of which is the work of Chelomei [75] on enhancing the stability of clastic systems. Recently discovered examples of the influence of vibrations upon stability are described in reference [76]. △

1.5. Averaging in Systems with Constant Frequencies. Systems with constant (i.e., independent on the slow variables) frequencies arise when considering a small nonlinear interaction of linear oscillatory systems (we note that in linear systems the frequency does not depend on the amplitude), the effect of quasi-periodic perturbations on linear oscillatory systems, or the action of fast quasi-periodic external forces on a nonlinear oscillatory system (for example, the influence of vibrations from two asynchronous engines on the motion of a ship or airplane).

We consider an analytic system of standard form (2) with constant frequencies. We shall assume that the components of the frequency vector ω are strongly incommensurable, i.e.,

$$|(k,\omega)|>c^{-1}|k|^{-\nu},\qquad c,\nu=\text{const}>0 \tag{11}$$

for all vectors $k\neq 0$ with integer components. It is known [27] that for $\nu>m-1$, the set of points ω for which there is no $c>0$ such that condition (11) is satisfied has measure zero.

Theorem 4. *If the frequencies of the unperturbed motion are constant and strongly incommensurable, then the difference between the slow motions in the exact system, $I(t)$, and in the averaged system, $J(t)$, remains small over time $1/\varepsilon$: if $I(0)=J(0)$, then*

$$|I(t)-J(t)|<c_1\varepsilon,\qquad 0\leqq t\leqq 1/\varepsilon.$$

---

[29] For $\delta=0$ and $K>1/\sqrt{2}$ the equilibrium is unstable. For $\delta=0$ and $K<1/\sqrt{2}$ the equilibrium is stable, but the proof of this fact requires new ideas (see §3.5 B). Elimination of fast variables leads, in every order of approximation, to a stable equilibrium with purely imaginary eigenvalues.

◁ The first-approximation change of coordinates in the procedure described in § 1.2 is given by the formula

$$I = J + \varepsilon u(J, \psi), \qquad u = \sum_{k \neq 0} \frac{f_k \exp(i(k, \psi))}{i(k, \omega)}. \tag{12}$$

The Fourier coefficients f_k of the analytic function f decay exponentially as the order of the harmonic increases: $|f_k| < c_3 \exp(-c_4^{-1}|k|)$. The denominators in formula (12) decay only like a power due to the strong incommensurability of the frequencies. It follows that series (12) converges and defines a coordinate change close to the identity. The proof is now completed exactly as for Theorem 1.1. ▷

Remarks. 1. It is clear from the proof that the theorem remains valid if the perturbation has only a finite, though sufficiently high, number of derivatives with respect to the phase; then the Fourier coefficients of the perturbation fall off like a power with sufficiently large exponent. It is readily verified that it suffices to assume the existence of $\nu + m + 1$ derivatives.

2. The proof shows also the validity of the result when the strong incommensurability condition is satisfied only for the vectors $k \in Z^m \setminus \{0\}$ figuring in the Fourier expansion of the perturbation f. △

If the weaker commensurability condition $(k, \omega) \neq 0$, $k \in Z^m \setminus \{0\}$, is satisfied instead of (11), then averaging can still be used to describe the motion. However, the accuracy may be less than ε (for example, $\sqrt{\varepsilon}$ or $1/|\ln \varepsilon|$); namely, the following result is valid:

Theorem 5. *For every $\eta > 0$ there is an $\varepsilon_0 = \varepsilon_0(\eta)$ such that for arbitrary ε with $0 < \varepsilon < \varepsilon_0$,*

$$|I(t) - J(t)| < \eta, \qquad 0 \leqq t \leqq 1/\varepsilon,$$

whenever $I(0) = J(0)$.

The proof can be found in [12].

For systems with constant incommensurable frequencies numerous results concerning the existence of invariant manifolds are available [140]. In particular, suppose that the averaged system has an equilibrium position or a periodic solution, and that the real parts of all (or, in the case of the periodic solution, of all but one) the characteristic exponents of the system linearized about it are different from zero. Then, close (relative to the slow variables) to this equilibrium or periodic solution, the exact system has an invariant torus of dimension m, respectively $m+1$. Even in the analytic case this torus is, as a rule, not analytic, neither in ε nor in the phase variables (see Proposition 1 and Example 6); however, the procedure for eliminating the fast variables permits us to construct for it an asymptotic expansion in ε.

In analyzing the motion on a torus born out of an equilibrium position of the averaged system one has to distinguish two cases, according to whether in the perturbed equations (2) $g = 0$ (i.e., the perturbation is quasi-periodic in time) or $g \neq 0$. If $g=0$, then the motion on the torus is conditionally-periodic with frequency vector ω [61]. If $g \neq 0$ the motion may be quite different. For instance, on a two-dimensional torus the motion is characterized by Poincaré's rotation number (in much the same way as in §1.4; see [120]). If the unperturbed frequencies are regarded as parameters of the problem, then, in an analytic system, for every vector $\omega_* \in R^m$ with strongly incommensurable components and every $\varepsilon > 0$ one can select unperturbed frequencies $\omega = \omega_* + \varepsilon \Delta(\varepsilon, \omega_*)$ so that the motion on the torus under consideration will be conditionally-periodic with frequency vector ω_* [61]. Here $\Delta(\varepsilon, \omega_*)$ is an analytic function of ε. The point is that for a suitable choice of the correction Δ the equation for the phases can be reduced to the form $\dot{\psi} = \omega_*$ by a change of variables $I, \varphi \mapsto J, \psi$ which is analytic, 2π-periodic in phases, and close to identity. The above assertion concerning the motion on the torus is a consequence of this fact.

The systems with constant frequencies are an important particular case of *systems in the standard form of Bogolyubov:*

$$\dot{x} = \varepsilon A(t, x, \varepsilon), \qquad x \in R^p,$$

in which it is assumed that function A satisfies the *condition of uniform average:* the limit

$$\lim_{T \to \infty} \frac{1}{T} \int_0^T A(t, x, 0)\, dt = A_0(x)$$

exists uniformly in x. In fact, upon introducing in the given system with constant frequencies the deviation $\xi = \varphi - \omega t$ from uniform rotation and denoting $x = (I, \xi)$, we obtain a system in standard form. The condition of uniform average is here fulfilled because $A(t, x, \varepsilon)$ is a quasi-periodic function of time t.

The *averaging principle of Bogolyubov* [11] consists of the replacement of the original system in standard form by the averaged system

$$\dot{y} = \varepsilon A_0(y).$$

Many results on averaging in systems with constant frequencies (among them Theorem 5 and the theorem on the birth of conditionally-periodic motions from equilibrium positions and periodic motions of the averaged system) can be generalized to the case of systems in standard form [11].

1.6. Averaging in Nonresonant Domains. Consider a multi-frequency system (2) in which the frequencies depend on the slow variables: $\omega = \omega(I)$, $I \in B$. We say that the domain B is *nonresonant in the first approximation of perturbation theory* (or simply *nonresonant*) if for every $I \in B$ the strong incommensurability condition

$$|(k, \omega(I))| > c^{-1} |k|^{-\nu} \tag{13}$$

holds, with some constants c and ν, for all vectors $k \neq 0$ with integer components such that the harmonic with phase (k, φ) appears in the Fourier expansion of the perturbation f in the right-hand side of system (2).

If the domain B is nonresonant then, in analogy with Theorem 4, the averaging method is applicable and guarantees an accuracy of order ε over times of order $1/\varepsilon$. If condition (13) is replaced by the weaker incommensurability condition $(k, \omega(I)) \neq 0$, then, in analogy with Theorem 5, the averaging method is also applicable, but with a possible loss in accuracy.

In a multi-frequency system in general position there are no nonresonant domains because the incommensurability condition (for the indicated vectors k) is violated, generally speaking, on an everywhere dense set. However, in applications one sometimes encounters problems in which such domains nevertheless exist. This is the case, for example, when the perturbation contains only finitely many harmonics and the frequencies are independent.

1.7. The Effect of a Single Resonance. The main peculiarities of multi-frequency systems are related to resonances. According to the averaging principle, in order to describe the motion near one selected resonance and far from the others, one has to average the equations of motion partially, taking into account the resonance selected. In many cases this approximation can be justified using the procedure of §1.3. Here we shall examine the resulting partially-averaged system. It has the form of a perturbed system, but the perturbation depends on the phases through one of their linear combinations with integer coefficients:

$$\dot{I} = \varepsilon f(I, \gamma), \quad \gamma = (k, \varphi),$$
$$\dot{\varphi} = \omega(I) + \varepsilon g(I, \gamma). \tag{14}$$

Here only one resonance is possible: $(k, \omega) = 0$. The main effects of a single resonance can be observed even on system (14), and for this reason its study is of considerable interest.

Let us assume that the solution of the system obtained by averaging (14) over γ intersects transversally the *resonant surface* (Fig. 31):

$$k \frac{\partial \omega}{\partial I} \langle f \rangle^{\gamma} \neq 0 \quad \text{for } (k, \omega) = 0.$$

The solution of the exact system (14) may behave quite differently. For some relations between phases a *capture into resonance* may occur: the point, falling in the neighborhood of the resonant surface, begins to move so that the commensurability relation that thus arises will be approximately preserved (Fig. 31). Averaging over γ cannot serve to describe such a motion: over time $1/\varepsilon$ the solutions of the exact and averaged systems diverge to a distance of order 1. However, capture into resonance and such a large averaging error can occur only for an exceptional set of initial conditions, the measure of which is bounded from above by a quantity of order $\sqrt{\varepsilon}$. For the remaining

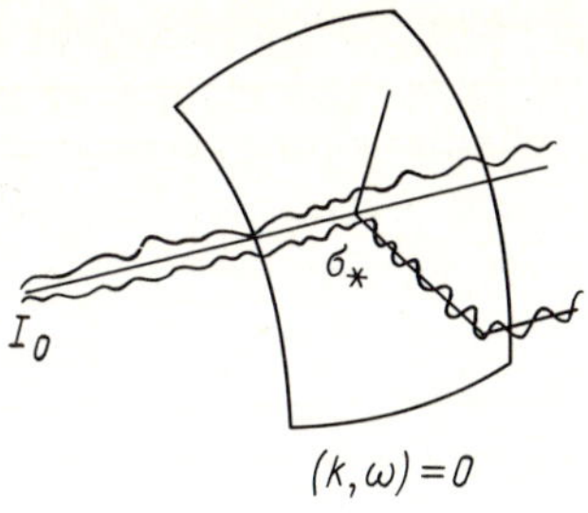

Fig. 31

initial conditions the averaged system describes the motion at least with accuracy $\sqrt{\varepsilon}\,|\ln\varepsilon|$ (under sufficiently general assumptions).

Further, it turns out that if capture into resonance does indeed occur then, as $\varepsilon\to 0$, the set of points whose fate is capture tends to fill the phase space densely: in every ball of diameter ε one can find both points undergoing capture and points avoiding it. If, as happens in practice, the initial conditions are known with an error exceeding ε, then one cannot determine with certainty whether or not the point will be captured by a resonance, and the problem becomes probabilistic in nature. However, one can show that the probability of capture into a resonance is small and tends to zero as $\sqrt{\varepsilon}$ for $\varepsilon\to 0$.

Let us illustrate the phenomena described above by an example. We first remark that in the variables I, γ, (14) becomes a single-frequency system. To the resonance there corresponds vanishing of the frequency, and to the passage through resonance there corresponds a change in the direction of rotation of γ.

Example 7. Suppose that the perturbed motion is described by the system of equations

$$\dot{I}=\varepsilon(1+a\sin\gamma-\tfrac{1}{4}I), \qquad \dot{\gamma}=I,\ a=\text{const}>0 \tag{15}$$

in the domain $|I|<2$. The corresponding averaged equation is

$$\dot{J}=\varepsilon(1-\tfrac{1}{4}J).$$

Differentiating the equation for γ with respect to time we get $\ddot{\gamma}=\varepsilon(1+a\sin\gamma-\frac{1}{4}\dot{\gamma})$. Introducing the slow time $\tau=\sqrt{\varepsilon}t$ and denoting differentiation with respect to τ by a prime, we obtain the equation

$$\gamma''=1+a\sin\gamma-\tfrac{1}{4}\sqrt{\varepsilon}\gamma', \tag{16}$$

describing the motion of a pendulum with constant torque and small friction. The phase portraits of the frictionless pendulum for $a<1$ and $a>1$ are shown in Fig. 32. When $a<1$ the phase portrait of the pendulum with small friction is the same as in the absence of friction. The pendulum shifts from reverse to direct rotation. The durations τ of the motion from the straight line $\gamma'=-1$

to the straight line $\gamma' = +1$ along two distinct phase curves may differ by a quantity of order 1. Returning to the original time and the original variable I we arrive at the following picture: all the points pass through the resonance $I=0$, i.e., here capture does not occur. The time spent to cross the $\sqrt{\varepsilon}$-neighborhood of the resonance may differ from trajectory to trajectory by a quantity of order $1/\sqrt{\varepsilon}$. Accordingly, on crossing this neighborhood an averaging error of order $\sqrt{\varepsilon}$ is accumulated.

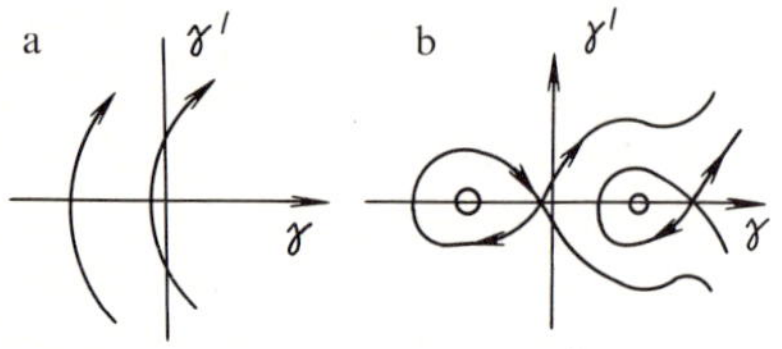

Fig. 32

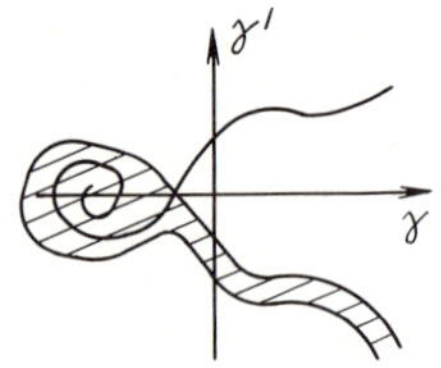

Fig. 33

For $a>1$ the phase portrait of the pendulum with friction is shown in Fig. 33. Along the *separatrix* a strip (of width of order $\sqrt{\varepsilon}$) is formed, consisting of the phase points for which the pendulum shifts from rotations to oscillations. In the original variables to the switch to oscillations there corresponds capture into resonance. In the unshaded region in Fig. 33 the pendulum shifts from reverse to direct rotation. For a trajectory passing at a distance $\xi>\varepsilon$ from the saddle point this shifting takes a time of order $|\ln \xi|$. Returning to the original variables, we see that a portion of order $\sqrt{\varepsilon}$ of phase points is captured into resonance. Consider the exceptional set, of measure of order $\sqrt{\varepsilon}$, consisting of the points which are captured into resonance, or lie in the resonance at the initial time, or pass to within ε of the saddle points. The points in the complement of this set cross the $\sqrt{\varepsilon}$-neighborhood of the resonance in a time t of order between $1/\sqrt{\varepsilon}$ and $|\ln \varepsilon|/\sqrt{\varepsilon}$. The averaging error accumulated at this crossing is of order between $\sqrt{\varepsilon}$ and $\sqrt{\varepsilon}\,|\ln \varepsilon|$.

The portrait of the problem in the original coordinates I, γ on the phase cylinder is shown in Fig. 34. At distance 1 from the resonance the loops of the separatrix lie apart at distance of order ε. Adjacent to the separatrix we see a shaded strip (of width of order $\varepsilon^{2/3}$), consisting of captured phase points. Therefore, when $\varepsilon \to 0$ the captured points tend to fill densely the phase space. △

The main phenomena connected with a single resonance occur in a $c\sqrt{\varepsilon}$-neighborhood, $c=\text{const}$, of the resonant surface. In such a neighborhood system (14) can be transformed to a "pendulum" form resembling equation (16). This transformation has been used in a number of works (see [79, 137, 143, 144, 150]). We shall now describe it. The points on the resonant surface

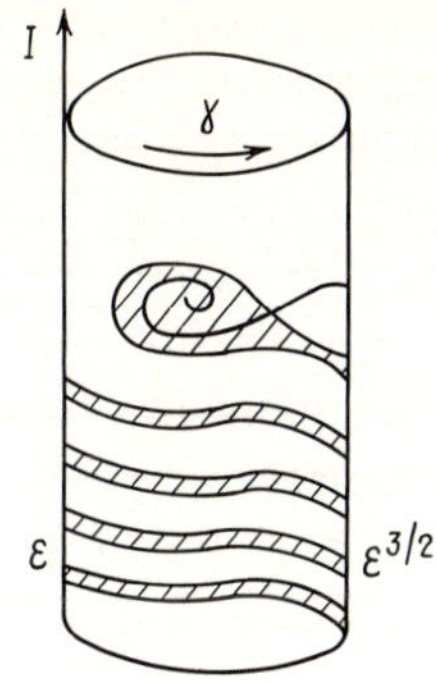

Fig. 34

will be denoted by $\sigma=(\sigma_1, \ldots, \sigma_{n-1})$. A point I in the neighborhood of the resonant surface will be specified by coordinates ρ, σ, where $\rho=(k, \omega(I))$, and σ is the projection of I on the surface. We introduce also the slow time $\tau=\sqrt{\varepsilon}\, t$ and the normalized distance to the resonance surface, $r=\rho/\sqrt{\varepsilon}$. Finally, we use a prime to denote differentiation with respect to τ. We thus get the system

$$\begin{aligned} \gamma' &= r+\sqrt{\varepsilon}\,\alpha_1(\gamma, \sigma, \sqrt{\varepsilon}\, r), \\ r' &= P(\gamma, \sigma)+\sqrt{\varepsilon}\, r\,\alpha_2(\gamma, \sigma, \sqrt{\varepsilon}\, r), \\ \sigma' &= \sqrt{\varepsilon}\,\alpha_3(\gamma, \sigma, \sqrt{\varepsilon}\, r), \end{aligned} \tag{17}$$

in which the functions P and α_i are 2π-periodic in γ. Letting $\varepsilon=0$ in (17) we obtain a Hamiltonian[30] system describing the motion of the pendulum in a potential field under the action of a constant torque:

$$\gamma'=r, \qquad r'=P(\gamma, \sigma), \qquad \sigma=\text{const}, \qquad \langle P\rangle^{\gamma}=k\,\frac{\partial\omega}{\partial I}\,\langle f\rangle^{\gamma} \neq 0. \tag{18}$$

This system is evidently integrable. The motion near resonance is described by means of a small perturbation (17) of this system.

Thus far, we have used averaging to describe the motion of points that pass through resonance without capture. Let us now show how averaging can be used to describe the motion of points that are captured into resonance. To this end, we shall use the equations of motion in form (17). Let us assume that the corresponding "intermediate" unperturbed system (18) (pendulum) satisfies the following two general position conditions: B) for every σ, the unstable singular points on its phase portrait are nondegenerate; B′) there are no separatrices connecting distinct singular points. Then the phase por-

[30] This should come as a surprise! The Hamiltonian nature of system (18) is discovered through calculation, and is by no means obvious beforehand.

trait resembles either Fig. 32b (though it may have more oscillation regions[31]) or Fig. 32a; here we are not interested in the second case, since it corresponds to absence of capture[32]. When σ changes, the existing oscillation regions do not vanish or coalesce with one another, and no new such regions appear. Pick one such region and consider in it the action-angle variables λ, χ of the unperturbed pendulum. The variation of σ, λ, χ in the perturbed system is described by the usual single-frequency system in which the roles of the phase, time, and the small parameter are played by χ, τ, and $\sqrt{\varepsilon}$, respectively. On averaging this system over χ we obtain equations which approximately describe the change of σ and λ over times τ of order $1/\sqrt{\varepsilon}$ (i.e., t of order $1/\varepsilon$)[33]. The change of σ characterizes the drift along the resonance surface, and that of λ – the amplitude of the oscillations about this surface. We say that the initial conditions σ_0, λ_0 for the solution of the system averaged in the oscillation region are taken on the separatrix if $2\pi\lambda_0$[34] is equal to the area of the oscillation region for $\sigma=\sigma_0$. Solutions with such initial conditions describe the motion following the moment of capture into resonance. Solutions terminating on the separatrix are defined in a similar manner; they describe the motion until the exit from resonance. The full trajectory of the point when capture into resonance occurs is approximated by a curve glued from several smooth pieces (Fig. 31). The first piece is a trajectory of the usual system averaged outside the resonance, up to the point, denoted σ_*, where it hits the resonant surface. The second piece is a curve on the resonant surface, determined by that oscillation region of the unperturbed pendulum in which the capture occurred. This curve is the σ-component of the solution of the system averaged in the required oscillation region with initial condition on the separatrix at $\sigma=\sigma_*$. If this solution hits the separatrix at some instant (suppose that this happens for $\sigma=\sigma_{**}$), then there is also a third piece – the solution of the ordinarily averaged system starting at the point σ_{**}. There may be several curves emanating from the point σ_* to meet the resonance surface; they correspond to capture into distinct oscillation regions (Fig. 31). One can show, under sufficiently general assumptions, that by glueing one of these curves to the first (nonresonant) piece one describes the motion with an error of order $O(\sqrt{\varepsilon}\,|\ln\varepsilon|)$ for most initial conditions. The exceptional initial conditions form a set whose measure tends to zero faster than any power of ε as $\varepsilon\to 0$; it consists of points which pass too close to the saddle points of the unperturbed pendulum.

---

[31] The oscillation regions of system (18) may contain unstable singular points. To simplify the exposition, we shall not consider this case here.

[32] It is nevertheless very important, being the case actually obtained for the majority of resonances after averaging (see § 1.8).

[33] These equations may have a nondegenerate equilibrium position with $\lambda\neq 0$. To it there corresponds, by Theorem 3, a limit cycle of the original system that lies inside the separatrix loop.

[34] Recall that $2\pi\lambda_0$ is the area of the domain bounded by a closed trajectory of the unperturbed system.

1.8. Averaging in Two-Frequency Systems. Consider the following *two-frequency perturbed system* with frequencies $\omega_1(I)$ and $\omega_2(I)$:

$$\dot{I}=\varepsilon f(I, \varphi, \varepsilon),$$
$$\dot{\varphi}_1=\omega_1(I)+\varepsilon g_1(I, \varphi, \varepsilon), \qquad \dot{\varphi}_2=\omega_2(I)+\varepsilon g_2(I, \varphi, \varepsilon). \tag{19}$$

We shall assume that its right-hand sides are analytic functions. We say that system (19) satisfies *condition A* if the rate at which the frequency ratio ω_1/ω_2 changes along its trajectories is different from zero:

$$\mathscr{L}(I, \varphi, \varepsilon)=\left(\omega_1 \frac{\partial \omega_2}{\partial I}-\omega_2 \frac{\partial \omega_1}{\partial I}\right) f>c_1^{-1}>0.$$

We say that system (19) satisfies *condition $\bar{A}$* if the rate of change of the frequency ratio ω_1/ω_2 along the trajectories of the corresponding averaged system is different from zero:

$$L(I)=\langle\mathscr{L}\rangle^{\varphi}=\left(\omega_1 \frac{\partial \omega_2}{\partial I}-\omega_2 \frac{\partial \omega_1}{\partial I}\right) F>c_1^{-1}>0, \qquad F=\langle f\rangle^{\varphi},\ \varepsilon=0.$$

Hereafter c_i and C_i will be positive constants.

Theorem 6 ([51]). *Suppose condition A is satisfied. Then the difference between the slow motion in the perturbed system, $I(t)$, and in the averaged system, $J(t)$, remains small over time $1/\varepsilon$: if $I(0)=J(0)$, then*

$$|I(t)-J(t)|<c_2\sqrt{\varepsilon}, \qquad 0 \leqq t \leqq 1/\varepsilon.$$

◁ We define the number $N=N(\varepsilon)$ by the following condition: in perturbation (19) the sum of the amplitudes of harmonics of order larger than N does not exceed ε^2. For an analytic function the amplitude of the harmonic decreases exponentially as the order grows; hence $N<C_1|\ln\varepsilon|$. We say that the resonance $k_1\omega_1+k_2\omega_2=0$, with k_1, k_2 coprime, is *essential* if its order $|k_1|+|k_2| \leqq N$. In the frequency plane resonances are represented by straight lines with rational slopes passing through the origin of coordinates (Fig. 35). The straight lines corresponding to essential resonances are distributed rather sparsely: one can shown that the angle between two such adjacent lines is not smaller than $C_2^{-1}|\ln\varepsilon|^{-2}$.

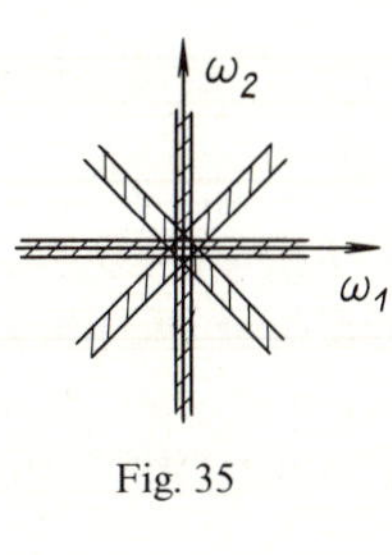

Fig. 35

If the resonance is not essential, then it practically has no effect on the motion over times $1/\varepsilon$. The effect of an essential resonance is felt in a narrow strip about the resonant line (Fig. 35). We call this strip a *resonant zone.* As in Example 7, the actual width of the resonant zone is of order $\sqrt{\varepsilon a_k}$, where a_k is an upper bound for the amplitudes of the resonant harmonics of the perturbation; a_k decays exponentially as $|k|$ grows.

Condition A shows that the point passes successively through the nonresonant and resonant zones portrayed in Fig. 35. In a nonresonant zone the change of variables indicated in §1.2 is defined, and the first approximation it reduces the exact system to the averaged one. The total difference between the solutions of the exact and averaged systems accumulated over the nonresonant zones due to the mismatch left after reduction does not exceed $C_3\sqrt{\varepsilon}$. In resonant zones the averaging in no way describes the motion. However, the total width of these zones is of order $\sqrt{\varepsilon}$. In view of condition A, a solution spends a time of order $1/\sqrt{\varepsilon}$ in these zones. Over this time the solutions of the exact and the averaged systems may diverge to a distance of at most $C_4\sqrt{\varepsilon}$. In the end, the total deviation accumulated over the nonresonant and resonant zones does not exceed $c_2\sqrt{\varepsilon}$. ▷

The condition of Theorem 6 can be weakened. We say that system (19) satisfied *condition* A' (see [103]) if it satisfies condition $\bar{A}$ and, in addition, on each resonance $k_1\omega_1+k_2\omega_2=0$ the rate of change of the frequency ratio along the trajectories of the system partially averaged with this resonance taken into account (see §1.1) is different from zero:

$$L(I)=\left(\omega_1\frac{\partial\omega_2}{\partial I}-\omega_2\frac{\partial\omega_1}{\partial I}\right)F_k(I,\gamma)>\tfrac{1}{2}c_1^{-1}>0, \tag{20}$$

where $\gamma=k_1\varphi_1+k_2\varphi_2$ and $F_k(I,\gamma)$ is the sum of those harmonics of function f which depend on the phase γ. The quantity $|F_k|$ decreases as $|k|$ is increased. Hence, there is a number N_0, which does not depend on ε, such that for $|k|>N_0$ inequality (20) is a consequence of condition $\bar{A}$. We call the resonances with $|k|\leqq N_0$ ($|k|>N_0$) *strong* (respectively, *weak*). It suffices to verify inequality (20) for strong resonances.

Theorem 6′. *If condition A' is satisfied, then the conclusion of Theorem* 6 *holds true.*

◁ As in the proof of Theorem 6, we must examine separately the motion in the resonant and nonresonant zones. In a nonresonant zone the motion is well approximated by the averaged system. Condition $\bar{A}$ guarantees that the phase point does not stick in such a zone. As in Theorem 6, the total error of averaging accumulated over the nonresonant zones does not exceed $C_5\sqrt{\varepsilon}$. Now in a resonant zone the coordinate change of §1.3 is defined, leading in the first approximation to the system partially averaged with this resonance taken into account. Hence, it follows from condition A' that the phase point cannot stick in a resonant zone either. The time spent by a solution in such a zone is of the order of the zone's width, divided by ε. The remaining estimates are identical to those given in the proof of Theorem 6. ▷

Condition $\bar{A}$ formulated in the beginning of this paragraph does not preclude capture into resonance. It turns out that if this condition is satisfied the total effect of the passage through resonances is the same as the effect of a single resonance, described in §1.7.

Theorem 7 ([150]). *Suppose that the given system satisfies condition $\bar{A}$ and also a certain condition B (which is actually almost always satisfied). Then for all initial points I_0, φ_0 outside a set of measure not exceeding $c_2\sqrt{\varepsilon}$, the difference between the slow motion $I(t)$ in the exact system and the motion $J(t)$ in the averaged system remains small over time $1/\varepsilon$: if $I(0)=J(0)$, then*

$$|I(t)-J(t)|<c_3\sqrt{\varepsilon}\,|\ln\varepsilon|, \qquad 0\leqq t\leqq 1/\varepsilon.$$

Given any $\kappa\geqq c_2\sqrt{\varepsilon}$, the following estimate holds for all initial points outside a set of measure not exceeding κ: if $I(0)=J(0)$, then

$$|I(t)-J(t)|<c_4\sqrt{\varepsilon}\,|\ln c_5^{-1}\kappa|, \qquad 0\leqq t\leqq 1/\varepsilon.$$

◁ In nonresonant zones and in resonant zones of weak resonances the analysis is carried out as above. The averaging error accumulated in these zones does not exceed $C_6\sqrt{\varepsilon}$. It remains to examine the strong resonances; as we know, there are not too many of them. The content of condition B is that for each strong resonance the system partially averaged with this resonance accounted for satisfies condition B of §1.7 (the nondegeneracy of the singular points of the corresponding "pendulum" system; see §1.7). Now in a neighborhood of a given strong resonance we perform the coordinate change indicated in §1.3 and thus, in the first approximation, we transform the original system into the corresponding partially averaged system. The latter differs only by a small perturbation from the single-resonance system considered in §1.7, where we also described the passage through resonance. For all initial conditions, except those in a set of measure of order $\sqrt{\varepsilon}$, the time necessary for crossing a resonant zone does not exceed $C_7|\ln\varepsilon|/\sqrt{\varepsilon}$. The averaging error accumulated over this time does not exceed $C_8\sqrt{\varepsilon}\,|\ln\varepsilon|$. Combining the estimates for all the zones, we conclude that the total error does not exceed $c_3\sqrt{\varepsilon}\,|\ln\varepsilon|$. For initial conditions in the complement of a set of measure $\kappa>C_9\sqrt{\varepsilon}$, the time needed for crossing a resonant zone does not exceed $C_{10}|\ln C_{11}^{-1}\kappa|/\sqrt{\varepsilon}$, and the corresponding averaging error does not exceed $c_4\sqrt{\varepsilon}\,|\ln c_5^{-1}\kappa|$. ▷

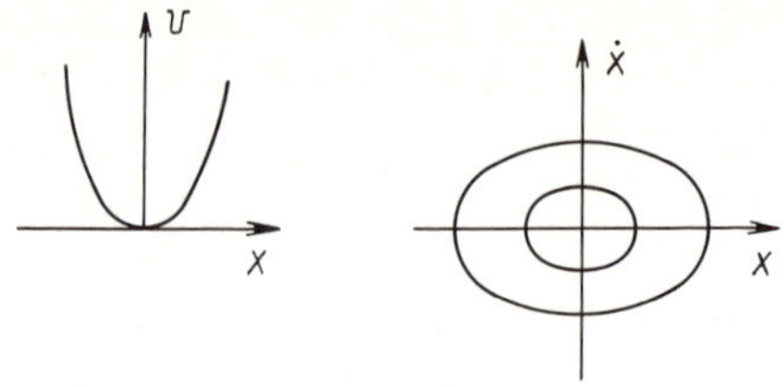

Fig. 36

Example 8. Consider the oscillations of a particle in a potential well, in the presence of a small periodic excitation and a small friction:

$$\ddot{x} = -\frac{\partial U}{\partial x} + \varepsilon a S(t) - \varepsilon \dot{x},$$

where S is a 2π-periodic function of t; the graph of the funtion $U(x)$ and the phase portrait of the unperturbed ($\varepsilon=0$) system are shown in Fig. 36. We shall assume that the unperturbed system is nonlinear, so that motions on distinct trajectories have distinct periods. Let $h=\frac{1}{2}\dot{x}^2+U(x)$ be the energy of the unperturbed motion, and φ the phase on the unperturbed trajectories. The perturbed system has two frequencies: the variables φ and t are fast, whereas h is slow. The equation for h is

$$\dot{h} = \varepsilon[a \dot{x} S(t) - \dot{x}^2].$$

Averaging over φ and t yields

$$\dot{h} = -\varepsilon \langle \dot{x}^2 \rangle.$$

We shall consider the motion in the domain $1<h<2$ (though the results are valid for the entire domain $h \geqq 0$; see [144]). Through averaging we can see that h is a decreasing function and all points eventually leave this domain. For sufficiently small values of a condition A is satisfied, and the averaging method describes the motion for all initial conditions. Condition $\bar{A}$ is satisfied for all a. It guarantees that averaging is applicable for most initial conditions. But a portion of order $\sqrt{\varepsilon}$ of the phase points may get stuck into resonances in the domain $1<h<2$. A detailed analysis of the motion in this problem for $U=\frac{x^2}{2}+\frac{\mu x^4}{4}$ (Duffing's problem) is contained in [144]. △

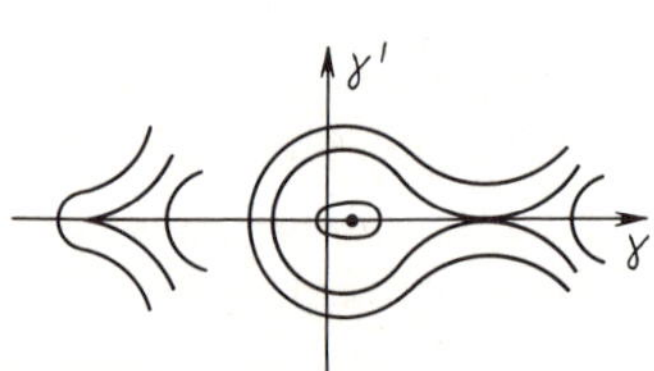

Fig. 37

Remarks. 1. For two-frequency systems, the case must still be investigated in which condition $\bar{A}$ is not fulfilled, i.e., the ratio of the frequencies of fast motion does not vary monotonically in the averaged system.

2. Also requiring further investigation is the case in which condition B fails, i.e., the phase portrait of the "pendulum" corresponding to resonance may be of the type shown in Fig. 37 (it is assumed that condition $\bar{A}$ is fulfilled). If there is a single slow variable, then the following estimate (which is the best that can be obtained) holds: in the complement of a set of measure $\kappa > c_2\sqrt{\varepsilon}$, the averaging error does not exceed $c_3\sqrt{\varepsilon}/\sqrt{\kappa}$; see [150]. If there are several slow variables, the same estimate of the error is know so far only in the complement of a set of measure $\kappa > c_2\varepsilon^{1/3}$; see [168]. △

Let us give some examples of nonconservative two-frequency systems:

– a pendulum under the action of a nonconservative force which is periodic in time [77];

– two weakly coupled nonlinear oscillators with a small friction;

– fast rotation of a heavy rigid body in a resisting medium [47];

– motion of an asteroid in the three-body problem (see Chapter 2) in a resisting medium, or under a small "pull".

In these problems most of the solutions can be described by means of an independent averaging over phases. However, for certain relations between parameters, capture into resonance may occur.

1.9. Averaging in Multi-Frequency Systems. The case where the number of frequencies is larger than two has been studied much less than the two-frequency case. A characteristic feature of the two-frequency systems is that resonant surfaces are simply arranged in the phase space (see Fig. 35). For a higher number of frequencies, the disposition of these surfaces is completely different.

Example 9. Consider the unperturbed three-frequency system

$$\dot{\varphi}_1 = I_1, \quad \dot{\varphi}_2 = I_2, \quad \dot{\varphi}_3 = 1, \quad \dot{I}_s = 0 \quad (s = 1, 2).$$

Here the resonant surfaces are all the straight lines given by rational equations in the plane (I_1, I_2) (Fig. 38). A curve in the plane I intersects many resonant lines: a) under small angles (since arbitrarily close to any linear element there is the linear element of a resonant line), and b) close to points of

Fig. 38

mutual intersection of resonant lines, i.e., multiple-resonance points. Therefore, whereas in the two-frequency case the main effect is passing through a single resonance, in the case of a higher number of frequencies it is necessary to take into account tangency to resonances and the joint effect of several (in the above case of two) resonances. △

An analysis that takes into account the detailed structure of these phenomena for multi-frequency systems is still needed. Nevertheless, some estimates which justify the applicability of the averaging method are available. They were derived on the basis of the following general considerations: If a set of points close to resonant surfaces has a small measure, then for most initial conditions the time spent by the phase curve in this set is short; hence, it is natural to expect that for most initial conditions the averaging method provides a correct description of the motion.

General results in this direction were obtained by Anosov [49] and Kasuga [102]. A theorem of Anosov asserts that *for every positive number* ρ, *the measure of the set of initial conditions (in a compact subset of the phase space) for which the error made in describing the exact motion by averaging exceeds* ρ, *i.e.*,

$$\operatorname{mes}\{I_0, \varphi_0 : \max_{0 \leq t \leq 1/\varepsilon} |I(t) - J(t)| > \rho \text{ if } I(0) = J(0) = I_0\},$$

tends to zero as $\varepsilon \to 0$. This result was proved for perturbed systems of a more general form than the standard form (2): it is not assumed that the common level sets of the integrals of the unperturbed problem are tori; rather, it is required that for almost all constant values of these integrals the unperturbed motion on the common level set be ergodic.

For systems of the standard form (2), the technique of [102] permits one to obtain the following estimate of the averaging error:

Theorem 8 [152]. *Suppose one of the following two nondegeneracy conditions is satisfied: the rank of the map* $I \mapsto \omega(I)$ *equals the number* m *of frequencies, or the rank of the map* $I \mapsto (\omega_1(I) : \omega_2(I) : \ldots : \omega_m(I))$ *(into the projective space) equals* $m-1$. *Then the mean error (over the initial conditions) of the averaging method does not exceed a quantity of order* $\sqrt{\varepsilon}$:

$$\int \max_{0 \leq t \leq 1/\varepsilon} |I(t) - J(t)| \, dI_0 \, d\varphi_0 < c_1 \sqrt{\varepsilon}. \tag{21}$$

Corollary. *Let* $E(\varepsilon, \rho)$ *denote the set of initial conditions, belonging to some fixed compactum, for which the error attains the value* ρ. *Then*

$$\operatorname{mes} E(\varepsilon, \rho) < c_1 \sqrt{\varepsilon}/\rho.$$

An equivalent formulation: *the following bound of the averaging error holds in the complement of a set of measure* κ:

$$|I(t) - J(t)| < c_1 \sqrt{\varepsilon}/\kappa. \tag{22}$$

Estimate (22) cannot be improved (in the class of power estimates) [152]. However, it seem plausible that it may be possible to improve it if we restrict ourselves to generic (general position) perturbations (compare with the estimate obtained for two-frequency systems under conditions $\bar{A}$, B: the error does not exceed $c_2\sqrt{\varepsilon}\ln|\kappa|$).

If $m \geqq n+2$ Theorem 8 is not applicable, but the estimates (21), (22) nevertheless hold for almost all members of a typical family of frequencies with a sufficiently large number of parameters λ (V.I. Bakhtin, Funct. Anal. Appl., *20*, No. 2, 1–7 (1986)). Here instead of the nondegeneracy of frequencies required in Theorem 8, one uses the inequality: $|(k, \omega)| + |\partial(k, \omega)/\partial I| > c^{-1}|k|^{-\nu}$ for $\nu > m-1$ and all $k \in Z^m \setminus \{0\}$. It holds for almost all values of λ with some constant $c > 0$.

Theorem 9 (V.I. Bakhtin). *For generic systems with m fast and n slow variables, the mean value of the error of the averaging method, calculated over the initial conditions, does not exceed a quantity of order* $\varepsilon^{1/(k+1)}$, *provided that* $m \leqq \binom{n+k}{k} - n$.

Accordingly, the right-hand side of estimate (22) takes the form $c_1\varepsilon^{1/(k+1)}/\rho$. The nongeneric systems lie on a certain hypersurface in the space of all systems.

§2. Averaging in Hamiltonian Systems

Poincaré himself called the problem of studying the effects of small Hamiltonian perturbations on an integrable system the *fundamental problem of dynamics*. This problem has numerous applications, and historically it gave rise to the first formulations of the averaging principle as well as to the first results of perturbation theory. The formal side of the theory is in principle the same here as in the case of general non-Hamiltonian systems. However, the nature of the evolution under the effects of Hamiltonian perturbations is completely different. Accordingly, the methods used to justify the procedures of perturbation theory are essentially different from those encountered in the non-Hamiltonian case.

2.1. Application of the Averaging Principle. Suppose that the given unperturbed Hamiltonian system is completely integrable, that some domain in its phase space is foliated into invariant tori, and that action-angle variables I, φ: $I=(I_1, \ldots, I_n) \in B \subset R^n$, $\varphi=(\varphi_1, \ldots, \varphi_n) \pmod{2\pi} \in T^n$ are introduced in this domain. The Hamiltonian H_0 of the unperturbed system depends only on the action variables: $H_0 = H_0(I)$. The equations governing the unperturbed motion have the usual form:

$$\dot{I}=0, \qquad \dot{\varphi}=\partial H_0/\partial I.$$

Suppose now that the system is subjected to a small Hamiltonian perturbation. The perturbed motion is described by the system with Hamiltonian $H=H_0(I)+\varepsilon H_1(I, \varphi, \varepsilon)$, which is 2π-periodic in φ:

$$\dot{I}=-\varepsilon \frac{\partial H_1}{\partial \varphi}, \quad \dot{\varphi}=\frac{\partial H_0}{\partial I}+\varepsilon \frac{\partial H_1}{\partial I}. \tag{23}$$

These equations are in the standard form to which the averaging principle applies. Unless otherwise stipulated, the functions H_0 and H_1 are assumed to be analytic.

Remark. One frequently encounters problems in which the perturbation is also periodic in the time variable t. This case is reduced to the one considered here by introducing the new phase $\varphi_{n+1}=t$ and the canonically-conjugate variable I_{n+1}. The evolution of the extended set of phase variables is described by the system of equations with the Hamiltonian

$$H'=I_{n+1}+H_0(I_1, \ldots, I_n)+\varepsilon H_1(I_1, \ldots, I_n, \varphi_1, \ldots, \varphi_n, \varphi_{n+1}, \varepsilon),$$

which has the same standard form (23). △

Let us assume that all the frequencies $\partial H/\partial I_j$ do not vanish identically and that they do not satisfy identically linear integral relations. In agreement with the principle of § 1.1, in order to describe approximately the variation of the I variables we average equations (23) over the phases φ.

Theorem 10. *In a Hamiltonian system with n degrees of freedom and n frequencies evolution of slow variables does not occur, in the sense that the averaged system has the form* $\dot{J}=0$.

◁ When calculating the integral of $\partial H_1/\partial \varphi_j$ over an n-dimensional torus one can first integrate with respect to the variable φ_j. This one-dimensional integral is equal to the increment of the periodic function H_1 over its period, i.e., it is zero. ▷

Remark. In order to preserve the Hamiltonian form of the equations we generalize the principle of § 1.1 slightly by averaging also the second equation (23), which governs the variation of the angles (phases) φ. The averaged system obtained in this way has the Hamiltonian $\mathscr{H}(J, \varepsilon)=H_0(J)+\varepsilon \mathscr{H}_1(J)$, $\mathscr{H}_1=\langle H_1(J, \varphi, 0)\rangle^{\varphi}$. We see that the phases rotate uniformly with frequencies $\partial \mathscr{H}/\partial J$. △

Example 10. Consider the *planar circular restricted three-body problem* (Ch. 2, § 5). We denote the mass of Jupiter by ε, and assume it to be small when compared with the mass of the Sun. This system has two and a half degrees of freedom (two degrees of freedom plus the explicit periodic dependence on the time variable). Let us pass to a uniformly rotating barycentric coordinate system (with origin at the center of mass of the Sun-Jupiter system), having one axis directed toward Jupiter, and the other perpendicular

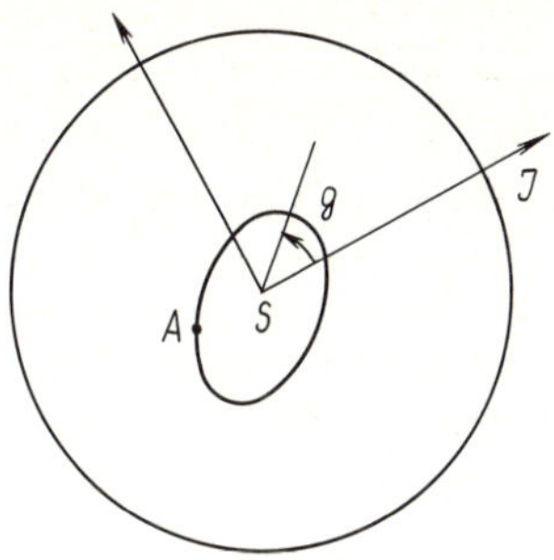

Fig. 39

to it and lying in the plane of Jupiter's orbit (Fig. 39). In this way we obtain a system with two degrees of freedom.

For $\varepsilon=0$ this becomes the unperturbed two-body problem in a rotating coordinate system. In the phase space of this system, the domain of elliptic motions is foliated into two-dimensional invariant tori. We can take the canonical Delaunay elements (see Chapter 2) L, G, l, g, with $L=\sqrt{a}$ and $G=\sqrt{a(1-e^2)}$, as action-angle variables; here a and e are respectively the major semiaxis and the eccentricity of the orbit of the asteroid, l is the mean anomaly of the asteroid, and g is the longitude of the pericenter, measured from the axis directed toward Jupiter (Fig. 39). The Delaunay elements are canonical variables in phase space. They can also be used for describing the perturbed motion. Let us average the perturbed equations for the Delaunay elements over the fast phases l and g. According to Theorem 10 and the related remark, in the averaged system L and G (and hence, a and e) are first integrals, while the phases l and g rotate uniformly with frequencies that differ from the unperturbed frequencies by quantities of order ε. Thus, the averaging principle leads to the following picture of the motion: the asteroid moves along an ellipse which rotates slowly about its focus (the center of mass of the Sun-Jupiter system). △

Example 11. Consider the rotation of a heavy rigid body around a fixed point. We denote by ε the distance from the fixed point to the center of mass of the body and regard it as a small quantity. For $\varepsilon=0$ we obtain the Euler-Poinsot problem (Ch. 4). The action-angle variables $I_1, I_2, \Theta, \varphi_1, \varphi_2, \vartheta$ for this problem are described in [20] (see also Ch. 3, §2.3). We recall that I_2 and Θ are the length of the angular momentum vector and its projection on the vertical axis, respectively, ϑ is the angle of rotation of this vector around the vertical axis and, for fixed I_2, the variables I_1, φ_1 and φ_2 determine the position of the body in the reference frame attached to the angular momentum vector and the vertical axis (Fig. 20). In these variables the Hamiltonian of the perturbed problem has the form

$$H=H_0(I_1, I_2)+\varepsilon H_1(I_1, I_2, \varphi_1, \varphi_2, \Theta).$$

Since Θ is a first integral of the problem, we obtain for I_j and φ_j a system with two degrees of freedom and two frequencies. Applying the averaging principle we conclude that the "actions" I_j are first integrals, while the phases φ_j undergo a uniform rotation close to the rotation in the Euler-Poinsot problem. An analysis of the equation $\dot{\vartheta}=\varepsilon\,\partial H_1/\partial\Theta$ readily shows that in the approximation under consideration the variation of the angle ϑ is close to uniform rotation with an angular velocity of order ε. We are thus led to the conclusion that in the coordinate system attached to the angular momentum vector and the vertical axis, the body moves "almost according to Euler-Poinsot", whereas the angular momentum vector itself undergoes a slow precession around the vertical axis. △

One often encounters *properly* (or *intrinsically*) *degenerate* problems (Ch. 4, §2.1), in which the unperturbed Hamiltonian depends only on a part of the "action" variables and, accordingly, some of the unperturbed frequencies vanish identically:

$$H=H_0(I_1, \ldots, I_r)+\varepsilon H_1(I, \varphi, \varepsilon), \qquad r<n.$$

The phases φ_j, $j>r$, are slow variables. In agreement with the averaging principle, to give an approximate description of the evolution one has to average the perturbed motion over the fast phases φ_i, $i\leqq r$. Proceeding in much the same way as above, one can prove the following statement:

Theorem 11. *In a Hamiltonian system with n degrees of freedom and $r<n$ frequencies the variables canonically conjugate to the fast phases are first integrals of the averaged system.*

In agreement with this theorem, we obtain through averaging a reduced Hamiltonian system with $n-r$ degrees of freedom for the slow phases and their canonically conjugate variables. If the number of fast phases is smaller by only one unit than the number of degrees of freedom (simple degeneracy), then the reduced system has only one degree of freedom. Therefore, in the case of simple degeneracy the averaging principle permits an approximate integration of the problem (as in the nondegenerate case).

Example 12 (Gauss' Problem). Consider the (nonplanar) *circular restricted three-body problem.* We consider that the mass of Jupiter is small in comparison with that of the Sun. The equations of motion in the rotating coordinate system introduced in Example 10, written in canonical Delaunay elements $L, G, \Theta, l, g, \vartheta$ (Ch. 2), are simply degenerate, that is, in the unperturbed motion the angle g (the argument of the latitude of the asteroid's pericenter) is constant. Averaging over the fast phases l, ϑ in this problem is called *Gauss averaging*. The averaged system is integrable (N.D. Moiseev). By Theorem 2.1, L and Θ are first integrals of the averaged system. The variation of G and g after averaging is governed by a Hamiltonian system with one degree of freedom, whose Hamiltonian depends on L and Θ as parameters. The phase portraits of this system for all values of L and Θ were constructed in [187] (using a computer, because the Hamiltonian is rather complicated). This problem was studied analytically in four limiting cases: for small inclinations, in the Hill case (where Jupiter lies considerably farther from the Sun than the asteroid) [127], in the exterior Hill case (where the asteroid lies considerably farther from the Sun than Jupiter) [193], and in the case where the orbits of Jupiter and the asteroid are uniformly close [128].

For small inclinations the motion is qualitatively the same as in the planar case.[35]

Investigation of the Hill case revealed a new phenomenon: for orbits with large inclinations the eccentricity undergoes strong oscillations. In particular, for orbits with 90° inclination which are originally almost circular, the eccentricity grows until it reaches the value 1, which results in the orbit becoming a segment and in collision with the Sun. This may perhaps explain why the Solar system is almost planar, and also why spherically-symmetric planets have no satellites with large inclinations relative to the plane of the Solar system.

In the exterior Hill problem a small precession of the asteroid's orbit occurs around the normal to the plane of Jupiter's orbit, and the orbit itself rotates slowly in its own plane as a rigid body. For vertical orbits precession does not occur. For orbits with inclination $\arccos\frac{1}{\sqrt{5}}$ rotation in the plane of the orbit does not occur.[36] △

Example 13 *(the Lagrange-Laplace theorem on the stability of the Solar system).* We consider the *n-body problem* under the assumption that the mass of one of the bodies (Sun) is considerably larger than the mass of the others (planets). The unperturbed system is here, by definition, the one in which the planets do not interact among themselves and the Sun is at rest. The unperturbed system decomposes into $n-1$ Kepler problems. We assume that the unperturbed orbits of the planets are Keplerian ellipses, and describe each of them using *Poincaré's canonical elements*[37] [15]. We thus get canonical variables for the perturbed system. In the present case there are $n-1$ fast phases, namely, the mean longitudes of the planets. The canonically

[35] If at the initial time the inclination is small, then in the averaged system it remains small for the entire duration of the motion [151].

[36] It is precisely in this way that the orbit of a distant satellite of an axially-symmetric planet evolves (see, e.g., [151]).

[37] These are canonical variables in which the problem is regular for small eccentricities and inclinations. The Poincaré elements Λ, ξ, p, λ, η, q are connected with the Delaunay elements as follows:

$$\Lambda=L, \qquad \xi=\sqrt{2(L-G)}\cos(g+\vartheta), \qquad p=\sqrt{2(G-\Theta)}\cos\vartheta,$$

$$\lambda=l+g+\vartheta, \qquad \eta=-\sqrt{2(L-G)}\sin(g+\vartheta), \qquad q=-\sqrt{2(G-\Theta)}\sin\vartheta.$$

For null eccentricities and inclinations, $\xi=\eta=p=q=0$. The variable λ is the *mean longitude* of the planet.

conjugate variables $\Lambda_j=\sqrt{\mu_j a_j}$, $j=1, \ldots, n-1$, are first integrals of the system averaged over the fast phases. Here a_j are the major semiaxes of the Keplerian planetary orbits, and μ_j are factors depending on the masses. Thus, in the averaged system there is no evolution of the major semiaxes of the planets. This important conclusion is known as Laplace's theorem on the absence of secular perturbations of semiaxes.

Moreover, it turns out that the averaged system has a stable equilibrium position, corresponding to all the planets moving in one plane and in the same direction on circular orbits. The motion of planets which corresponds to small oscillations in the averaged system linearized about this equilibrium position is known as *Lagrangian motion.* It admits the following simple geometric interpretation. Consider the vector with the tail at the focus, directed toward the perihelion of planet, and having length proportional to the planet's eccentricity (the *Laplace vector*). The projection of this vector on the base plane of the coordinate system is a sum of $n-1$ uniformly rotating vectors. The set of angular velocities of these vectors is the same for all the planets. The vector whose length is proportional to the inclination of the planet, and which is directed along the line of intersection of the plane of the planet's orbit and the base plane (line of nodes), is a sum of $n-2$ uniformly rotating vectors.[38] If at some moment of time the eccentricities and inclinations are sufficiently small, then in the averaged system they will remain small for the entire duration of the motion. In particular, collisions of planets and escape at infinity are not possible. This assertion is known as the Lagrange-Laplace theorem on the stability of the Solar system. From the moment this theorem was proved (1784), a mathematical problem central to celestial mechanics became that of establishing this stability result, obtained for the averaged system, for the exact system. Many branches of the theory of dynamical systems, among them perturbation theory and ergodic theory, have arisen in the attempt of solving this problem. To the present day considerable progress in this direction has been achieved. Thus, it was discovered that for sufficiently small planetary masses, a large portion of the domain of phase space corresponding to unperturbed motion on identically directed Keplerian orbits of small eccentricities and inclinations is filled up by conditionally-periodic motions close to Lagrangian motions (see §3). Therefore, "stability" holds for most initial conditions. For an exceptional set of initial conditions the evolution of major semiaxes, if it occurs, is very slow: its mean velocity decays exponentially as the strength of the perturbation is decreased linearly [31] (see §3.4 below). However, it is not yet known whether such an evolution actually occurs, let alone whether it may lead to the destruction of the Solar system. △

Questions concerning the correspondence between solutions of the exact and averaged systems in all the above examples are of a general nature,

[38] The fact that one of the frequencies is missing is connected with the existence of the angular momentum integral of the system.

and can be settled in the framework of the Kolmogorov-Arnold-Moser (KAM) theory; see §3.

For an approximate description of the evolution when the frequencies of unperturbed motion are close to integral commensurability one resorts to partial averaging with the resonances taken into account (see §1.1). For the Hamiltonian systems considered here, this simply means omitting in the Fourier expansion of the perturbed Hamiltonian all the harmonics whose phases vary fastly for the commensurability relations in question. This procedure also leads to first integrals of the averaged system.

Theorem 12. *A Hamiltonian system partially averaged with r independent resonances taken into account has $n-r$ first integrals in involution which are linear integral combinations of the original slow variables I_j.*

◁ We make the symplectic change of variables $I, \varphi \mapsto p, q$ with generating function $W=(p, R\varphi)$, where R is a unimodular integral matrix whose first r rows form a basis in the subgroup of Z^m generated by the coefficient vectors of the resonance relations; R exists according to [73]. In the new variables, averaging amounts to omitting those harmonics in the Hamiltonian which contain the phases $q_{r+1}, \ldots, q_n$. The canonically-conjugate variables $p_{r+1}, \ldots, p_n$ are first integrals of the averaged system. ▷

If there is only one resonance, then the system partially averaged with this resonance taken into account has $n-1$ first integrals in involution (different from the energy integral), and hence it is integrable.

Example 14. Suppose that in the *planar circular restricted three-body problem* the period of the unperturbed motion of the asteroid is approximately half the period of revolution of Jupiter. We use the canonical variables L, G, l, g of Example 10. The quantity $l+2g$ is a slow variable. The generating function W and the new variables p_j, q_j, introduced in the proof of Theorem 12 are given by the formulas

$$W=p_1(l+2g)+p_2(-l-g),$$

and respectively

$$p_1=G-L, \qquad q_1=l+2g,$$
$$p_2=G-2L, \qquad q_2=-l-g.$$

Following averaging with the resonance taken into account, p_2 becomes a first integral, and for p_1, q_1 we obtain a Hamiltonian system with one degree of freedom. Its phase portrait for various values of p_2 and small p_1 is shown in Fig. 40, in which we took $-p_1=-\sqrt{L}(\sqrt{1-e^2}-1)\approx\sqrt{L}e^2/2$ and q_1 as polar coordinates. △

In the single-resonance case the phase portrait of the averaged system is similar to that of the pendulum in a conservative field (under sufficiently general assumptions) [33]. In fact, in the variables p, q introduced in the

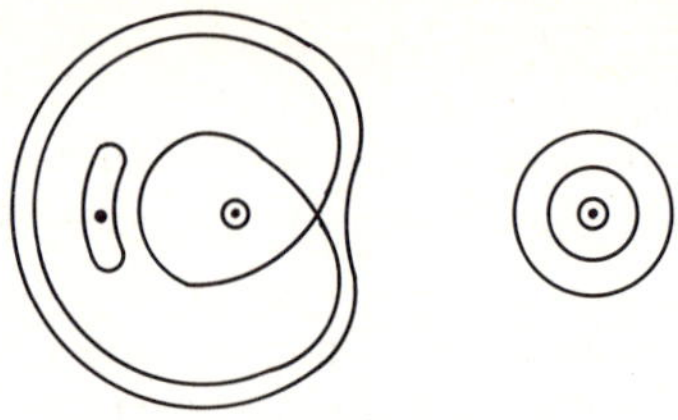

Fig. 40

proof of Theorem 12, the averaged Hamiltonian takes the form

$$F = F_0(p_1, p_2, \ldots, p_n) + \varepsilon F_1(p_1, \ldots, p_n, q_1).$$

Let $p_1^* = p_1^*(p_2, \ldots, p_n)$ be a simple resonance value of p_1, i.e.,

$$\left(\frac{\partial F_0}{\partial p}\right)_{p_1^*} = 0, \quad \left(\frac{\partial^2 F_0}{\partial p_1^2}\right)_{p_1^*} = a \neq 0.$$

In the $\sqrt{\varepsilon}$-neighborhood of this value we introduce the new variable $P_1 = (p_1 - p_1^*)/\sqrt{\varepsilon}$ and, accordingly, the new Hamiltonian $\Phi = F/\sqrt{\varepsilon}$. If the original Hamiltonian F is regular in the neighborhood of the resonance, then

$$\begin{aligned} \Phi &= \sqrt{\varepsilon}(\tfrac{1}{2} a P_1^2 + V(q_1)) + O(\varepsilon), \\ V(q_1) &= F_1(p_1^*, p_2, \ldots, p_n, q_1). \end{aligned} \tag{24}$$

We shall omit the explicit dependence of a and V on the parameters $p_2, \ldots, p_n$. Ignoring the terms $O(\varepsilon)$ in (24), we obtain the Hamiltonian of the system describing the motion of a pendulum in a conservative field, whose phase portrait is shown in Fig. 41. On this portrait we distinguish regions of oscillatory and rotational motions of the pendulum, separated by separatrices. The typical dimension of a region of oscillations in the variable p_1 and the typical amplitude of the oscillations of p_1 are each of order $\sqrt{\varepsilon}$; the typical period of the oscillations is of order $1/\sqrt{\varepsilon}$. The equilibrium positions in Fig. 41 are called stationary resonance modes (regimes). If we also consider the variables $q_2, \ldots, q_n$, to these modes there correspond conditionally-periodic motions (and if the number of degrees of freedom in $n = 2$, periodic motions).

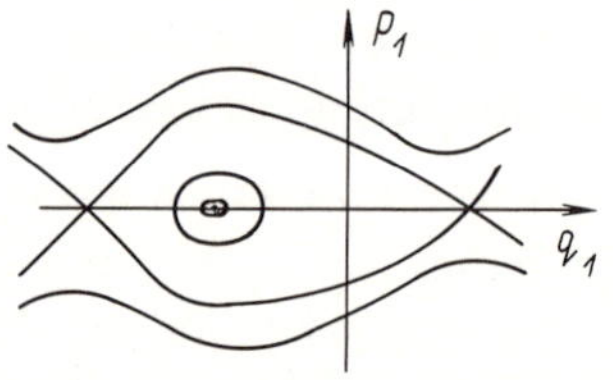

Fig. 41

Remarks. 1. If in the domain of variables under consideration the Hamiltonian is not a regular function, the phase portrait of the averaged system may look different from that shown in Fig. 41. This happens, for instance, in the problem of Example 14, if small values of the eccentricity are considered (cf. Figs. 40 and 41).

2. Effects connected with resonances are often encountered unexpectedly in nature. The large perturbations of Saturn's motion produced by Jupiter ("the large inequality") are related to the fact that the ratio of their Keplerian frequencies is 2/5. Three resonance relations are known for the satellites of Saturn: the frequency ratio of Mimas and Thethys is (approximately) 2/1, of Enceladus and Dione is also 2/1, and of Titan and Hyperion is 4/3. The frequency of the axial rotation of Mercury is 3/2 of its orbital frequency. Tables of commensurability relations found in the Solar system are given in [56]. In most cases the causes of these commensurability relations are not known. The partial averaging procedure discussed above can be successfully used to describe the motion in the vicinity of resonances. △

2.2. Procedures for Eliminating Fast Variables. In §1.2 we described remarkable changes of variables by which we can formally eliminate the fast phases in the right-hand sides of the equations of motion. These changes of variables play a central role in all problems related to averaging. In the case of the Hamiltonian system presently under consideration, they can be chosen to be symplectic. We shall now describe the basic procedure for symplectic elimination of fast variables.

A. Lindstedt's Method. This is one of the earliest methods for eliminating fast phases. We owe its contemporary form to Poincaré [33].

Suppose given a perturbed Hamiltonian system (23) with n degrees of freedom and n frequencies. We assume that there are no identical resonance relations between frequencies. Let us try to find a symplectic change of variables $I, \varphi \mapsto J, \psi$, close to the identity, such that the new Hamiltonian $\mathscr{H}$ will depend only on the slow variables: $\mathscr{H} = \mathscr{H}(J, \varepsilon)$. We seek the generating function of this change of variables and the new Hamiltonian as formal power series in ε:

$$I = J + \varepsilon \frac{\partial S}{\partial \varphi}, \qquad \psi = \varphi + \varepsilon \frac{\partial S}{\partial J}, \tag{25}$$

$$S(J, \varphi, \varepsilon) = S_1(J, \varphi) + \varepsilon S_2(J, \varphi) + \dots$$

$$\mathscr{H}(J, \varepsilon) = \mathscr{H}_0(J) + \varepsilon \mathscr{H}_1(J) + \dots .$$

The functions S_i must be 2π-periodic in φ. The original and new Hamiltonians are related as follows:

$$\mathscr{H}(J, \varepsilon) = H_0\left(J + \varepsilon \frac{\partial S}{\partial \varphi}\right) + \varepsilon H_1\left(J + \varepsilon \frac{\partial S}{\partial \varphi}, \varphi, \varepsilon\right).$$

Equating here the terms of the same order in ε we get the system of equations

$$\begin{aligned} \mathscr{H}_0(J) &= H_0(J), \\ \mathscr{H}_1(J) &= \frac{\partial H_0}{\partial J}\frac{\partial S_1}{\partial \varphi} + H_1(J, \varphi, 0), \\ \mathscr{H}_i(J) &= \frac{\partial H_0}{\partial J}\frac{\partial S_i}{\partial \varphi} + F_i(J, \varphi), \quad i \geqq 2. \end{aligned} \tag{26}$$

The functions F_i are polynomials in $\partial S_1/\partial\varphi, \ldots, \partial S_{i-1}/\partial\varphi$. If the notations $\langle\cdot\rangle^\varphi$ and $\{\cdot\}^\varphi$ for the averaging and integration operators introduced in §1.2 are used, the solution of system (26) is given by the formulas

$$\begin{aligned} \mathscr{H}_1 &= \langle H_1\rangle^\varphi, \qquad S_1 = -\{H_1\}^\varphi + S_1^0(J), \\ \mathscr{H}_i &= \langle F_i\rangle^\varphi, \qquad S_i = -\{F_i\}^\varphi + S_i^0(J), \quad i \geqq 2. \end{aligned}$$

Here S_i^0 are arbitrary functions of J; one often takes $S_i^0 \equiv 0$.

The expression for the integration operator (see §1.2) contains denominators $(k, \omega(J)) = (k, \partial H_0/\partial J)$ with integer-valued vectors $k \neq 0$. Consequently, functions S_i are generally not defined on the dense set of points J at which these denominators vanish or are abnormally small.

Let us ignore small denominators for the moment and assume that the first m functions S_i are defined and smooth. We truncate the series for S at the terms of order ε^m and consider the change of variables defined by the truncated generating function $J\varphi + \varepsilon S_1(J, \varphi) + \ldots + \varepsilon^m S_m(J, \varphi)$. In the new variables we obtain a Hamiltonian in which only the terms of order $m+1$ or greater in ε depend on phases. Omitting these terms we obtain an integrable system of equations in which $J = \text{const}$, whereas the phase ψ rotates uniformly with a frequency that depends on J. Substituting this solution in the formulas for the change of variables we obtain an approximate solution of the original system. Its accuracy and the time interval during which it applies increase as the order m of the approximation increases. On the time interval $(0, T)$ this solution guarantees the accuracy $O(\varepsilon^{m+1}T)$ for the slow variables, and $O(\varepsilon^{m+1}T^2)$ for the fast ones. Should the series for the change of variables converge, the procedure described above would allow us to integrate the original perturbed system.

To attach a realistic meaning to these considerations in the presence of *small denominators*, we write the perturbation as

$$\varepsilon H_1(I, \varphi, \varepsilon) = H^{(1)}(I, \varphi, \varepsilon) + H^{(2)}(I, \varphi, \varepsilon) + \ldots,$$

where each $H^{(i)}$ is a trigonometric polynomial in φ which is bounded in modulus by a quantity of order ε^i. We shall carry out the procedure described above for eliminating fast variables, regarding $H^{(i)}$ as the term of order ε^i in the expansion of the perturbation in a series in ε ("ignoring" that the function $H^{(i)}/\varepsilon^i$ itself depends on ε). Then in each order of approximation of the procedure only finitely many small denominators occur and, accordingly, there will be only a finite collection of surfaces $(k, \omega(J)) = 0$ (their number depends on ε and on the order m of the approximation), on which the functions S_i, $1 \leqq i \leqq m$, are not defined. Outside a small neighborhood of the union of these surfaces, the truncated generating function introduced above defines a coordinate change that approximately integrates the original system of equations. By the foregoing discussion, the system of functions S_i, $1 \leqq i < \infty$, is, as a whole, not defined on a dense set of values of J.

Lindstedt's method is very effective because it provides a simple procedure for integrating approximately the perturbed Hamiltonian system. This method has played a large part in the development of perturbation theory, since it led to an expansion of the general solution of the perturbed Hamiltonian system in a formal series containing only terms periodic in time. Poincaré called the methods yielding such expansions "new", as opposed to the "old" methods which led to *secular terms* of the form t^m and $t^m \sin lt$, $t^m \cos lt$ [33]. The discovery of the "new" methods has completely changed the formulation of the stability problem for perturbed Hamiltonian systems (including the Solar system). The occurrence of secular terms in the "old" methods, although actually caused by the method of expansion itself (similar to the way in which a secular term arises in the expansion $\sin(1+\varepsilon)t = \sin t + \varepsilon t \cos t + \ldots$), was regarded as a test for instability of motion[39]. Efforts were directed toward proving, for concrete perturbations, that no such terms are present in the leading orders of the expansion. For the Solar system Laplace showed that secular terms are absent to the first order in the strength of the perturbation. Poisson found that to the second order in the strength of the perturbation there are no purely secular terms (of the form t^m), whereas mixed terms (of the form $t^m \sin lt$, $t^m \cos lt$) do indeed arise. The "new" methods made it possible to construct formal theories free of secular terms in which the problem is the convergence of the resulting expansions.

As Poincaré noted, in general Lindstedt's series diverge.

Example 15. Consider the following system in which the frequencies of the unperturbed motion are constant and incommensurable:

$$H = \omega_1 I_1 + \omega_2 I_2 + \varepsilon(I_1 + \textstyle\sum a_k \sin(k, \varphi)), \qquad k \neq 0$$
$$|(\omega, k)| > c\,|k|^{-\nu}, \qquad c, \nu = \text{const} > 0, \; a_k = \exp(-|k|).$$

Here one can calculate all the approximations of Lindstedt's method. If the Lindstedt series were convergent, then along the motions the quantity $|I|$ would undergo only bounded oscillations.

On the other hand, the perturbed system can be readily integrated. The phases rotate uniformly with frequencies $\omega_1 + \varepsilon$ and ω_2, and the equation governing the variation of I is solved by a quadrature. For a rational frequency ratio $(\omega_1 + \varepsilon)/\omega_2$ the quantity $|I|$ increases unboundedly along the

[39] Here we have in mind *Lagrange stability*. A motion is called Lagrange stable if its trajectory remains forever in a bounded domain of the phase space.

motions (as can be readily verified). We conclude that Lindstedt's series diverge. △

As is often the case, divergence is connected with the fact that one constructs certain objects that actually do not exist. Here the situation is roughly as follows: If for some J Lindstedt's series converge, then the perturbed system has an invariant torus $J=\text{const}$, on which the phase rotates with frequency vector $\hat{\omega}(J, \varepsilon)=\frac{\partial H_0(J)}{\partial J}+\varepsilon\frac{\partial \mathscr{H}_1(J)}{\partial J}+\dots$. The unperturbed frequencies are incommensurable (the value of J was indeed chosen to guarantee this). However, the frequencies of the perturbed motion become commensurable for certain values of ε. Then the invariant torus is foliated into tori of lower dimension. Such a situation is highly degenerate and does not arise in generic systems. For this reason Lindstedt's series generally diverge.

Remark. There exists a variant of Lindstedt's method in which one seeks invariant tori with incommensurable frequencies fixed beforehand. The corresponding series converge [26]. The proof of the convergence of these series rests upon their identification as convergent series of the KAM theory (see §3). A direct proof of the convergence is not yet known. △

B. von Zeipel's Method. This method extends Lindstedt's procedure to the case in which one eliminates from the Hamiltonian only some of the phases. It permits us to consider systems with proper degeneration and resonances. Von Zeipel's method reveals the potentialities of the methods of Delaunay and Bohlin developed earlier for the same purpose.

Suppose again that system (23) is subjected to a symplectic change of coordinates of form (25), close to the identity. We seek the new Hamiltonian as a formal series $\mathscr{H}(J, \psi, \varepsilon)=\mathscr{H}_0(J)+\varepsilon\mathscr{H}_1(J, \psi)+\dots$. The old and new Hamiltonians are related as follows:

$$\mathscr{H}_0(J)+\varepsilon\mathscr{H}_1\left(J, \varphi+\varepsilon\frac{\partial S}{\partial J}\right)+\dots=H_0\left(J+\varepsilon\frac{\partial S}{\partial\varphi}\right)+\varepsilon H_1\left(J+\varepsilon\frac{\partial S}{\partial\varphi}, \varphi, \varepsilon\right).$$

Equating here the terms of the same order in ε, we again obtain relations (26), but now the functions $\mathscr{H}_i$ depend on the phases, and the F_i's are calculated differently.

Let $\chi\in T^{n-r}$ be the set of phases that we wish to eliminate from the Hamiltonian, and let $\gamma\in T^r$ be those that remain. Then we can take

$$\begin{aligned} &\mathscr{H}_1=\langle H_1(J, \varphi, 0)\rangle^{\chi}, && S_1=-\{H_1-\mathscr{H}_1\}^{\varphi}+S_1^0(J), \\ &\mathscr{H}_i=\langle F_i(J, \varphi)\rangle^{\chi}, && S_i=-\{F_i-\mathscr{H}_i\}^{\varphi}+S_i^0(J), \quad i\geqq 2. \end{aligned} \tag{27}$$

Here S_i^0 are arbitrary functions of J; for instance, we can choose $S_i^0\equiv 0$.

If, as above, we truncate the series for S at the terms of order ε^m, then consider the "truncated" coordinate change, and finally ignore the terms of order ε^{m+1} in the transformed Hamiltonian, we obtain the Hamiltonian of the m-th approximation. It does not depend on the phases χ, and accordingly the resulting approximate system possesses $n-r$ first integrals and reduces to a system with r degrees of freedom. Here, as in Lindstedt's method, small denominators appear. To ensure that in each order of approximation will appear only finitely many small denominators, the described procedure must be modified by analogy with §2.2.A.

Suppose that the system in question is *properly degenerate*, i.e., the unperturbed Hamiltonian does not depend on some of the "action" variables. Then both slow and fast phases are present and the procedure discussed above enables us to eliminate the fast phases from the Hamiltonian. The first-approximation system for the new variables coincides with the system averaged over the fast phases.

Suppose now that the frequencies of the unperturbed system satisfy r independent resonance relations. We relabel the phases so that when these relations are fulfilled the first r (last $n-r$) phases are semifast (respectively, fast). Then the described procedure permits us to formally eliminate the fast phases from the Hamiltonian. The first-approximation system for the new variables is identical with the system partially averaged with the resonances taken into account. In practice, here one can avoid introducing resonant combinations of phases as new variables, and instead work in the original variables. Then in formulas (27) averaging over the fast phases is replaced by the following operation: in the Fourier expansion one omits those harmonics which in the unperturbed motion oscillate in the presence of the given resonance relations.

In general, von Zeipel's series, like Lindstedt's series, diverge.

Instead of von Zeipel's method, one often resorts to its modified version proposed by G. Hori and A. Deprit (see, e.g., [94]). In this version the symplectic change of coordinates that eliminates the phases is given not by a generating function, but rather by a *generator* – a function $W(J, \psi, \varepsilon) = W_1 + \varepsilon W_2 + \dots$ such that the required transformation $I, \varphi \mapsto J, \psi$ is the shift by time ε along the trajectories of the Hamiltonian system with the Hamiltonian W. This is more convenient, since a generating function contains simultaneously old (φ) and new (J) variables, whereas the generator depends only on the new variables. Hence, by using the generator one is spared the additional task of solving functional equations in order to insure that all expressions involved will contain only the old or only the new variables. Simple recursion relations are available, by which the coefficients in the power series expansion of the new variables and Hamiltonian in ε may be expressed in terms of the old variables and Hamiltonian, and the generator. The coefficients in the expansion of the generator are determined successively from a system of relations equivalent to (27). For example, the first approximation for the generator is simply identical with the first approximation for the generating function: $W_1(J, \psi) \equiv S_1(J, \psi)$. Computer programs are available which implement the Hori-Deprit procedure in character form. One of these programs was used to improve Delaunay's classical theory of Lunar motion [84].

C. Methods of the KAM Theory. The Kolmogorov-Arnold-Moser (KAM) theory provides convergent methods for integrating perturbed Hamiltonian systems. The basis of these methods is the construction of successive changes of coordinates which kill in the Hamiltonian the terms depending on the fast phases in increasingly higher orders in the small parameter. The procedure of successive coordinate changes was proposed by Newcomb. We owe its present form to Poincaré, who, however, considered Newcomb's procedure to be equivalent to that of Lindstedt.

As the works of A.N. Kolmogorov [22] and V.I. Arnold [6, 7] revealed, this procedure of successive coordinate transformations actually possesses the remarkable property of quadratic convergence: following m transformations, the mismatch in the Hamiltonian depending on the phases has order ε^{2^m} (ignoring small denominators). This *"superconvergence"* annihilates the effects of small denominators and ensures the convergence of the full procedure on a certain "nonresonance" set.

The procedure of successive coordinate changes can be implemented in different ways. Below we describe Arnol'd construction, which is similar to Newcomb's original method.

Consider the unperturbed Hamiltonian system with Hamiltonian

$$H(I, \varphi, \varepsilon) = H_0(I) + \varepsilon H_1(I, \varphi, \varepsilon). \tag{28}$$

We make a symplectic coordinate transformation $I, \varphi \mapsto J, \psi$, such that in the new coordinates the terms of order ε in the Hamiltonian will not depend on the phases. Such a coordinate transformation has already been constructed in §2.2.A, while considering the first approximation in Lindstedt's method. It is given by the generating function

$$J\varphi + \varepsilon S(J, \varphi), \qquad S = -\{H_{1N}(J, \varphi, \varepsilon)\}^{\varphi}. \tag{29}$$

Here $\{\cdot\}^{\varphi}$ designates the integration operator and $H_{1N}(J, \varphi, \varepsilon)$ is the sum of the harmonics of order less than or equal to N in the Fourier expansion of function H_1, where N is an integer. We select N so that the modulus of the remainder $R_{1N} = H_1 - H_{1N}$ of the Fourier series will not exceed ε. The new Hamiltonian has the form

$$\begin{aligned}
\mathscr{H}(J, \psi, \varepsilon) &= \mathscr{H}_0(J, \varepsilon) + \varepsilon^2 \mathscr{H}_1(J, \psi, \varepsilon),\\
\mathscr{H}_0(J, \varepsilon) &= H_0(J) + \varepsilon \langle H_1 \rangle^{\varphi},\\
\varepsilon^2 \mathscr{H}_1(J, \psi, \varepsilon) &= \left[H_0\left(J + \varepsilon\frac{\partial S}{\partial \varphi}\right) - H_0(J) - \varepsilon \frac{\partial H_0}{\partial J}\frac{\partial S}{\partial \varphi}\right] +\\
&\quad + \varepsilon\left[H_1\left(J + \varepsilon\frac{\partial S}{\partial \varphi}, \varphi, \varepsilon\right) - H_1(J, \varphi, \varepsilon)\right] + \varepsilon R_{1N}(J, \varphi, \varepsilon).
\end{aligned} \tag{30}$$

In the right-hand side of the last of these equalities one has to express φ in terms of ψ, J using the formulas giving the coordinate transformation.

The new Hamiltonian has the same form as the old one, but the phases appear only in the terms of order ε^2. We now subject the system we obtained to a similar coordinate transformation. This "pushes" the phases further to the terms of order ε^4 (see (30)). After m such coordinate transformations the dependence of phases is preserved only in the terms of order ε^{2^m}. We remind the reader that in Lindstedt's method, after the m-th approximation coordinate transformation, the dependence on the phases is "pushed" to the terms of order ε^{m+1} only.

The estimate ε^{2^m} indicates the formal order in ε of the mismatch in the Hamiltonian. Actually the mismatch can be considerably larger due to the influence of small denominators.

Example 16. Consider the first coordinate transformation described above in the domain in which the frequencies satisfy the usual incommensurability condition $|(k, \omega(J))| \geqq \kappa |k|^{-\nu}$, $0 \leqq |k| \leqq N$. Then, obviously,

$$|I-J| = \left|\varepsilon \frac{\partial S}{\partial \varphi}\right| \sim \frac{\varepsilon}{\kappa}, \qquad |\varphi-\psi| = \left|\varepsilon \frac{\partial S}{\partial J}\right| \sim \frac{\varepsilon}{\kappa^2}, \qquad |\varepsilon^2 \mathscr{H}_1| \sim \frac{\varepsilon^2}{\kappa^2}.$$

The value of κ may vary from a quantity of order 1 to one of order $\sqrt{\varepsilon}$ (for $\kappa \sim \sqrt{\varepsilon}$ we have $\left|\varepsilon \dfrac{\partial S}{\partial J}\right| \sim 1$) and the generating function does not necessarily define a bijective correspondence $I, \varphi \leftrightarrow J, \psi)$. For $\kappa \sim \sqrt{\varepsilon}$ we get $|\varepsilon^2 \mathscr{H}_1| \sim \varepsilon$ instead of the formal estimate ε^2. △

We shall consider the whole sequence of coordinate changes in the nonresonant set on which the resulting small denominators are bounded from below by $c\sqrt{\varepsilon}|k|^{-\nu}$, where $c = \text{const} > 0$, $\nu = \text{const} > n-1$ and $|k|$ is the order of the harmonic corresponding to the denominator. It turns out that on this set the fast increase of the order of the mismatch with ε counteracts the influence of the small denominators, and the composition of successive coordinate transformations converges. This assertion is central to the KAM theory. Its implications are formulated in §3.

The procedure of successive coordinate transformations guarantees superconvergence also in the case of those degenerate systems for which the degeneracy is "removable". By this we mean that the Hamiltonian of the problem has the form

$$H = H_{00}(I) + \varepsilon H_{01}(I) + \varepsilon^2 H_1(I, \varphi, \varepsilon),$$

where H_{00} depends only on $r < n$ action variables, while $H_{01}(I)$ depends on all n variables. One takes $H_{00} + \varepsilon H_{01}$ to be the unperturbed Hamiltonian. The unperturbed problem has n frequencies, as in the nondegenerate case, but r of them are of order 1, whereas the remaining $n-r$ are of order ε. The perturbation is $1/\varepsilon$ times smaller than the minimal frequency. The setup of the procedure of successive coordinate transformation is exactly the same as in the nondegenerate case. It turns out that it converges on the corresponding nonresonant set [7].

In the foregoing analysis we assumed that the perturbation H_1 is an analytic function. If H_1 has only finite smoothness (i.e., a finite number of derivatives), then the procedure of successive coordinate transformations described above leads to a "loss of derivatives": in each order of approximation the perturbation has fewer derivatives than in the preceding one. For this reason the procedure breaks down after a finite number of steps. For the case of perturbations with finite smoothness Moser proposed a modified method that uses a smoothing technique invented by Nash [149]. As is known, a smooth function can be approximated arbitrarily well by analytic functions. If the given function is periodic in some of its variables, then one can choose for the approximant a trigonometric polynomial in these variables. Now in the expression for the generating function (29) of the first-approximation coordinate transformation let H_{1N} be an analytic function which is a trigonometric polynomial in the phases and approximates H_1 to within ε. Such a coordinate transformation eliminates the phases from the Hamiltonian up to terms of order ε^2. In the subsequent approximations we proceed in the same manner. Under such a procedure the smoothness of the perturbation is preserved. It follows from the results of [146] and [147] that for a sufficiently smooth perturbation the successive approximations converge on the nonresonant set.[40] Following Moser's initial results, the number of derivatives required for the perturbation was gradually reduced in the works of Moser, Rüssmann and Pöschel. In [166] it is shown that in the nondegenerate case it suffices to assume that the perturbation is of class C^r with $r>2n$ (here r is not necessarily an integer, so that C^r designates the Hölder space).

§3. The KAM Theory

The Kolmogorov-Arnold-Moser (KAM) theory is a perturbation theory for conditionally-periodic motions of Hamiltonian and related systems that works over infinite time intervals. In particular, it provides a rigorous justification of the fundamental conclusion, drawn from the heuristic averaging principle and the formal integration procedures, about absence of evolution in such systems.

3.1. Unperturbed Motion. Nondegeneracy Conditions. We recall the basic notions underlying integrable systems. Consider an unperturbed integrable Hamiltonian system with Hamiltonian $H_0(I)$. Its phase space is foliated by invariant tori $I=\text{const}$. The motion on each torus is conditionally-periodic

[40] The technique of successive approximations with smoothing has led also to new implicit function theorems of Nash-Moser type in nonlinear functional analysis [147, 192, 97].

with frequency vector $\omega(I)=\partial H_0/\partial I$. A torus on which the frequencies are rationally independent is said to be *nonresonant*. Each phase trajectory on such a torus fills it densely (and is called a *winding* of the torus). The remaining tori $I=\text{const}$ are said to be *resonant*. They are foliated by tori of smaller dimension. The unperturbed system is said to be *nondegenerate* if the frequencies are functionally independent:

$$\det\left(\frac{\partial \omega}{\partial I}\right)=\det\left(\frac{\partial^2 H_0}{\partial I^2}\right)\neq 0.$$

In a nondegenerate system the nonresonant tori form a dense set of full measure. The resonant tori form a set of measure zero which, however, is also dense. Moreover, the sets of resonant tori with any number of independent frequencies from 1 to $n-1$ are each dense; in particular, the set of tori on which all phase trajectories are closed is dense.

The unperturbed system is called *isoenergetically nondegenerate* if one of the frequencies does not vanish and the ratios of the remaining $n-1$ frequencies to it are functionally independent on the energy level $H_0=\text{const}$. The condition for isoenergetic nondegeneracy has the form

$$\det\begin{bmatrix} \dfrac{\partial^2 H_0}{\partial I^2} & \dfrac{\partial H_0}{\partial I} \\ \dfrac{\partial H_0}{\partial I} & 0 \end{bmatrix}\neq 0$$

(as can be readily verified). In an isoenergetically nondegenerate system the sets of nonresonant and resonant tori are both dense on each energy level, but, as above, the former has full measure, whereas the latter has measure zero.

3.2. Invariant Tori of the Perturbed System. Consider now the perturbed system with Hamiltonian

$$H(I,\varphi,\varepsilon)=H_0(I)+\varepsilon H_1(I,\varphi,\varepsilon). \tag{31}$$

The theorem of Kolmogorov [22, 6] formulated below describes the fate of nonresonant tori under perturbation.

Theorem 13 (Kolmogorov's theorem). *If the unperturbed system is nondegenerate or isogenetically nondegenerate, then for a sufficiently small Hamiltonian perturbation most nonresonant invariant tori do not vanish but are only slightly deformed, so that in the phase space of the perturbed system there are invariant tori densely filled with conditionally-periodic phase curves winding around them, with a number of independent frequencies equal to the number of degrees of freedom. These invariant tori form a majority in the sense that the measure of the complement of their union is small when the perturbation is small. In the case of isoenergetic nondegeneracy the invariant tori form a majority on each level manifold of the energy.*

The invariant tori constructed in this theorem are called *Kolmogorov tori*, and their union is the *Kolmogorov set*. The proof of the theorem is based on the convergent procedure for eliminating fast phases described in §2.2.C.

The following facts can be added to the general formulation of Kolmogorov's theorem:

1°. The theorem is valid if the unperturbed Hamiltonian is analytic while the perturbation is of class C^r, $r>2n$ [166]. (In the original formulation it was assumed that the perturbation is also analytic [6].)

2°. Suppose we are given a number ν such that $n-1<\nu<\frac{1}{2}r-1$. For a sufficiently small C^r-perturbation the frequencies of motion on the Kolmogorov tori form the following Cantor set:

$$\Omega_\kappa=\{\xi\colon \xi\in\Omega-\kappa\subset R^n,\ |(k,\xi)|>\kappa|k|^{-\nu}\ \forall k\in Z^n\setminus\{0\}\}.$$

Here Ω denotes the set of values of the unperturbed frequencies $\omega(I)$, $\Omega-\kappa$ the set of points whose κ-neighborhood is contained in Ω, and κ is a quantity of order $\sqrt{\varepsilon}$ [167]. We recall that the Lebesgue measure of the complement of Ω_κ in Ω does not exceed a quantity of order κ.

3°. The measure of the complement of the Kolmogorov set does not exceed a quantity of order $\sqrt{\varepsilon}$. The deformation of the preserved torus, i.e., its deviation from the unperturbed torus carrying conditionally-periodic motions with the same frequencies, depends on the arithmetic properties of the frequencies. If the frequencies belong to Ω_δ with $\delta>\kappa$ (see 2°), then the deformation does not exceed a quantity of order $\varepsilon/\delta\lesssim\sqrt{\varepsilon}$ [122, 156, 167, 183].

4°. The Kolmogorov tori form a smooth family [122, 167, 183]. To elaborate on this consider first the nondegenerate case. For simplicity we shall assume that the frequency mapping $I\mapsto\omega(I)$ is a diffeomorphism. Then for perturbations of class C^r, with $r>3\nu+2>3n-1$, there exists a diffeomorphism

$$\Psi\colon\ \Omega\{\xi\}\times T^n\{\vartheta\}\to R^n\{I\}\times T^n\{\varphi\},$$

whose restriction to the Cantor set $\Omega_\kappa\times T^n$ of standard tori maps this set into the Kolmogorov set. In the variables ξ, ϑ, the equations of motion take for $\xi\in\Omega_\kappa$ the form

$$\dot{\xi}=0,\quad \dot{\vartheta}=\xi.$$

The diffeomorphism Ψ has anisotropic smoothness in the sense that it has more derivatives with respect to the phase ϑ than with respect to the frequency ξ. The smoothness with respect to ϑ (respectively, ξ) is the smoothness of an individual invariant torus (respectively, the intrinsic smoothness of the family of tori).

The smoothness class of the diffeomorphism Ψ is estimated from above by the smoothness class of the perturbation. In particular, if the perturbation is analytic, then Ψ is analytic in ϑ (i.e., each torus is analytic) and infinitely differentiable in ξ (i.e., the tori form an infinitely differentiable family). This is better than we could expect, since if Ψ were analytic in both variables, then the perturbed system would be completely integrable.

In the case of isoenergetic nondegeneracy the family of tori on each energy level is parametrized analogously by the frequency ratios (and depends smoothly on energy).

5°. The perturbed system is completely integrable on a Cantor set [167]. This means that, in the nondegenerate case, for a sufficiently smooth perturbation there is a symplectic coordinate transformation $I, \varphi \mapsto J, \psi$ with generating function $J\varphi + \varepsilon S(J, \varphi, \varepsilon)$, and a nondegenerate Hamiltonian $\mathscr{H}(J, \varepsilon)$, such that

$$H_0\left(J+\varepsilon \frac{\partial S}{\partial \varphi}\right)+\varepsilon H_1\left(J+\varepsilon \frac{\partial S}{\partial \varphi}, \varphi, \varepsilon\right)\bigg|_{B_{\kappa,\varepsilon}\times T^n} = \mathscr{H}(J, \varepsilon);$$

moreover, this equality can be differentiated sufficiently many times. Here $B_{\kappa,\varepsilon}$ is the preimage of the standard Cantor set Ω_κ (see 2°) under the mapping $J \mapsto \partial \mathscr{H}/\partial J$. An analogous statement holds true in the case of isoenergetic nondegeneracy. The functions S and $\mathscr{H}$ are obtained from the corresponding functions, given by the procedure of §2.2.C, by smoothing in the "gaps" between the Kolmogorov tori.

In problems of celestial mechanics one frequently encounters situations of *proper degeneracy*, in which the unperturbed Hamiltonian is independent of some of the "action" variables. We say that the perturbation *removes* the degeneracy if the perturbed Hamiltonian can be written in the form

$$H = H_{00}(I) + \varepsilon H_{01}(I) + \varepsilon^2 H_{11}(I, \varphi, \varepsilon), \tag{32}$$

where H_{00} depends only on the first r "action" variables and is nondegenerate or isoenergetically nondegenerate in these variables, while H_{01} depends, generally speaking, on all the "action" variables and is nondegenerate in the last $n-r$ of them (i.e., the Hessian of H_{01} with respect to $I_{r+1}, \ldots, I_n$ is different from zero). We call the system with Hamiltonian $H_{00}+\varepsilon H_{01}$ the intermediate system.

Theorem 14 ([7]). *Suppose the unperturbed system is degenerate, but the perturbation removes the degeneracy. Then a large subset of the phase space is filled by invariant tori which are close to the invariant tori* $I=\text{const}$ *of the intermediate system. The phase curves are conditionally-periodic windings on these tori with a number of frequencies equal to the number of degrees of freedom. Among these frequencies,* r *(respectively,* $n-r$*) correspond to the fast (respectively, slow) variables. If the unperturbed Hamiltonian is isoenergetically nondegenerate in the* r *variables on which it depends, then the invariant tori described above form a majority on each energy level manifold of the perturbed system.*

The proof of this theorem, too, rests on the procedure of §2.2.C.

Remark. In many problems the perturbation is periodic not only in the phases, but also in the time variable: $H = H_0(I) + \varepsilon H_1(I, \varphi, t, \varepsilon)$. This case can be reduced, as we have already mentioned, to the autonomous case by introducing time as a new phase. If $\det(\partial^2 H_0/\partial I^2) \neq 0$, then the resulting autonomous system is isoenergetically nondegenerate, and according to Theorem 13 it has many n-dimensional invariant tori. If the resulting unperturbed system is properly degenerate, but the perturbation removes the degeneracy, then the existence of invariant tori is guaranteed by Theorem 14. △

Suppose that the Hamiltonian has the form $H = H_0(I, p, q) + \varepsilon H_1(I, \varphi, p, q)$, where H_0 has for every value of I the equilibrium $(p, q) = 0 \in R^{2m}$, whose eigenvalues are not purely imaginary, and is nondegenerate with respect to I. Then for small ε most of the unperturbed invariant tori $p = q = 0$, $I = \text{const}$, are not destroyed [95] (tori with a number of frequencies smaller than the number of degrees of freedom were investigated also by V.K. Melnikov and J. Moser; see [26]). If $m = 1$ the tori are conserved also for purely imaginary eigenvalues [26].

In [26] Moser constructed a perturbation theory for quasi-periodic motions of non-Hamiltonian systems. In particular, he proved the conservation of invariant tori in reversible system.

3.3. Systems with Two Degrees of Freedom

A. Absence of Evolution. In systems with two degrees of freedom a consequence of the presence of a large number of invariant tori is that for all (and not merely the majority of) initial conditions evolution does not occur:

Theorem 15 ([6]). *In an isoenergetically-nondegenerate system with two degrees of freedom, for all initial conditions the values of the "action" variables remain forever near their initial values.*

◁ In the present case, the phase space is four-dimensional, the energy level sets are three-dimensional, and the Kolmogorov tori are two-dimensional, filling a large part of each energy level set. Each such two-torus divides the three-dimensional energy level manifold (Fig. 42 shows these tori positioned on the energy level set). A phase curve starting in a gap between two invariant tori of the perturbed system remains forever trapped between these tori. The corresponding "action" variables remain forever near their initial values. The oscillations of the "action" variables do not exceed a

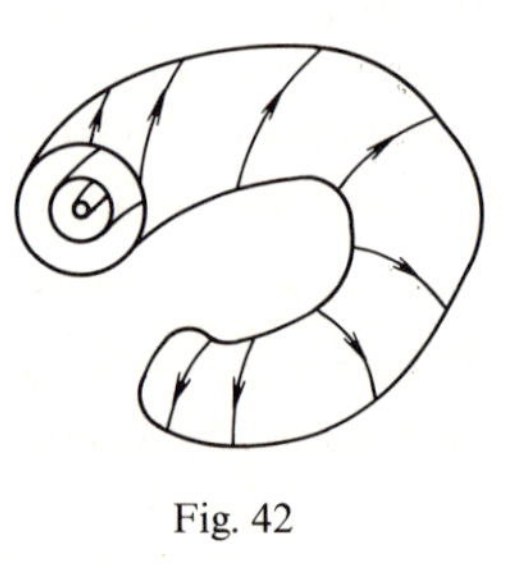

Fig. 42

quantity of order $\sqrt{\varepsilon}$, because the measure of the gap and the difference between a torus and its unperturbed counterpart ($I=\text{const}$) are bounded by quantities of this order. ▷

Evolution is also absent in the case of *proper degeneracy* if the perturbation removes the degeneracy and on the energy level set there are many invariant tori. For a Hamiltonian of form (32) the condition for removing the degeneracy is written as

$$\frac{\partial H_{00}}{\partial I_1} \neq 0, \qquad \frac{\partial^2 H_{01}}{\partial I_2^2} \neq 0. \tag{33}$$

Here I_1, I_2 are the "action" variables in our problem.

Theorem 16 ([7]). *If a properly degenerate system with two degrees of freedom satisfies condition* (33), *then for all initial conditions the values of the "action" variables remain forever near their initial values.*

A degenerate system with two degrees of freedom is in the following sense "more integrable" than the usual perturbed system:

Theorem 17 ([156]). *Suppose the degenerate system* (32) *satisfies condition* (33) *and also the condition* $\partial H_{01}/\partial I_2 \neq 0$ *(the latter means that the "slow" frequency does not vanish). Then the measure of the set of tori that disappear under perturbation is exponentially small* ($O(\exp(-\text{const}/\varepsilon))$ *instead of* $O(\sqrt{\varepsilon})$ *in the nondegenerate case), while the deviation of a perturbed torus from the torus* $I=\text{const}$ *is of order* ε.

◁ The cause of this phenomenon is that the "fast" and "slow" variables differ by a factor of $1/\varepsilon$, and to resonances between them there correspond in the perturbation harmonics of high order $1/\varepsilon$ and, accordingly, of small amplitude $O(\exp(-\text{const}/\varepsilon))$. ▷

Remark. Analogous statements are valid if the system has one and a half degrees of freedom, i.e., if a time-periodic perturbation is imposed on a system with one degree of freedom. In this case the appropriate nondegeneracy condition is $\partial^2 H_0/\partial I^2 \neq 0$. In the case of a properly degenerate Hamiltonian system with one and a half degrees of freedom the Hamiltonian has the form

$$H = \varepsilon H_{01}(I) + \varepsilon^2 H_{11}(I, \varphi, t, \varepsilon). \tag{34}$$

The condition for removing the degeneracy is written $\partial^2 H_{01}/\partial I^2 \neq 0$. The measure of the vanishing tori is exponentially small, provided the condition $\partial H_{01}/\partial I \neq 0$ is also satisfied. △

In nondegenerate systems with two degrees of freedom which are not isoenergetically nondegenerate, an evolution of the "action" variables may sometimes occur outside the invariant tori.

Example 17 [31]. The system with Hamiltonian $H=\frac{1}{2}(I_1^2-I_2^2)+\varepsilon\sin(\varphi_1 - \varphi_2)$ has the "fast" solution $I_1=-\varepsilon t$, $I_2=\varepsilon t$, $\varphi_1=-\frac{1}{2}\varepsilon t^2$, $\varphi_2=-\frac{1}{2}\varepsilon t^2$. The fact is that on the unperturbed energy level set there lies the ray $I_1 = -I_2$, along which the frequency ratio remains constant and equal to 1. This ray is a "superconductivity channel". △

B. Gaps Between the Kolmogorov Tori. We next describe the structure of the gap between Kolmogorov's tori which arises near a given resonance. For simplicity we consider the case of one and a half degrees of freedom (see the Remark in § 3.3.A). To give an approximate description of the motion we proceed in accordance with § 2.1 and average the perturbation taking into account the resonance $k_1 \omega + k_2 = 0$. For $I, q = \varphi + (k_2/k_1)t$ we obtain a Hamiltonian system with one degree of freedom. In the generic case, its phase portrait in the vicinity of a resonance value of I is similar to that of the pendulum (see § 2.1 and Fig. 43a). This portrait may be thought of as the section of the phase space I, φ, t of the averaged system by the plane $t = 0$. To equilibrium positions and to the separatrices shown in Fig. 43a there correspond, respectively, periodic solutions and surfaces asymptotic to these solutions for $t \to \pm\infty$; to the closed curves there correspond two-dimensional invariant tori.

Fig. 43

We can sharpen this picture slightly by performing several steps of the procedure for eliminating the fast variables (§ 2.2) and subsequently omitting in the Hamiltonian the supplementary terms of higher order in ε. This again yields a system with one degree of freedom whose phase portrait resembles closely that shown in Fig. 43a.

What is the effect of the omitted small terms on the motion? The periodic solutions are preserved (this follows from the implicit function theorem). The surfaces asymptotic to these solutions are also preserved. However, the surfaces asymptotic for $t \to +\infty$ and $t \to -\infty$ to distinct solutions (or even to the same solution) are no longer necessarily identical (see Fig. 43b which depicts the section of the phase space by the plane $t = 0$). This is the phenomenon of *splitting* (or *bifurcation*) *of separatrices* discovered by Poincaré [33].

The fate of invariant tori is settled by Theorems 14 and 17. The motion near a resonance can be described by a system with removable degeneracy of form (34), only now the role of the small parameter is played by $\sqrt{\varepsilon}$; the phase variables in this system are time and the action-angle variables on the phase portrait shown in Fig. 43a. It follows that far from the separatrices one finds all the tori obtained through averaging, except for a portion of order $O(\exp(-c^{-1}/\sqrt{\varepsilon}))$, with $c=\text{const}>0$. Near separatrices a special analysis is needed. It shows that there exist tori exponentially close to the separatrices, so that the separatrices are confined in a zone of exponentially small width [157].

As one moves away from the resonance the tori found above become the ordinary Kolmogorov tori.

Remarks. 1. The problem of calculating the exponentially small angle between the split separatrices has been considered recently by V.F. Lazutkin.

2. It is natural to expect that for systems with two degrees of freedom in general position the total measure of the "torus-free" set corresponding to all resonances will be exponentially small. However, this has not yet been proved. △

3.4. Diffusion of Slow Variables in Higher-Dimensional Systems, and its Exponential Estimate. In investigating the perturbed motion outside the invariant tori, we must distinguish between the case of two degrees of freedom and that of a higher number.

For two degrees of freedom the invariant tori are two-dimensional and divide the three-dimensional energy level set, thereby precluding evolution (§3.3.A).

If the number n of degrees of freedom is greater than two, the n-dimensional invariant tori do not divide the $(2n-1)$-dimensional energy level manifold, but are arranged in it like points in a plane or lines in space. In this case the "gaps" corresponding to different resonances are connected to one another, so that the tori do not prevent a phase curve which originates near a resonance from wandering far away.

Conjecture ([7]). The typical case in higher-dimensional problems is *topological instability:* through an arbitrarily small neighborhood of any point there pass phase trajectories along which the slow variables drift away from the initial value by a quantity of order 1.

The KAM theory proves the metric stability[41], i.e., stability for most initial conditions. Thus, according to the conjecture formulated above, the typical situation in higher-dimensional problems is a combination of metric stability and topological instability.

There exist many examples of evolution of slow variables in metrically stable problems. Most of them are concerned with highly degenerate situations in which the unperturbed Hamiltonian is not a steep function (the definition of steep functions is given below). Only one example of evolution in a system with steep unperturbed Hamiltonian has been investigated analyt-

[41] Here this word is not used with its formal meaning, but as synonymous with "absence of evolution": $\sup_{-\infty<t<\infty} |I(t)-I(0)|\to 0$ as $\varepsilon\to 0$.

ically [50]. The average velocity of the evolution in this example is exponentially small ($O(\exp(-1/\sqrt{\varepsilon}))$.

Numerical experiments show that the evolution of the "action" variables apparently has no directional character; rather, it represents a more or less random wandering in the resonance regions between the invariant tori. This process is called *"diffusion"* [191].[42] Discussions of the questions arising here can be found in [80, 130, 191].

In generic systems the average velocity of the "diffusion" is exponentially small. The needed genericity condition is called the *steepness condition.* An analytic function is said to be *steep* if it has no stationary points and if its restriction to any plane of arbitrary dimension has only complex-isolated stationary points.[43]

Theorem 18 (N.N. Nehoroshev [31, 180]). *Suppose that the unperturbed Hamiltonian $H_0(I)$ is a steep function. Then for a sufficiently small perturbation*

$$|I(t)-I(0)|<\varepsilon^b \quad \text{for } 0 \leqq t \leqq \frac{1}{\varepsilon}\exp\left(\frac{1}{\varepsilon^a}\right)$$

in the perturbed Hamiltonian system. Here a, b are positive constants that depend on the geometric properties of the unperturbed Hamiltonian.

◁ The proof is based on the following considerations: In the region in which the frequencies of the unperturbed motion satisfy no resonance relations of order up to $1/\varepsilon$, the procedure for eliminating phases described in §2.2.A (Lindstedt's method) permits us to confine the dependence on the phases in the Hamiltonian to exponentially small terms. Consequently, in this region the evolution may be only exponentially slow.

A fast evolution (with velocity of order ε) is possible only at resonance. Near resonance, the procedure of §2.2.B (von Zeipel's method) permits us to confine the dependence on the nonresonant combination of phases to exponentially small terms. Ignoring these terms, we obtain a system which, according to Theorem 12, possesses linear first integrals. The fast evolution occurs in the plane defined by these integrals. The condition for exact resonance is, as one can readily check, that the gradient of the restriction of H_0 to this plane vanishes. Since H_0 is a steep function, exact resonance occurs at an isolated point. Consequently, under evolution the resonance is violated. It follows that fast evolution lasts only a short time; thus, an exponentially small upper estimate of the average velocity of evolution is indeed obtained. ▷

---

[42] The term "Arnold diffusion" is frequently used [Translator's note].

[43] The notion of steepness was introduced by N.N. Nehoroshev [159]. Here we take as definition a necessary condition for steepness found by Yu.S. Il'yashenko (Usp. Mat. Nauk *41*, No. 1, 193–194 (1986)).

If the steepness conditions are not fulfilled, then, as Example 17 shows, evolution may have an average velocity of order ε, causing the slow variables to drift away from their initial values at a distance of 1 during time $1/\varepsilon$.

3.5. Variants of the Theorem on Invariant Tori

A. Invariant Tori of Symplectic Mappings. Consider a mapping of the $2n$-dimensional "annulus" to itself, which is close to an n-dimensional rotation:

$$\begin{aligned} I' &= I + \varepsilon f(I, \varphi, \varepsilon), & I &\in B \subset R^n, \\ \varphi' &= \varphi + h(I) + \varepsilon g(I, \varphi, \varepsilon), & \varphi &\bmod 2\pi \in T^n. \end{aligned} \tag{35}$$

We assume that the mapping is exact-symplectic, i.e., it preserves the integral of the 1-form $I\,d\varphi$ on closed contours. The unperturbed mapping ($\varepsilon = 0$) is said to be *nondegenerate* if $\det(\partial h/\partial I) \neq 0$.

Theorem 19 ([85]). *Suppose the unperturbed mapping is analytic and nondegenerate. Then for any sufficiently small perturbation of class C^r with $r > 2n+1$, in the $2n$-annulus $B \times T^n$ there exist invariant tori close to the tori $I = $ const, and the measure of the complement of their union is small when the perturbation is small. The images of any point of an invariant torus under the iterations of the mapping densely fill the torus.*

If $n = 1$, we obtain an area-preserving mapping of the ordinary annulus into itself (Fig. 44). The restriction of the unperturbed mapping ($\varepsilon = 0$) to each circle $I = $ const is a rotation. The nondegeneracy condition means that the angle of rotation changes from one circle to another (Fig. 44a). A circle for which the rotation angle is not commensurable with 2π is said to be nonresonant. The images of any of its points under the iterations of the mapping densely fill the circle. A circle for which the rotation angle is commensurable with 2π is said to be resonant. It consists of periodic points of the unperturbed mapping.

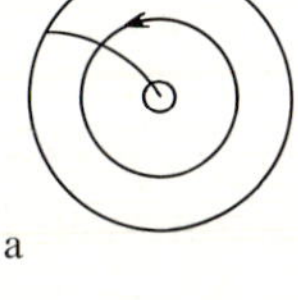
a

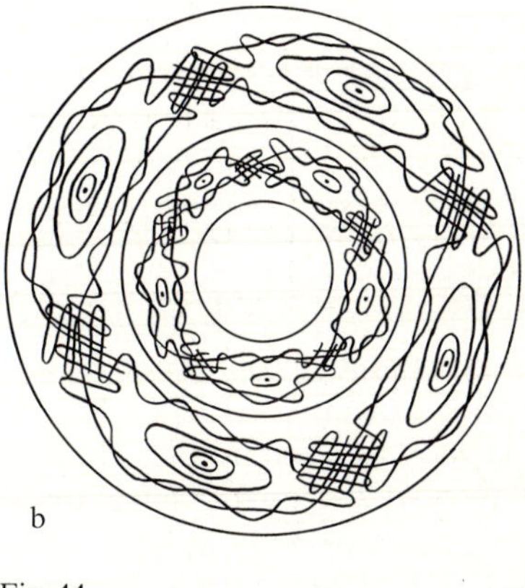
b

Fig. 44

Under perturbation the nonresonant circles (and especially those for which the rotation angle α is not too well approximated by 2π-rational numbers: $\left|\alpha-2\pi\frac{p}{q}\right|>c\sqrt{\varepsilon}q^{-\nu}$, $n+1<\nu<\frac{1}{2}(r+1)$) do not vanish, but are only slightly deformed. The resonant circles are destroyed.

The theorems on invariant tori for Hamiltonian systems and symplectic mappings were first proved independently (though with almost identical methods). In the cases of finite and C^∞-smoothness, these theorems can be deduced from one another, since Poincaré mappings for Hamiltonian systems have the form (35), and conversely, every mapping of the form (35) can be obtained as such a Poincaré mapping [85]. According to [85], this holds true in the analytic case too, but the proof was not published.

B. Invariant Tori in the Theory of Small Oscillations. Other situations in which Kolmogorov tori are present are connected with the theory of small oscillations. In particular, let us consider a Hamiltonian system with n degrees of freedom in a neighborhood of one of its equilibrium positions. Suppose that this equilibrium is stable in the linear approximation, so that n characteristic frequencies $\omega_1, \ldots, \omega_n$ are defined. We assume also that among these frequencies there are no resonance relations of order $\leqq 4$, i.e., $k_1\omega_1+\ldots+k_n\omega_n \neq 0$ for all integers k_i such that $0<\sum|k_i|\leqq 4$.

Then the Hamiltonian can be reduced to the Birkhoff normal form (see Ch. 7, §3)

$$H=H_0(\tau)+\ldots,$$
$$H_0(\tau)=\sum\omega_i\tau_i+\tfrac{1}{2}\sum\omega_{ij}\tau_i\tau_j, \qquad \tau_i=\tfrac{1}{2}(p_i^2+q_i^2).$$

Here the dots denote terms of order higher than four in the distance from the equilibrium position. The system in the neighborhood of the equilibrium position is said to be *nondegenerate* if

$$\det\left(\frac{\partial^2 H_0}{\partial\tau^2}\right)=\det(\omega_{ij})\neq 0,$$

and *isoenergetically nondegenerate* if

$$\det\begin{bmatrix}\dfrac{\partial^2 H_0}{\partial\tau^2} & \dfrac{\partial H_0}{\partial\tau}\\ \dfrac{\partial H_0}{\partial t} & 0\end{bmatrix}_0=\det\begin{bmatrix}\omega_{ij} & \omega_i\\ \omega_j & 0\end{bmatrix}\neq 0.$$

If the system is nondegenerate or isoenergetically nondegenerate we say that the Hamiltonian is of *general elliptic type*.

The system with Hamiltonian H_0 is integrable and in it the motion occurs on invariant tori $I=\text{const}$. Hence, the system with Hamiltonian H is nearly integrable in a sufficiently small neighborhood of the equilibrium position. The situation is similar to that treated by Kolmogorov's theorem.

Theorem 20 ([7, 27]). *For a Hamiltonian of general elliptic type, in a neighborhood of the equilibrium position there exist invariant tori close to the tori of the linearized system. They form a set whose relative measure in the polydisc* $|\tau|<\varepsilon$ *tends to* 1 *as* $\varepsilon\to 0$. *In an isoenergetically nondegenerate system such tori fill a large subset of each energy level set near the equilibrium position.*

Remark ([167]). The relative measure of the set of invariant tori in the polydisc $|\tau|<\varepsilon$ is not smaller than $1-O(\varepsilon^{1/4})$. If among the frequencies there are no resonances of order less or equal to $l\geqq 4$, then this measure is even not smaller than $1-O(\varepsilon^{(l-3)/4})$. △

In the case $n=2$ isoenergetic nondegeneracy guarantees Lyapunov stability of the equilibrium position. For $n=2$ the condition for isoenergetic nondegeneracy says that the quadratic part of the function H_0 is not divisible by the linear part. Actually, even if the quadratic part is divisible by the linear part the equilibrium is, as a rule, stable. In fact, let us assume that the frequencies ω_1 and ω_2 satisfy no resonance relation of order less or equal to $l\geqq 4$. Then the Hamiltonian can be reduced to the normal form

$$H=H_0(\tau_1,\tau_2)+\dots,\qquad H_0=\sum_{1\leqq i+j\leqq [l/2]} A_{ij}\tau_1^i\tau_2^j,$$

where the dots denote terms of order higher than l in the distance from the equilibrium position. Consider the function $h_0(\varepsilon)=H_0(\varepsilon\omega_2,-\varepsilon\omega_1)$. If $h_0(\varepsilon)$ does not vanish identically, then the equilibrium position is stable.

Other cases in which analogous assertions concerning invariant tori and stability hold true arise in connection with the theory of small oscillations in the neighborhood of equilibrium positions of systems with periodic and conditionally-periodic coefficients, or in the neighborhood of periodic solutions of autonomous Hamiltonian systems, and also in the neighborhood of fixed points of symplectic mappings. The corresponding assertions are given in [8].

C. Quasi-Periodic Invariant Manifolds. Invariant manifolds connected with Kolmogorov tori exist in the case of quasi-periodic Hamiltonian perturbations (such problems were indicated by J. Tennyson). In this case the Hamiltonian has the form

$$H=H_0(p)+\varepsilon H_1(p,Aq,\varepsilon),\qquad (p,q)\in R^{2n},$$

where p, q are canonically conjugate variables, $A\colon R^n\to R^m$ is a linear operator, and the perturbation H_1 is 2π-periodic in $\varphi=Aq$. It is assumed that rank A $=n\leqq m$ (otherwise the order of the system can be reduced) and the adjoint operator $A^*\colon R^m\to R^n$ does not map vectors with integer components into zero (otherwise the number of angles φ can be reduced). Suppose that the unperturbed Hamiltonian H_0 is nondegenerate or isoenergetically nondegenerate and that A^* satisfies the strong incommensurability condition

$$|A^*k|>c^{-1}|k|^{-\nu}\quad \text{for all } k\in Z^m\setminus\{0\}\quad (\text{with } c,\nu=\text{const}).$$

Then a large subset of the phase space (and, in the case of isoenergetic degeneracy, a large subset of every energy level, too) is filled by n-dimensional invariant manifolds which are "quasi-periodic in q" and close to the "planes" $p=\text{const}$. These manifolds are obtained from the m-dimensional Kolmogorov tori of the auxiliary system with the Hamiltonian

$$H=H_0(A^* I)+\varepsilon H_1(A^* I, \varphi, \varepsilon), \qquad I\in R^m, \ \varphi \bmod 2\pi\in T^m,$$

through the formulas $p=A^* I$, $\varphi=Aq$. The proof of the existence of the indicated tori is based on the procedure for eliminating fast variables (§2.2 C) and is organized in the usual manner, since on every step of the procedure everything depends $A^* I$ rather than I. The strong incommensurability condition imposed above allows us to estimate the small denominators arising in the procedure.

If $n=2$ and H_0 is isoenergetically nondegenerate, the invariant manifolds found above divide the energy levels and there is no evolution of the slow variables p.

3.6. A Variational Principle for Invariant Tori. Cantori. An invariant torus of a Hamiltonian system carrying conditionally periodic motions with a given set of frequencies is the extremal of a certain variational principle, found by I.C. Percival [163, 164]. We now formulate this principle.

To this end it is convenient to pass from the Hamiltonian to the Lagrangian description of motion. Let $H(p, q)$ be the Hamiltonian of a system with n degrees of freedom. Suppose that from the relation $r=\partial H(p, q)/\partial p$ we can express p as a function $p=p(r, q)$. Then the change of the quantities $q, r=\dot{q}$ with time is described by Lagrange's equations with the Lagrangian

$$L(q, r)=p\cdot r-H(p, q).$$

Let $\omega\in R^n$ be the frequency vector of the sought-for conditionally-periodic motions. For every smooth function $f\colon T^n\{\vartheta\}\to R$ we denote

$$D_\omega f=\frac{d}{dt} f(\omega t+\vartheta)|_{t=0}=\frac{\partial f}{\partial \vartheta}\omega.$$

Let Σ be a smooth n-dimensional torus in the phase space $\{q, r\}$, given by the parametric relations $q=q_\Sigma(\vartheta)$, $r=D_\omega q_\Sigma(\vartheta)$, $\vartheta \bmod 2\pi\in T^n$. Any torus close to Σ, given by relations of the form

$$q=q_\Sigma(\vartheta)+\delta q(\vartheta), \qquad r=D_\omega q_\Sigma(\vartheta)+D_\omega \delta q(\vartheta),$$

will be called a variation of Σ. Let us introduce the functional

$$\Phi_\omega(\Sigma)=\langle L(q_\Sigma(\vartheta), D_\omega q_\Sigma(\vartheta))\rangle^\vartheta,$$

where the angle brackets stand for averaging over ϑ.

Theorem 21 (Variational Principle [163]). *The smooth torus Σ is an invariant torus of the given system carrying conditionally-periodic motions with frequency vector ω if and only if it is a stationary point of the functional Φ_ω.*

◁ Using integration by parts and the fact that the functions q_Σ and δq are 2π-periodic in ϑ, the first variation of the functional Φ_ω can written as

$$\delta\Phi_\omega=\left\langle\frac{\partial L}{\partial q}\,\delta q+\frac{\partial L}{\partial r}\,D_\omega\,\delta q\right\rangle^{\vartheta}=\left\langle\left(\frac{\partial L}{\partial q}-D_\omega\,\frac{\partial L}{\partial r}\right)\delta q\right\rangle^{\vartheta}.$$

If the torus Σ is invariant and filled by conditionally-periodic motions $q_\Sigma(\omega t+\vartheta)$, $D_\omega q_\Sigma(\omega t+\vartheta)$, then by virtue of Lagrange's equations $\delta\Phi_\omega=0$, i.e., Σ is a stationary point of the functional Φ_ω.

Conversely, if $\delta\Phi_\omega=0$, then $D_\omega\dfrac{\partial L}{\partial r}-\dfrac{\partial L}{\partial q}=0$. Hence, $q_\Sigma(\omega t+\vartheta)$, $D_\omega q_\Sigma(\omega t+\vartheta)$ is a conditionally-periodic solution of the given system. ▷

This variational principle permits to seek invariant tori as stationary points of the functional Φ_ω.

According to a famous dictum of David Hilbert, "every problem of calculus of variations has a solution, provided the word "solution" is suitably understood" [190]. The Kolmogorov tori are extremals of the variational principle formulated above for nearly integrable system and frequency vectors ω with strongly incommensurable components. What "solution" does the variational problem posed above have for systems far from integrable, or for abnormally commensurable frequencies? The answer is presently known only in the case of two degrees of freedom (Mather [133, 134] and Aubry [53]). It turned out that the solution is a *cantorus* (this terms was proposed by Percival): an invariant set obtained by embedding in phase space a Cantor subset of the standard two-dimensional torus. More precise formulations are given below.

For simplicity we consider a system with one an a half (rather than two) degrees of freedom, whose Hamiltonian $H(p, q, t)$ is 2π-periodic in time t and in the coordinate q. We assume that the system has two invariant tori, given by the equations $p=p_0$ and $p=p_1>p_0$. Consider the Poincaré mapping at time 2π for this system:

$$f=(f_p, f_q \bmod 2\pi)\colon R\times S^1\to R\times S^1.$$

Mapping f preserves area and leaves invariant the circles $p=p_0$ and $p=p_1$, as well as the annulus Π that they bound.

We shall assume that $\partial f_p/\partial p>0$ in Π; in this case f is referred to as a *twist mapping*. We let ν_0 and ν_1 denote the Poincaré rotation numbers of f on the boundary circles. Since f is a twist mapping, $\nu_0<\nu_1$.

Theorem 22 ([133]). *For every $\nu\in(\nu_0, \nu_1)$ there exists a (not necessarily continuous) mapping h of the standard circle S^1 into the annulus Π, $h=(h_p, h_q \bmod 2\pi)\colon S^1\{\vartheta\}\to\Pi\{p, q\}$, which takes the rotation of S^1 by the angle $2\pi\nu$ into the restriction of the given transformation f to the image of S^1 under h: $f(h(\vartheta))=h(\vartheta+2\pi\nu)$. Moreover, h has the following properties:*

a) h_q *is a nondecreasing function;*

b) *if* ϑ *is a continuity point of* h_q, *then so are* $\vartheta+2\pi\nu$ *and* $\vartheta-2\pi\nu$;

c) *the function* h_p *can be calculated by the formula* $h_p(\vartheta)=g(h_q(\vartheta), h_q(\vartheta+2\pi\nu))$, *where g is a smooth function;*

d) *if* ν *is irrational, then there are no intervals on which* h_q *is constant.*

◁ The required function h_q is sought as a minimum point of a functional which is the discrete analog of the functional Φ_ω introduced above. For details we refer the reader to [133, 134]. ▷

Let us examine some corollaries of this result. If $\nu=m/n$ is rational, then for every ϑ the point $h(\vartheta)\in\Pi$ is taken into itself by n iterations of the mapping f. Correspondingly, on the universal covering of the annulus (i.e., the strip $p_0<p<q_1$, $-\infty<q<\infty$) the coordinate q of the point increases by $2\pi m$. The existence of such periodic points is one of the known consequences of Poincaré's geometric theorem (proved by Birkhoff [10]). In this case the original Hamiltonian system has a periodic solution of period $2\pi n$, which makes m complete rotations in the angle q over a period.

If ν is irrational and the mapping h is continuous, then the original Poincaré mapping has an invariant curve homeomorphic to a circle, and its restriction to this curve is topologically conjugate to the rotation of the circle through the angle $2\pi\nu$. The original Hamiltonian system has a two-dimensional invariant torus filled by conditionally-periodic motions with frequency ratio ν.

Now suppose that ν is irrational and h_q has discontinuities. Then according to assertion b) of the theorem, the discontinuity points are dense in the circle. Since, according to a), h_q is nondecreasing, there also exist continuity points, and they too are dense. We let Ξ and Σ denote the closure of the sets of points $h_q(\vartheta) \bmod 2\pi\in S^1$ and $h(\vartheta)\in\Pi$, respectively, such that ϑ is a continuity point of h_q. Then according to assertions b)–d) of the theorem, Ξ is a Cantor set on the circle, while Σ is a "one-dimensional" invariant cantorus; motion on Σ is characterized by the rotation number ν. To Σ there corresponds an invariant cantorus in the original Hamiltonian system. There exist examples of mappings having no continuous invariant curves which are not null-homotopic [53]. All invariant sets of these mappings corresponding to irrational rotation numbers are one-dimensional cantori.

The one-dimensional cantori found above are probably unstable. This conjecture is based on the following argument: Let ϑ_* be a discontinuity point of h_q. Then the intervals $(h_q(\vartheta_*+2\pi\nu j-0), h_q(\vartheta_*+2\pi\nu j+0))$, $j=0, \pm1, \dots$ are "holes" in the Cantor set Ξ on the circle; the endpoints of these intervals belong to Ξ. Consequently, they are disjoint, and their lengths tend to zero as $j\to\pm\infty$. It follows that the distance between the points $h(\vartheta_*+2\pi\nu j-0)$ and $h(\vartheta_*+2\pi\nu j+0)$ on the one-dimensional cantorus Σ also tends to zero as $j\to\pm\infty$. Therefore, motion along (transverse to) the cantorus Σ is accompanied by contraction (respectively, dilation), as follows from preservation of area. But this is a criterion for instability.

Instability of cantori for mappings

$$p' = p + V(q), \quad q' = q + p', \quad \langle V \rangle^q = 0,$$

can be proved under the additional assumption that for at least one trajectory on the cantorus the spectrum of the Schrödinger operator defining the variational equation is separated from zero (the nonnegativity of the spectrum follows from the variational principle, the spectrum arising in the case of a smooth invariant curve obviously begins at zero). Specifically, for each hole in the cantorus one can construct two curves which connect its endpoints in such a way that under the action of the given mapping the length of one of them tends to zero as time approaches ∞, while the length of the other behaves similarly as time approaches $-\infty$ (Ya.G. Sinai, to be published). These curves are the stable and unstable manifolds of the cantorus.

The stable manifold of a cantorus may intersect the unstable manifold of another cantorus. It follows from the Poincaré recurrence theorem that the set of these intersection points has measure zero. Hence, it seems plausible that the set of all cantori, too, has measure zero (Sinai).

It seems that the discovery of cantori can explain the following result of numerical analysis of the mappings considered above: during a large number of iterations the point may move in a domain bounded, seemingly, by an invariant curve, and then, after a relatively small number of iterations, cross this curve and start moving in the region lying on the other side relative to this curve. The point is that although a one-dimensional cantorus does not divide the plane, it may nevertheless serve as a kind of partial "fence" which the phase point cannot cross too easily. For this reason the point is forced to move along the fence before slipping through some gap. A detailed numerical investigation of this process can be found in [130].

3.7. Applications of the KAM Theory. In many classical problems of mechanics and physics the KAM theory has provided the rigorous justification of conclusions obtained earlier by using the heuristic averaging principle and formal perturbation theory. We list the best known examples (see [8, 27]):

– In the problem of the motion of a heavy rigid body fixed at a point (Example 11), suppose that the kinetic energy of the body is sufficiently large in comparison with its potential energy (at the initial moment of time). Then the length of the angular momentum vector and its inclination to the horizon remain forever near their initial values (provided that the initial values of the energy and angular momentum differ sufficiently from values for which the body can rotate around its medium principal axis). For most initial conditions the motion of the body is close to the superposition of the Euler-Poinsot motion and a slow azimutal precession;

– In the *planar circular restricted three-body problem* (Example 10), if the mass of Jupiter is sufficiently small, then the magnitude of the major semiaxis and the eccentricity of the Keplerian ellipse of the asteroid will remain forever near their initial values (provided that initially this ellipse does not intersect Jupiter's orbit). For most initial conditions the motion remains forever close to Keplerian motion on a ellipse which rotates slowly around its focus.

– In the *n-body problem* (Example 13), if the masses of the planets are sufficiently small, then a large portion of that region of phase space corresponding to unperturbed motion of all planets on identically oriented Keplerian ellipses of small eccentricities and inclinations is filled up by conditionally-periodic motions close to the Lagrangian motions (see Example 13).

– Most geodesics on a surface close to a triaxial ellipsoid oscillate between two closed curves close to lines of curvature of the surface, densely filling the ring between them; this ring is the projection on the configuration space (i.e., on the surface in question) of an invariant torus in phase space filled by trajectories of the geodesic flow.

– There exists a magnetic field such that, in a neighborhood of a given circle, most lines of force wind around two toroidal surfaces which are embedded into one another, and which surround the circle. The remaining lines of force are confined forever between these toroidal surfaces. Under a small perturbation of the field, most of these surfaces do not vanish, but are only slightly deformed (such a field configuration is used for plasma confinement in toroidal chambers).

The statement on the stability of equilibrium positions of systems with two degrees of freedom in the general elliptic case has also many applications.

Example 18 (Stability of the equilateral libration points). The planar restricted circular three-body problem in the rotating coordinate system of Example 10 has two degrees of freedom. The *equilateral* (or *triangular*) *libration points* are its equilibrium positions [131]. As was already known to Lagrange, these equilibria are stable in the linear approximation if $\mu<\mu_1 =\frac{1}{2}(1-\frac{1}{9}\sqrt{69})\approx 0.03852$, and unstable otherwise (here $\mu/(1-\mu)$ is the ratio of the masses of Jupiter and the Sun, and we assume that $\mu<1/2$). It was found that to resonances of order $\leqq 4$ there correspond the values

$$\mu=\mu_2=\tfrac{1}{2}(1-\tfrac{1}{45}\sqrt{1883})\approx 0.02429,$$

and

$$\mu=\mu_3=\tfrac{1}{2}(1-\tfrac{1}{15}\sqrt{213})\approx 0.01352$$

(see [83]). Moreover, the condition for isoenergetic nondegeneracy is violated only for the value $\mu=\mu_c\approx 0.0109$ (see [83]; first it was established that the problem is degenerate only at finitely many points [125], and subsequently the critical value μ_c has been calculated).

By a result of §3.3, if $0<\mu<\mu_1$ and $\mu\neq\mu_2, \mu_3, \mu_c$, the equilateral libration points are stable. In [131] it is shown that for $\mu=\mu_2$ and $\mu=\mu_3$ there is instability, and for $\mu=\mu_c$, stability. Finally, in [181] it is shown that for $\mu=\mu_1$ there is stability. △

The stability of stationary rotations of a heavy rigid body about a fixed point has been investigated in a similar manner in [108].

We next formulate a number of results that follow from the assertions on the stability of equilibrium positions of time-periodic systems with one degree of freedom, of periodic motions of systems with two degrees of freedom, and of periodic points of symplectic mappings of the plane.

– If the equilibrium position of a pendulum in a periodically varying external field is stable in the linear approximation and if among its characteristic multipliers there are no points of the circle with argument $\frac{\pi}{3}j, \frac{\pi}{4}j, j=0, 1, \ldots, 7$, then this position is stable.

– The closed geodesics on a general-position surface in three-dimensional space which are stable in the linear approximation are stable.

– If the trajectory of a little ball bouncing between two concave walls in general position (or, equivalently, a light ray reflected from two mirrors (Fig. 45)), is stable in the linear approximation, then it is stable.

Fig. 45

Applications of the KAM theory to the problem of conservation of adiabatic invariants over infinite time intervals are described in §4.

Less traditional applications concern calculation of the short-wave approximation of the eigenvalues and eigenfunctions of the Schrödinger, Laplace, and Laplace-Beltrami operators [123]. For the sake of definiteness we shall focus here on the Schrödinger operator. The short-wave approximation formulas permit us to construct, in terms of solutions of the equations of motion of a classical mechanical system, approximate solutions to the Schrödinger equation describing the behavior of the corresponding quantum system. In particular, if the classical system has an invariant torus in phase space satisfying arithmetic quantization conditions, then the short-wave approximation formulas permit us to construct, in terms of this torus, the asymptotics of an eigenvalue of the Schrödinger operator and of the corresponding almost-eigenfunction.[44] In a nearly integrable system there are many invariant tori, which actually form a smooth family (§2.2). Thus, in general there are many tori satisfying the quantization conditions, thereby enabling us to approximate a large part of the spectrum of the corresponding Schrödinger operator.

[44] An almost-eigenfunction satisfies the equation for the eigenfunction approximately, but is not necessarily close to a genuine eigenfunction.

§4. Adiabatic Invariants

In this section we describe the effect of a sufficiently slow (gentle) variation of the parameters upon the motion in an integrable Hamiltonian system. An *adiabatic invariant* of such a system is defined as a function of phase variables and of parameters which changes very little under a considerable change of the parameters. The main result of the theory concern single-frequency systems.

4.1. Adiabatic Invariance of the Action Variable in Single-Frequency Systems. Consider a Hamiltonian system with one degree of freedom whose parameters change slowly, i.e., the Hamiltonian of the system has the form $E=E(p, q, \lambda)$, with $\lambda=\lambda(\tau)$, $\tau=\varepsilon t$, and $0<\varepsilon\ll 1$ (for example, a pendulum whose length changes slowly). The function $\lambda(\tau)$ will be assumed to be sufficiently smooth.

Definition. A function $I(p, q, \lambda)$ is called an *adiabatic invariant* if for every $\kappa>0$ there is an $\varepsilon_0=\varepsilon_0(\kappa)$ such that for $\varepsilon<\varepsilon_0$ the variation of $I(p(t), q(t), \lambda(\varepsilon t))$ for $0\leqq t\leqq 1/\varepsilon$ does not exceed κ.[45]

Let us assume that for each fixed value of λ the Hamiltonian $E(p, q, \lambda)$ has closed phase curves (for instance, those encircling the equilibrium position of the pendulum (Fig. 46)), on which the frequency of motion is different from zero. Then we can introduce action-angle variables for the system with fixed λ:

$$I=I(p, q, \lambda), \qquad \varphi=\varphi(p, q, \lambda) \bmod 2\pi.$$

Our next objective is to prove the adiabatic invariance of the quantity I.

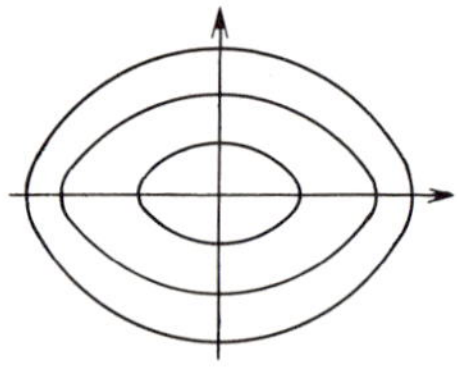

Fig. 46

Proposition 1. *The change of the variables I, φ in a system with a slowly varying parameter is described by the Hamiltonian $H=H_0(I, \lambda)+\varepsilon H_1(I, \varphi, \varepsilon t)$:*

$$\dot I=-\varepsilon\frac{\partial H_1}{\partial\varphi}, \qquad \dot\varphi=\frac{\partial H_0}{\partial I}+\varepsilon\frac{\partial H_1}{\partial I}, \tag{36}$$

[45] Such almost-conserved quantities were first discovered by Boltzmann in his study of adiabatic processes in thermodynamics. The term "adiabatic invariant" was introduced by Ehrenfest. Many different definitions of adiabatic invariants have been proposed. The one given here, which seems to be generally accepted at present, was formulated by A.A. Andronov, M.A. Leontovich, and L.I. Mandel'shtam. A detailed exposition of the history of this problem is given in [55].

where H_0 is the Hamiltonian E expressed in terms of I, λ, and H_1 is 2π-periodic in φ.

Notice that equations (36) are in the standard form to which the averaging principle of §1.1 applies.

Proposition 2. *The "action" variable I is a first integral of the system averaged over the phase.*

◁ The right-hand side of the equation for I is the derivative of a periodic function, and consequently its mean value is equal to zero. ▷

Theorem 23. *If the frequency $\omega(I, \lambda)$ of the given system with one degree of freedom does not vanish, then the "action" variable $I(p, q, \lambda)$ is an adiabatic invariant:*

$$|I(p(t), q(t), \lambda(\varepsilon t)) - I(p(0), q(0), \lambda(0))| < c\varepsilon \quad \text{for } 0 \leqq t \leqq \frac{1}{\varepsilon},$$

with $c = \text{const} > 0$.

◁ By Theorem 1 of §1, the averaging principle describes the solution of the given single-frequency system over time $1/\varepsilon$ with accuracy of order ε, and by Proposition 1, I is a first integral of the averaged system. ▷

Example 19. For a harmonic oscillator the energy-to-frequency ratio $I = h/\omega$ is an adiabatic invariant. △

Suppose that in a Hamiltonian system with $n \geqq 2$ degrees of freedom the variation of the Hamiltonian E in all but one of the coordinates (denoted q) is slow: $E = E(p, q, y, \varepsilon x)$. Here q, x are coordinates, and p, y are the canonically conjugate momenta.[46] An adiabatic invariant for such a system is a function of the phase variables which changes very little over time $1/\varepsilon$. We call the system with one degree of freedom obtained by setting $\varepsilon x = \text{const}$, $y = \text{const}$, the *unperturbed* system. Now suppose that in the phase portrait of the unperturbed system there are closed trajectories (Fig. 46) on which the motion has nonzero frequency, so that we can introduce action-angle variables

$$I = I(p, q, y, \varepsilon x), \qquad \varphi = \varphi(p, q, y, \varepsilon x) \bmod 2\pi.$$

We let $H_0(I, y, \varepsilon x)$ denote the Hamiltonian E, expressed through I, y, εx.

Theorem 24. *The "action" variable I is an adiabatic invariant of the system with Hamiltonian $E(p, q, y, \varepsilon x)$. The variation of the variables y, εx with time is described (with accuracy of order ε over the time interval $1/\varepsilon$) by the system with Hamiltonian $H_0(I, y, \varepsilon x)$, which contains I as a parameter (this approximation is called adiabatic).*

---

[46] The case of a Hamiltonian that varies slowly with time is reduced to this case by taking time as a new coordinate and introducing the canonically conjugate momentum.

◁ The transformation $(p, q) \mapsto (I, \varphi)$ can be completed by a transformation $(y, x) \mapsto (Y, X) = (y + O(\varepsilon), x + O(1))$ so that the resulting full transformation will be symplectic. The variation of the new variables is described by the Hamiltonian

$$H = H_0(I, Y, \varepsilon X) + \varepsilon H_1(I, \varphi, Y, \varepsilon X, \varepsilon).$$

Averaging over the phase φ we obtain a system which describes the change of the solutions over the time interval $1/\varepsilon$ up to an error of order ε. The action variable I is a first integral of this system. ▷

Remarks. 1. If the original system has two degrees of freedom, then Theorem 2 leads to a system with one degree of freedom, and the averaged equations can be integrated.

2. The same quantity I is also an adiabatic invariant of any system with a Hamiltonian $F = E(p, q, y, \varepsilon x) + \varepsilon E_1(p, q, y, x, \varepsilon)$ that differs from E by a small term which does not vary slowly with x. △

Example 20. For the motion in a quadratic *potential fosse* streched along the x-axis (Fig. 47) we have

$$E = \frac{p^2 + y^2 + \omega^2(\varepsilon x)\, q^2}{2}, \qquad I = \frac{p^2 + \omega^2(\varepsilon x)\, q^2}{2\,\omega(\varepsilon x)}, \qquad H_0 = \frac{y^2}{2} + I\,\omega(\varepsilon x).$$

Suppose, for instance, that the function $\omega(\varepsilon x)$ is even and, as $|x|$ grows, it first increases and then decreases (Fig. 48a). The phase portrait of the system with Hamiltonian H_0 is shown in Fig. 48b. We see that if the value of the initial longitudinal velocity is not too large, the point will be confined to the middle of the fosse (Fig. 48c). The corresponding condition on the velocity (the *"confinement condition"*) is usually written in the form

$$\frac{E_{\|}}{E_{\perp}} < \frac{\omega_m - \omega_0}{\omega_0}, \tag{37}$$

where $E_{\|} = y^2/2$ and $E_{\perp} = (p^2 + \omega^2 q^2)/2$ are respectively the energies of the longitudinal and the transverse motion of the point at the middle of the fosse (for $x = 0$), and ω_0 and ω_m designate the minimum and maximum values of the function $\omega(\varepsilon x)$.

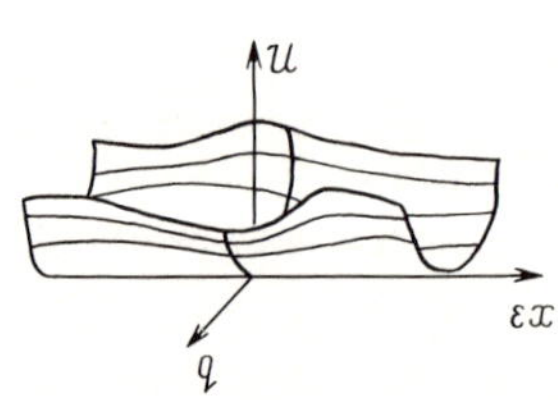

Fig. 47

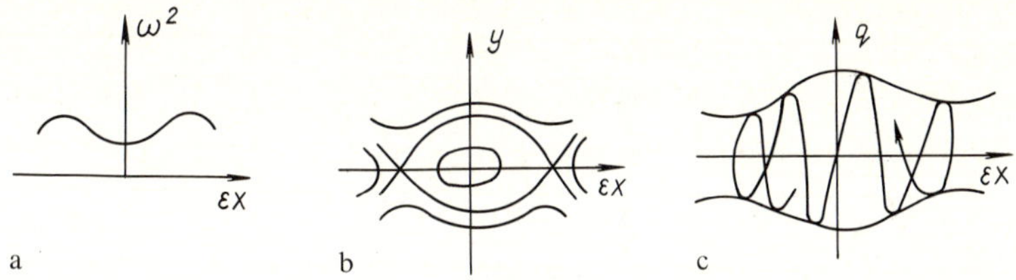

Fig. 48

The validity of these conclusions over times $1/\varepsilon$ follows from Theorem 24. It is extended to infinite time intervals using the KAM theory (see §4.5). △

Example 21. Short-wave excitations propagate along rays. A *wave-guide* is a potential fosse for rays. A medium whose refraction index changes slowly along a curve (the wave guide's axis) and rapidly in the direction transverse to it is named a *slowly-irregular* wave-guide; the refraction index has a maximum on the wave-guide axis. Suppose, for example, that the wave-guide axis is a straight line and the medium is two-dimensional (Fig. 49). Then the propagation of rays is described by the system with the Hamiltonian

$$E = p^2 + y^2 - n^2(q, \varepsilon x),$$

where x and q are the coordinates along the wave-guide axis and the axis perpendicular to it, respectively, the momenta y, p specify the direction of the ray, and n^2 is the refraction index [119]. We are interested in the solutions of this system on the energy level $E = 0$.

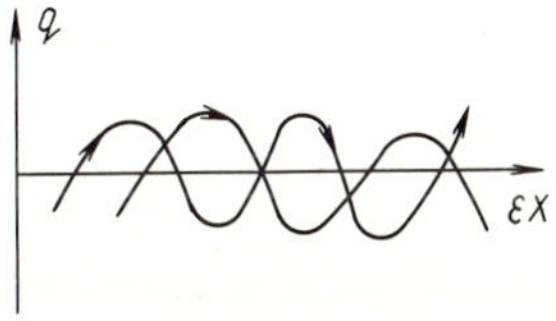

Fig. 49

We may assume that near the axis the refraction index is a quadratic function: $n^2 = a^2(\varepsilon x) - b^2(\varepsilon x) q^2$. Then in the notations of Theorem 24

$$I = \frac{p^2 + b^2 q^2}{2b}, \qquad H_0 = y^2 - a^2(\varepsilon x) + 2 I b(\varepsilon x).$$

These relations permit us to describe the behavior of rays. The ray portrait is shown in Fig. 49 (in the case in which there is no ray "supression"). In this way one can describe the propagation of light in light-guides and the propagation of short radiowaves and of sound in layered media [119]. △

Example 22. In a constant magnetic field a charged particle moves on a spiral around a line of force of the field. This motion is the superposition

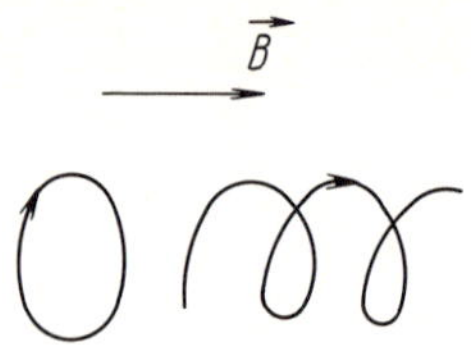

Fig. 50

of gyration on a circle around the line of force, called Larmor gyration, and a drift of this circle (Fig. 50). The Hamiltonian system describing this motion has three degrees of freedom.

Using the invariance of the Hamiltonian under translations along the field and under rotations around the direction of the field, the number of degrees of freedom can be reduced to one. All trajectories of the reduced system are periodic, and its "action" variable is the *magnetic moment* $I = v_\perp^2/(2B)$, where $v_\perp$ is the component of the particle velocity perpendicular to the field, and B the field intensity.[47] If the field is slowly-nonuniform (i.e., it changes only slightly over the length of the Larmor radius), then the magnetic moment is an adiabatic invariant [161]. The theory of motion in a slowly-nonuniform field is described in [161] without resorting to the Hamiltonian formalism; the Hamiltonian theory is constructed in [93, 129].

We now consider the particular case of a slowly-nonuniform axially-symmetric field whose lines of force lie in planes passing through the symmetry axis (Fig. 51). In cylindrical coordinates r, ϑ, z the Hamiltonian does not depend on the angle ϑ. The problem reduces to one with two degrees of freedom and, by Theorem 4, its investigation can be carried to its conclusion. It turns out (see [7]) that the motion in the (r, z)-plane is described by the Hamiltonian

$$E = \frac{y^2 + p^2 + B^2(\varepsilon x) q^2}{2} + \varepsilon E_1(p, q, y, \varepsilon x, \varepsilon).$$

Here x and q are coordinates along and respectively transverse to the line of force of the magnetic field around which the point moves (this line is singled out by the condition of conservation of the momentum canonically-conjugate to the angle ϑ), and $B(\varepsilon x)$ is the value of the field intensity on this line. Discarding the terms of order ε in the Hamiltonian we obtain the potential fosse of Example 20. The confinement condition (37) shows which particles are locked in this fosse. On this principle of confinement of charged particles is based the construction of traps *(mirrors)* for plasma known as adiabatic traps [161] (or traps with magnetic stoppers). A gigantic natural adiabatic trap is the Earth's magnetic field (Fig. 52). △

Adiabatic invariance is also observed in systems with collisions.

[47] The magnetic moment is proportional to the flux of the field across the Larmor circle.

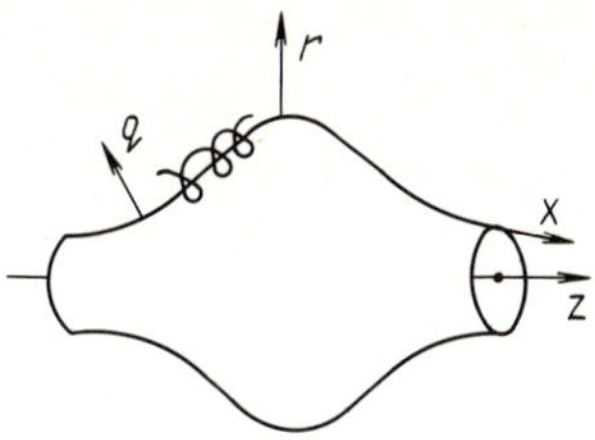

Fig. 51

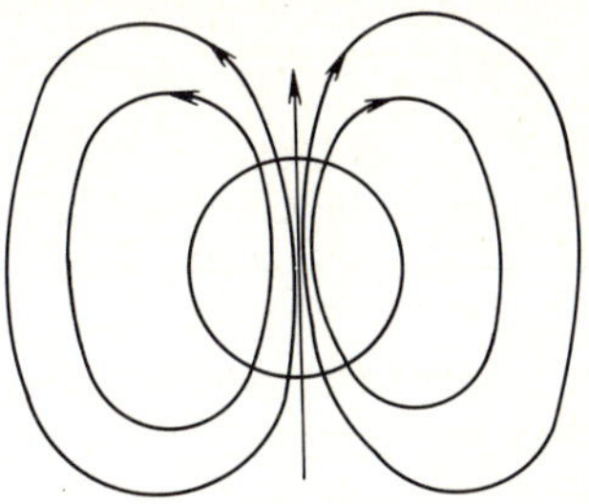

Fig. 52

Example 23. For the motion of a small elastic ball between two slowly moving walls (Fig. 53), an adiabatic invariant is the product of the velocity of the ball and the distance between the walls. This fact can be established by direct calculation; alternately, it can be deduced from Theorem 23 by passing to a limit.

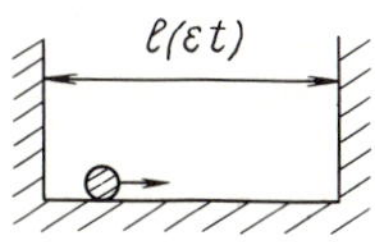

Fig. 53

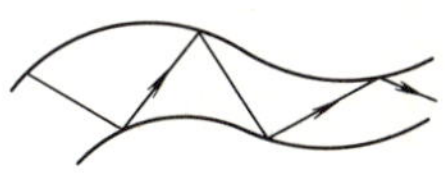

Fig. 54

Example 24 ([119]). For the propagation of rays in a planar slowly-irregular[48] *light-guide* with mirror walls (Fig. 54) an adiabatic invariant is the product of the distance between the walls and the sine of the angle that the ray makes with the wall.

4.2. Adiabatic Invariants of Multi-Frequency Hamiltonian Systems

Definition. A function of phase coordinates and of parameters is called an *almost adiabatic invariant* if for every $\rho>0$ the measure of the set of initial conditions for which the variation of this function along a solution exceeds ρ in time $1/\varepsilon$ tends to zero as $\varepsilon\to 0$.

We consider a Hamiltonian system with $n\geqq 2$ degrees of freedom whose Hamiltonian depends on a slowly varying parameter λ. We assume that for each fixed λ the system is completely integrable, so that it admits action-angle coordinates I, φ. Propositions 1 and 2 of §4.1 are valid for this system (and the proof is exactly the same): the behavior of the variables I, φ is described by a system of form (36), and the "action" I is a first integral of the averaged system. How do the variables I behave in the exact (nonaveraged) system?

[48] A light-guide is said to be slowly-irregular if its width and the direction of the walls vary slowly along the guide.

Suppose that $\det(\partial^2 H_0/\partial I^2) \neq 0$. Then by Theorem 8 the variations of the variables I remain smaller than ρ over time $1/\varepsilon$ if we neglect a set of initial conditions of measure $c\sqrt{\varepsilon}/\rho$, $c = \text{const} > 0$, in the phase space (which is assumed here to be compact). Therefore, the "action" variables I are almost-adiabatic invariants of the given nondegenerate multi-frequency Hamiltonian system.

If the system has two frequencies, then the estimate for the variation of the variables I can be sharpened using the results of § 1.6.

There are examples showing that even in two-frequency systems there may exist a set of initial conditions of measure $\sqrt{\varepsilon}$ for which an almost adiabatic invariant can vary by a quantity of order 1 over time $1/\varepsilon$ due to capture into a resonance [54]. Adiabatic invariants in single-frequency systems are preserved over time intervals considerably larger than $1/\varepsilon$, and under a periodic change of the parameter-even indefinitely. In multi-frequency systems the picture changes drastically. Examples show that for a set of initial conditions of measure of order 1, an almost adiabatic invariant can undergo a variation of order 1 over time $1/\varepsilon^{3/2}$ due to temporary captures into resonances.

In the foregoing discussion it was assumed that for each fixed λ the system is completely integrable. Almost adiabatic invariants exist also in the opposite, far more frequent case, in which for each fixed λ the motion on almost all energy levels $E(p, q, \lambda) = \text{const}$ is ergodic. Suppose that the surface $E(p, q, \lambda) = h$ is smooth and bounds a domain of finite phase volume (for an open set of values of h and λ). We denote this volume by $I(h, \lambda)$. Then the function $I(E(p, q, \lambda), \lambda)$ is an almost adiabatic invariant [102].

The adiabatic invariants of linear multi-frequency systems are investigated in detail in [126]. The theory developed therein is concerned with linear Hamiltonian systems with coefficients periodic in time which also depend on a parameter $\lambda = \lambda(\varepsilon t)$ that varies slowly with time. It is assumed that for each fixed λ the system is strongly stable, i.e., it is stable and an arbitrary, sufficiently small change in the coefficients does not destroy the stability. All multipliers of a strongly stable system lie on the unit circle and are different from ± 1 (see, e.g., [8]). Consequently, under the variation of λ the multipliers move on the upper and lower semicircles without passing from one half plane to the other.[49]

Definition. We say that a group of multipliers of the unperturbed ($\lambda = \text{const}$) system, arranged successively on the unit circle, forms a *cluster* if under the variation of λ according to the rule $\lambda = \lambda(\varepsilon t)$ these multipliers collide with one another, but not with the other multipliers (Fig. 55).

Theorem 25 ([126]). *The given linear Hamiltonian system has at least as many independent adiabatic invariants, varying by a quantity of order ε over time $1/\varepsilon$, as there are clusters of multipliers of the unperturbed system on*

[49] Since the system is real the multipliers lie symmetrically with respect to the real axis.

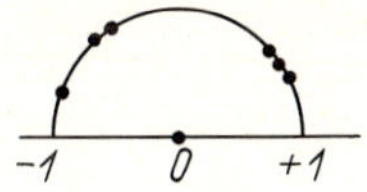

Fig. 55

the upper unit semicircle. These adiabatic invariants are quadratic forms in the phase variables whose coefficients depend (periodically) on time and on the parameter λ.

Corollaries. 1. *If for all values of λ the multipliers are distinct, then the number of independent adiabatic invariants is equal to the number of degrees of freedom.*

2. *The given linear system has at least one adiabatic invariant.*

If in the given system collisions of multipliers occur only at isolated moments of the slow time εt, then the number of adiabatic invariants is equal to the number of degrees of freedom. Far away from the collision moments the adiabatic invariants undergo oscillations of order ε. Near a collision moment the adiabatic invariants corresponding to the colliding multipliers may change by a quantity of order $\gg \varepsilon$ (and by a quantity of order $\sqrt{\varepsilon}$ in the case of collision of multipliers with nonzero velocity).

4.3. Procedure for Eliminating Fast Variables. Conservation Time of Adiabatic Invariants. For single-frequency Hamiltonian systems with slowly varying parameters the fast variables can be eliminated by means of symplectic transformations of coordinates and one can thereby obtain quantities that are preserved with a high degree of accuracy. In the variables I, φ, Y, X introduced in §4.1 the Hamiltonian of the problem has the form

$$H = H_0(I, Y, \varepsilon X) + \varepsilon H_1(I, \varphi, Y, \varepsilon X, \varepsilon). \tag{38}$$

Theorem 26. *For a C^∞-smooth single-frequency Hamiltonian* (38) *the fast phase φ can be eliminated by means of a symplectic transformation of coordinates $I, \varphi, Y, X \mapsto J, \psi, \eta, \xi$.*

◁ The new Hamiltonian $\mathscr{H} = \mathscr{H}(J, \eta, \varepsilon\xi, \varepsilon)$ and the generating function $J\varphi + \varepsilon S(J, \varphi, \eta, \varepsilon X, \varepsilon)$ are connected by the relation

$$H\left(J + \varepsilon \frac{\partial S}{\partial \varphi}, \varphi, \eta + \varepsilon^2 \frac{\partial S}{\partial \varepsilon X}, \varepsilon X, \varepsilon\right) = \mathscr{H}\left(J, \eta, \varepsilon X + \varepsilon^2 \frac{\partial S}{\partial \eta}, \varepsilon\right). \tag{39}$$

We seek $\mathscr{H}$ and S as formal series in ε:

$$\mathscr{H} = H_0 + \varepsilon \mathscr{H}_1 + \varepsilon^2 \mathscr{H}_2 + \ldots, \qquad S = S_1 + \varepsilon S_2 + \ldots. \tag{40}$$

Substituting these series in (39) and equating the terms of the same order in ε we get the system of equations

$$\frac{\partial H_0}{\partial J}\frac{\partial S_1}{\partial \varphi}+H_1(J,\varphi,\eta,\varepsilon X,0)=\mathscr{H}_1(J,\eta,\varepsilon X),$$

$$\frac{\partial H_0}{\partial J}\frac{\partial S_i}{\partial \varphi}+G_i(J,\varphi,\eta,\varepsilon X)=\mathscr{H}_i(J,\eta,\varepsilon X),\qquad i\geqq 2.$$

The function G_i is determined by the terms $S_1, \mathscr{H}_1, \ldots, S_{i-1}, \mathscr{H}_{i-1}$ in expansions (40). In the notations $\langle\cdot\rangle^\varphi$ and $\{\cdot\}^\varphi$, introduced in § 1.2 for the averaging and integration operators, respectively, the solution of the last system is written:

$$\mathscr{H}_1=\langle H_1\rangle^\varphi,\qquad S_1=-\{H_1\}^\varphi+S_1^0,$$
$$\mathscr{H}_i=\langle G_i\rangle^\varphi,\qquad S_i=-\{G_i\}^\varphi+S_i^0,\quad i\geqq 2,$$

where S_i^0 are arbitrary functions of $J, \eta, \varepsilon X$. One often takes $S_i^0\equiv 0$. ▷

Corollaries. 1. *The "function" J is a formal first integral of the problem.*

2. *Truncate the series for the coordinate transformation at the terms of order ε^m. Then in the new coordinates introduced by the truncated transformation the Hamiltonian will depend on the phase only through terms of order ε^{m+1}. Moreover, the function J introduced by the truncated transformation undergoes over time $1/\varepsilon$ only oscillations of order ε^{m+1}.*

Series (40) can diverge even in analytic systems (see [157] for an example). The next proposition describes the limiting attainable accuracy with which the phase can be eliminated from a Hamiltonian.

Proposition 3 ([157], cf. Theorem 2). *The Hamiltonian (38) of an analytic single-frequency system with slowly varying parameters can be transformed, by means of a symplectic transformation of coordinates that differs from the identity by $O(\varepsilon)$, into a sum of two terms, the first of which does not depend on the phase, while the second is exponentially small ($O(\exp(-c_1^{-1}/\varepsilon))$, with $c_1=\text{const}>0$).*

Corollary. *In an analytic single-frequency system the "action" I undergoes over time $T=\exp(\frac{1}{2}c_1^{-1}/\varepsilon)$ only oscillations of order ε. Therefore, adiabatic invariance holds on an exponentially large time interval.*[50]

◁ The coordinate transformation of Proposition 3 takes I into a quantity $J=I+O(\varepsilon)$ which varies exponentially slowly and hence changes over time T by an exponentially small quantity. ▷

4.4. Accuracy of the Conservation of Adiabatic Invariants. Suppose given a system with one degree of freedom depending on a slow parameter λ which

[50] Here it is assumed that over this time the trajectories of the system do not leave a given domain of size of order 1.

varies so that it tends sufficiently fast to definite limits as $\varepsilon t \to \pm\infty$. Then there exist the limit values $I(+\infty)$ and $I(-\infty)$ of the adiabatic invariant I, and one can speak about the increment of I

$$\Delta I = I(+\infty) - I(-\infty)$$

over the infinite time interval $(-\infty, +\infty)$. Although for finite t the quantity I undergoes oscillations of order ε, the increment ΔI is, as it turns out, much smaller than ε.

If the parameter varies smoothly (i.e., $\lambda \in C^\infty$), then $\Delta I = O(\varepsilon^\infty)$, i.e., ΔI decays faster than any power of ε as $\varepsilon \to 0$ [124]. In fact, the procedure of §4.3 permits us, for every m, to introduce a quantity J which undergoes only oscillations of order $O(\varepsilon^m)$ along the motion, and which coincides asymptotically with I for $t \to \pm\infty$.

If λ depends analytically on εt, then $\Delta I = O(\exp(-c^{-1}/\varepsilon))$, $c = \text{const} > 0$ [155, 178]. In fact, by Proposition 3, we can introduce the quantity J which undergoes only exponentially small oscillations along the motion, and which coincides with I for $t \to \pm\infty$.

For the linear oscillator

$$\ddot{x} = -\omega^2(\varepsilon t)\, x, \qquad \omega(\pm\infty) = \omega_\pm,$$

with analytic frequency $\omega(\varepsilon t) > \text{const} > 0$, the asymptotics of ΔI is known [129, 91]. Its calculation is based on the analytic continuation of the solution for complex values of the time variable t.

In [178] a method was found for analytic continuation of solutions of nonlinear perturbed systems, and was used to calculate the constant c in the formula $\Delta I = O(\exp(-c^{-1}/\varepsilon))$ and, for a number of cases, the factor that multiplies the exponential. It was found that c is the minimum of the imaginary parts of the increments of phases in the unperturbed motion

$$\dot{\varphi} = \frac{\partial H_0(I_-, \lambda(\varepsilon t))}{\partial I}, \qquad I_- = I(-\infty),$$

calculated along the paths in the complex t-plane which go from the real axis to the singular points of the Hamiltonian and to the zeros of the unperturbed frequency in the upper half plane (Fig. 56).

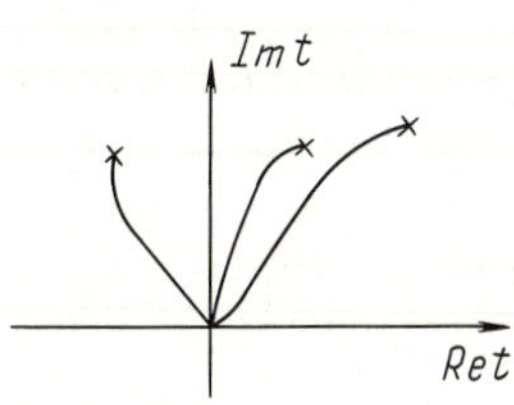

Fig. 56

4.5. Perpetual Conservation of Adiabatic Invariants. Over an infinite time interval adiabatic invariants may undergo a strong evolution due to accumulation of small perturbations.

Example 2.5. We consider the linear oscillator

$$\ddot{x} = -\omega^2(1+\alpha \cos \varepsilon t)\,x, \qquad \alpha = \text{const} < 1.$$

For arbitrarily small ε the equilibrium position $x=0$ may become unstable (the phenomenon of parametric resonance [8]). The adiabatic invariant of this oscillator changes unboundedly. △

It turns out, however, that under a periodic variation of the parameter such a failure of conservation of the adiabatic invariant is directly related to the linearity of the system (more precisely, to the fact that the period of oscillations does not depend on the amplitude). In a nonlinear system, as the amplitude increases the frequency changes, and the oscillations do not have enough time to grow before the resonance condition is violated.

Definition. An adiabatic invariant is called *perpetual* if for $-\infty < t < \infty$ its value deviates little from the initial value, provided ε is sufficiently small.

Theorem 27 ([7]). *Under slow periodic variations of the Hamiltonian of a nonlinear oscillatory system with one degree of freedom, the "action" variable I is a perpetual adiabatic invariant. A large subset of the phase space is filled out by invariant tori close to the tori $I = \text{const}$.*

◁ First, let us formulate the required nonlinearity condition for the given system. In action-angle variables the Hamiltonian has the form

$$H = H_0(I, \tau) + \varepsilon H_1(I, \varphi, \tau), \qquad \tau = \varepsilon t, \tag{41}$$

and is 2π-periodic in φ and τ. We let $\bar{\omega}(I)$ denote the mean value over τ of the "adiabatic" frequency of variation of the phase, $\omega(I, \tau) = \partial H_0/\partial I$. In the statement of the theorem, the system is considered to be nonlinear if the mean frequency $\bar{\omega}(I)$ depends on I:

$$\frac{\partial \bar{\omega}}{\partial I} \neq 0. \tag{42}$$

The proof of the theorem rests on the results of the KAM theory (§3). To reduce the equations of motion to the standard form of the KAM theory ("the fundamental problem of dynamics") some transformations must be effected.

Firstly, we replace φ by the new phase

$$\psi = \varphi - \frac{1}{\varepsilon} \int_0^\tau \omega(I, \xi)\, d\xi + \frac{1}{\varepsilon}\, \bar{\omega}(I)\, \tau.$$

As a result, the principal term of the Hamiltonian will no longer depend on the phase τ: the Hamiltonian takes the form

$$\mathscr{H}=\mathscr{H}_0(I)+\varepsilon\mathscr{H}_1(I,\psi,\tau),$$

where $\mathscr{H}_0$ is the mean value of $H_0(I,\tau)$ over τ, and $\mathscr{H}_1$ is 2π-periodic in ψ and τ.

Secondly, we take the phase ψ as a new time variable. Then the Hamiltonian of the problem takes the form (see Ch. 4, §1.1):

$$F=-\varepsilon I=-\varepsilon F_0(-h)+\varepsilon^2 F_1(h,\psi,\tau,\varepsilon),$$

where h is the variable canonically conjugate to the phase τ (the old Hamiltonian $\mathscr{H}$ with opposite sign) and F_0 is the function inverse to $\mathscr{H}_0$.

We thus obtain the system with one and a half degrees of freedom and *proper degeneracy* considered in §3.3. Condition (42) guarantees that the degeneracy is removable. According to the results of §3.3, in this system there are many invariant tori close to the tori $h=\text{const}$; the value of h oscillates perpetually with an amplitude of order ε. Consequently, in the phase space I,φ,τ of the original system there are many invariant tori close to the tori $I=\text{const}$; the value of I oscillates perpetually with an amplitude of order ε. ◁

Remark. If the parameters of the system depend in conditionally-periodic manner on time and the frequency vector $\varepsilon\Omega$ satisfies the usual incommensurability conditions: $|(k,\Omega)|>c^{-1}|k|^{-\nu}$, $k\in Z^m\setminus\{0\}$, then I is also a perpetual adiabatic invariant [7]. △

A perpetual adiabatic invariant also exists (under certain assumptions) in any conservative system with two degrees of freedom whose Hamiltonian depends slowly on one of the coordinates [7]. By Theorem 24, the motion in such a system is approximately described by the Hamiltonian $H_0(I,y,\varepsilon x)$. Suppose that the phase curves of this Hamiltonian for fixed I are closed (as in Fig. 46). Then in the approximation considered, the motion in phase space occurs on the two-dimensional tori specified by the conditions $I=\text{const}$, $H_0=\text{const}$. This motion has two frequencies, one of which is smaller than the other by a factor of $1/\varepsilon$. If for given $H_0=\text{const}$ the frequency ratio varies with I, then it follows from the results of the KAM theory that in the exact system there exist many invariant tori close to the tori of the approximate system. As always in the case of two degrees of freedom, this implies stability: the "action" variable I remains perpetually close to its initial value; for details see [7]. From this conclusion it follows, in particular, that charged particles are held indefinitely in the axially-symmetric magnetic trap of Example 22.

Chapter 6
Nonintegrable Systems

A common feature of the various approaches to the problem of integration of Hamiltonian systems indicated in Chapter 4 is the existence of a sufficiently large number of independent first integrals, or "conservation laws". Unfortunately, in typical situations, not only do we fail to find such integrals, but they simply do not exist, since the trajectories of Hamiltonian systems, generally speaking, do not lie on low-dimensional invariant manifolds. We have in mind, of course, integrals which exist on the entire phase space: a complete set of independent first integrals always exists in a small neighborhood of a nonsingular point.

The earliest rigorous results on nonintegrability of Hamiltonian systems are due to Poincaré. The essence of Poincaré's idea is that an intricate behavior of solutions (for example, creation of nondegenerate periodic solutions, splitting of asymptotic surfaces, etc.) is incompatible with the complete integrability of Hamilton's equations. In this chapter we present the main methods for proving the nonintegrability of Hamiltonian systems, based on the isolation of various nontrivial dynamical effects which are not peculiar to completely integrable systems. A more detailed exposition is given in [21].

§1. Near-Integrable Hamiltonian Systems

Here we describe a method, attributable mostly to Poincaré, for proving the nonexistence of additional analytic integrals, based on the investigation of bifurcations of long-periodic solutions.

Suppose that the direct product $M = D \times T^n\{\varphi \bmod 2\pi\}$ (where D is a domain in $R^n = \{I\}$) is endowed with the standard symplectic structure $dI \wedge d\varphi$, and $H\colon M \times (-\varepsilon_0, \varepsilon_0) \to R$ is an analytic function, such that $H(I, \varphi, 0) = H_0(I)$. For $\varepsilon = 0$ we have a completely integrable Hamiltonian system with the Hamiltonian H_0. Consider the full system

$$\dot{I} = -\frac{\partial H}{\partial \varphi}, \quad \dot{\varphi} = \frac{\partial H}{\partial I}, \quad H = H_0(I) + \varepsilon H_1(I, \varphi) + \ldots \tag{1}$$

for small values of the parameter ε. Numerous examples are given in Chapter 5. In this section we investigate the problem of existence of first integrals of equations (1) which are independent of function H and analytic (or, more precisely, formally analytic) in the parameter ε.[51] We recall that the existence of a complete commutative set of such integrals is connected with the feasibility of the construction and the convergence of different variants of perturbation theory for Hamiltonian systems (see Ch. 5, §2.2).

1.1. Poincaré's Method. We first give a number of definitions. Let

$$H_1 = \sum_{m \in Z^n} h_m(I) \exp(i\langle m, \varphi\rangle).$$

The *Poincaré set* is the set of values $I \in D$ for which there exist $n-1$ linearly independent vectors $k_1, \ldots, k_{n-1} \in Z^n$ such that

1) $\langle k_s, \omega(I)\rangle = 0$, $1 \leqq s \leqq n-1$, $\omega = H_0'$;

and

2) $h_{k_s}(I) \neq 0$.

We let $\mathscr{A}(V)$ denote the class of functions that are analytic in a domain $V \subset R^n$. The set $\Lambda \subset V$ is called a *key set* (or *set of uniqueness*) for $\mathscr{A}(V)$ if any analytic function that vanishes on Λ vanishes identically on V. Thus, if two analytic functions coincide on Λ, then they coincide on the whole domain V. For example, a set of points of an interval $\Delta \subset R$ is a key set for the class $\mathscr{A}(\Delta)$ if and only if it has a limit point in the interior of Δ. The sufficiency of this condition is obvious; necessity follows from Weierstrass's theorem on infinite products. We note that if Λ is a key set for the function class $C^\infty(V)$, then Λ is dense in V.

Theorem 1. *Suppose that the unperturbed system is nondegenerate:* $\det(\partial^2 H_0/\partial I^2) \neq 0$ *in the domain D; that $I^0 \in D$ is a noncritical point of H_0; and that in every neighborhood U of I^0 the Poincaré set is a key set for $\mathscr{A}(U)$. Then the Hamiltonian equations* (1) *do not admit a formal first integral F, independent of H, representable as a formal power series $\sum_{s \geqq 0} F_s(I, \varphi)\, \varepsilon^s$ with coefficients analytic in $D \times T^n$* (cf. [33, 20]).

A formal series $\sum f_s \varepsilon^s$ is regarded as being equal to zero if all $f_s = 0$. The series F is a formal integral of the canonical equations with Hamiltonian H if the formal series $\{H, F\}$ is equal to zero. Two formal series, F and G, are regarded as dependent if all the second-order minors of their Jacobian matrix are identically zero as formal power series in ε.

To prove Theorem 1 we need the following lemma.

Lemma 1. *Suppose that the functions F_s: $D \times T^n \to R$ are continuously differentiable and that the series $\sum F_s(I, \varphi)\, \varepsilon^s$ is a formal integral of equations* (1) *with a nondegenerate function H_0. Then*

[51] A formal series in ε is said to be formally analytic if its coefficients are analytic functions of the phase variables.

1) *$F_0(I, \varphi)$ does not depend on φ,*

and

2) *H_0 and F_0 are dependent at the points of the Poincaré set.*

◁ The condition $\{H, F\}=0$ is equivalent to the sequence of equations

$$\{H_0, F_0\}=0, \quad \{H_0, F_1\}+\{H_1, F_0\}=0, \ldots . \tag{2}$$

From the first equation it follows that F_0 is an integral of the unperturbed equations with Hamiltonian H_0. Suppose the torus $I=I^*$ is nonresonant. Then $F_0(I^*, \varphi)$ does not depend on φ, because every trajectory on a nonresonant torus fills it densely. To complete the proof of 1) it remains to use the continuity of F_0 and the fact that the set of nonresonant tori of a nondegenerate integrable system is dense in the phase space.

From the second equation of sequence (2) we derive a series of equalities for the Fourier coefficients h_m and f_m of the functions H_1 and $F_1 = \sum f_m(i) \exp(i\langle m, \varphi\rangle)$:

$$\left\langle m, \frac{\partial H_0}{\partial I}\right\rangle f_m(I)=\left\langle m, \frac{\partial F_0}{\partial I}\right\rangle h_m(I), \qquad m\in Z^n.$$

The condition for the solvability of these equations with respect to f_m at a point of the Poincaré set is the linear dependence of the vectors $\partial H_0/\partial I$ and $\partial F_0/\partial I$. ▷

◁ *Proof of Theorem 1.* Since at the point $I^0 \in D$ at least one of the derivatives $\partial H_0/\partial I_1, \ldots, \partial H_0/\partial I_n$ is different zero, it follows that in a small neighborhood U of I^0 one can take, say, $H_0, I_2, \ldots, I_n$ as local coordinates (provided that $\partial H_0/\partial I_1 \neq 0$).

By Lemma 1, the functions H_0 and F_0 are dependent on the Poincaré set. Since the minors of the Jacobian matrix

$$\begin{bmatrix} \dfrac{\partial H_0}{\partial I_1} \cdots \dfrac{\partial H_0}{\partial I_n} \\[2ex] \dfrac{\partial F_0}{\partial I_1} \cdots \dfrac{\partial F_0}{\partial I_n} \end{bmatrix}$$

are analytic in U and the Poincaré set is a key set, H_0 and F_0 are dependent throughout U and, consequently, in the new coordinates, in a neighborhood of the value $H_0(I^0)$ we have $F_0=F_0(H_0)$.

We put $F-F_0(H)=\varepsilon\Phi$. Then Φ is a formal integral of the canonical equations (1). If we write $\Phi=\sum_{s\geq 0} \Phi_s \varepsilon^s$, then by Lemma 1, the function Φ_0 does not depend on the angle variables φ, and Φ_0 is dependent on H_0 in the domain U. Consequently, $\Phi_0=\Phi_0(H_0)$, and again we put $\Phi-\Phi_0(H)=\varepsilon\Psi$. But then $F=F_0(H)+\varepsilon\Phi_0(H)+\varepsilon^2\Psi$. Repeating this process as long as needed, we find that the expansion of all second-order minors of the Jacobian matrix

$$\begin{bmatrix} \frac{\partial H}{\partial I} & \frac{\partial H}{\partial \varphi} \\ \frac{\partial F}{\partial I} & \frac{\partial F}{\partial \varphi} \end{bmatrix}$$

in series of powers of ε begins with terms of arbitrarily high order. This implies the dependence of functions H and F. ▷

Theorem 2. *Suppose that H_0 is nondegenerate in the domain D, and that the Poincaré set is dense in D. Then equations* (1) *have no formal integral $\sum F_s \varepsilon^s$, independent of H, with infinitely differentiable coefficients F_s: $D \times T^n \to R$.*

This statement is readily proved by the same method as used for Theorem 1.

Remark. For $n=2$ it follows from Kolmogorov's theorem on conservation of conditionally-periodic motions that there exists a first integral analytic in ε with nonconstant continuous coefficients. By way of contrast, in the higher-dimensional case, it seems that for systems of general form even a continuous integral cannot exist (see [50]).

1.2. Creation of Isolated Periodic Solutions is an Obstruction to Integrability. We first recall some facts from the theory of periodic solutions of differential equations. The eigenvalues λ of the *monodromy operator* of a T-periodic solution are called *multipliers*, and the numbers α defined by $\lambda = \exp(\alpha T)$ are called *characteristic exponents*. The multipliers λ may be complex, and so the characteristic exponents are not uniquely determined. In the autonomous case one of the multipliers λ is always equal to 1 (the corresponding eigenvector is tangent to the periodic trajectory).

Proposition 1 (Poincaré-Lyapunov). *For a Hamiltonian system with n degrees of freedom, the characteristic polynomial $p(\lambda)$ of the monodromy operator is equal to its reciprocal: $p(\lambda^{-1}) = \lambda^{-2n} p(\lambda)$.*

For the proof see, for example [8].

Theorem 3 (Poincaré [33]). *Suppose that the Hamiltonian system with Hamiltonian function H has p first integrals $F_1 = H, F_2, \ldots, F_p$ which are independent at the points of the trajectory of a periodic solution. Then $p+1$ characteristic exponents of this solution vanish. If the integrals F_k pairwise commute, then at least $2p$ exponents vanish.*

Corollary 1. *Periodic solutions of autonomous Hamiltonian systems always have two null characteric exponents.*

One of the exponents vanishes because the system is autonomous, and the second due to the presence of the first integral H (which has no critical points on the trajectories of periodic solutions). If the remaining characteristic exponents are different from zero the periodic solution is said to be *nondegenerate*. Nondegenerate solutions are *isolated*: on the corresponding $(2n-1)$-dimensional level manifold of the energy H, in a small neighborhood of the T-periodic trajectory in question, there are no other periodic trajectories with period close to T. For systems with two degrees of freedom, a nondegenerate solution with real (purely imaginary) characteristic exponents is usually said to be *hyperbolic* (respectively, *elliptic*). Hyperbolic (elliptic) periodic solutions are unstable (respectively, stable).

Corollary 2. *If a Hamiltonian system has a complete set of first integrals in involution which are independent on the trajectory of a periodic solution, then the spectrum of its monodromy operator reduces to the point* $\lambda=1$.

Poincaré's theorem gives us a method for proving nonintegrability: if the trajectories of nondegenerate periodic solutions fill densely the phase space, or at least form a key set, then the Hamiltonian system has no additional analytic first integral. Apparently, in generic Hamiltonian systems the periodic trajectories are indeed everywhere dense (Poincaré [33], no. 36). This conjecture is not yet proven. In connection with it, we mention the following result concerning geodesic flows on Riemannian manifolds of negative curvature: all periodic solutions are hyperbolic, and the set of their trajectories is dense in the phase space [3].

For Hamiltonian systems close to integrable ones (near-integrable), one can establish the existence of a large number of nondegenerate periodic solutions and from this derive the results of § 1.1. For simplicity we confine ourselves to systems with two degrees of freedom.

Suppose that for $I=I^0$ the frequencies ω_1 and ω_2 of the unperturbed integrable problem are commensurable and that $\omega_1 \neq 0$. Then the perturbing function $H_1(I^0, \omega_1 t, \omega_2 t+\lambda)$ is periodic in t with a certain period T. Consider its mean value

$$\bar{H}_1(I^0, \lambda)=\lim_{S\to\infty} \frac{1}{S}\int_0^S H_1(I^0, \omega_1 t, \omega_2 t+\lambda)\, dt=\frac{1}{T}\int_0^T H_1\, dt.$$

Theorem 4 (Poincaré). *Suppose that the following conditions are satisfied:*

1) $\det(\partial^2 H_0/\partial I^2)\neq 0$ *at the point* $I=I^0$;

2) *for some* $\lambda=\lambda^*$ *the derivative* $\partial \bar{H}_1/\partial\lambda=0$, *but* $\partial^2 \bar{H}_1/\partial\lambda^2\neq 0$.

Then for small $\varepsilon\neq 0$ *the perturbed Hamiltonian system* (1) *has a T-periodic solution; this solution depends analytically on the parameter* ε, *and for* $\varepsilon=0$ *coincides with the periodic solution*

$$I=I^0, \quad \varphi_1=\omega_1 t, \quad \varphi_2=\omega_2 t+\lambda^*$$

of the unperturbed system. The two characteristic exponents $\pm\alpha$ *of this solution admit a convergent series expansion in powers of* $\sqrt{\varepsilon}$:

$$\alpha=\alpha_1\sqrt{\varepsilon}+\alpha_2\varepsilon+\alpha_3\varepsilon\sqrt{\varepsilon}+\dots,$$

where

$$\omega_1^2 \alpha_1^2 = \frac{\partial^2 \bar{H}_1}{\partial \lambda^2}(\lambda^*)\left(\omega_1^2 \frac{\partial^2 H_0}{\partial I_2^2} - 2\omega_1 \omega_2 \frac{\partial^2 H_0}{\partial I_1 \partial I_2} + \omega_2^2 \frac{\partial^2 H_0}{\partial I_1^2}\right). \tag{4}$$

A proof can be found in [33, 20].

The function $\bar{H}_1(I^0, \lambda)$ is 2π-periodic in λ. Hence, there are at least two values of λ for which $\partial \bar{H}_1/\partial \lambda = 0$. In general, these critical points are nondegenerate, and then there are as many local minima (where $\partial^2 \bar{H}_1/\partial \lambda^2 > 0$) as local maxima (where $\partial^2 H_1/\partial \lambda^2 < 0$). In the typical situation, for $I = I^0$ the quadratic form

$$\omega_1^2 \frac{\partial^2 H_0}{\partial I_2^2} - 2\omega_1 \omega_2 \frac{\partial^2 H_0}{\partial I_1 \partial I_2} + \omega_2^2 \frac{\partial^2 H_0}{\partial I_1^2} \neq 0. \tag{5}$$

We note that geometrically this means that the curve $H_0(I) = h$ does not have an inflection point at $I = I_0$. Therefore, the equation $d\bar{H}_1 = 0$ has as many roots for which $\alpha_1^2 > 0$ as roots for which $\alpha_1^2 < 0$. Equivalently, for small $\varepsilon \neq 0$ the perturbed system has as many periodic solutions of elliptic type as of hyperbolic type. In this situation one usually says that pairs of isolated periodic solutions are created at the disintegration of the unperturbed invariant torus $I = I^0$. According to results of the KAM theory, the trajectories of typical elliptic periodic solutions are "surrounded" by invariant tori. Hyperbolic periodic solutions have two invariant surfaces *(separatrices)*, filled out by solutions approaching asymptotically the periodic trajectory for $t \to \pm\infty$. Distinct *asymptotic surfaces* may intersect, forming a rather tangled network (see Fig. 44). The behaviour of asymptotic surfaces will be discussed in detail in the next section.

The main theoretical tool in proving the nonintegrability of perturbed equations is Lemma 1: if $F = F_0(I, \varphi) + \varepsilon F_1(I, \varphi) + \dots$ is a first integral of the canonical equations (1), then F_0 does not depend on φ, and the functions H_0 and F_0 are dependent in the Poincaré set. The first part of the lemma follows from the nondegeneracy of the unperturbed problem. Using Theorem 3, we prove the dependence of functions H_0 and F_0 on the set of unperturbed tori $I = I^0$ that satisfy the conditions of Theorem 4 and inequality (5).

◁ In fact, the periodic solutions $\Gamma(\varepsilon)$ that are created from the family of periodic solutions on the resonant torus $I = I^0$ satisfying the conditions of Theorem 4 are nondegenerate. Consequently, the functions H and F are dependent at all the points of the trajectory $\Gamma(\varepsilon)$ (Theorem 4). Now let $\varepsilon \to 0$. Then the periodic solution $\Gamma(\varepsilon)$ goes over into one of the periodic solutions $\Gamma(0)$ of the unperturbed system lying on the torus $I = I^0$, while the functions H and F become H_0 and F_0, respectively. By continuity, H_0 and F_0 are dependent at all the points of the trajectory $\Gamma(0)$. Consequently,

$$\operatorname{rank}\begin{bmatrix} \frac{\partial H_0}{\partial I} & \frac{\partial H_0}{\partial \varphi} \\ \frac{\partial F_0}{\partial I} & \frac{\partial F_0}{\partial I} \end{bmatrix} \leqq 1$$

at all points $(I, \varphi) \in \Gamma(0)$. In particular, at these points the Jacobian

$$\frac{\partial(H_0, F_0)}{\partial(I_1, I_2)} = 0.$$

To complete the proof we remark that the functions H_0 and F_0 are independent of φ. ▷

For small fixed values of the parameter $\varepsilon \neq 0$ Theorem 4 guarantees the existence of a large (but finite) number of distinct isolated periodic solutions. Hence, from this theorem one cannot deduce the nonintegrability of the perturbed system for fixed values $\varepsilon \neq 0$. It is true, however, that in the case of two degrees of freedom, which is our present concern, the following statement is valid: *if the unperturbed system is nondegenerate, then for small fixed* $\varepsilon \neq 0$ *the perturbed Hamiltonian system has infinitely many distinct periodic solutions.* Unfortunately, they are not necessarily isolated. The existence of infinitely many periodic solutions follows from Kolmogorov's theorem on the conservation of conditionally-periodic motions (Ch. 5, §3.2), and from the *Poincaré-Birkhoff geometric theorem* (see [8]).

1.3. Applications of Poincaré's Method

a) We return to the restricted three-body problem considered in §5 of Ch. 2. We assume first that the mass of Jupiter μ is equal to zero. Then in "fixed" space the asteroid will rotate around the unit-mass Sun on Keplerian orbits. Suppose that the orbits are elliptic. Then it is convenient to pass from orthogonal coordinates to the canonical Delaunay elements L, G, l, g (see Ch. 4, §2.1, Example 4). In the new coordinates the equations of motion of the asteroid are canonical with the Hamiltonian $F_0 = -\frac{1}{2}L^{-2}$. If $\mu \neq 0$, then the full Hamiltonian F can be expanded in a series of increasing powers of μ: $F = F_0 + \mu F_1 + \dots$. Since in the moving coordinate system attached to the Sun-Jupiter system Keplerian orbits rotate with unit angular velocity, the Hamiltonian F depends on L, G, l, and $g-t$. We put $x_1 = L$, $x_2 = G$, $y_1 = l$, $y_2 = g - t$, and $H = F - G$. Function H depends now on x_i and y_i only, and it is 2π-periodic in the angular variables y_1 and y_2. We thus represent the equations of motion of the asteroid as the following Hamiltonian system:

$$\dot{x}_i = \frac{\partial H}{\partial y_i}, \quad \dot{y}_i = -\frac{\partial H}{\partial x_i}; \qquad H = H_0 + \mu H_1 + \dots, \quad H_0 = -\frac{1}{2x_1^2} - x_2. \tag{6}$$

The expansion of the perturbing function in a multiple trigonometric series in the angles y_1 and y_2 was already studied by J. Leverrier (see, e.g. [15]). It has the form

$$H_1 = \sum_{u=-\infty}^{\infty} \sum_{v=-\infty}^{\infty} h_{u,v} \cos[u y_1 - v(y_1 + y_2)].$$

The coefficients $h_{u,v}$, which depend on x_1 and x_2, are in general different from zero.

The Poincaré set of this problem consists of straight lines parallel to the x_2-axis: $u/x_1^3 - v = 0$, with $h_{u,v} \neq 0$. It is dense in the half-plane $x_1 > 0$. However, Theorem 1 on the absence of new analytic integrals does not apply directly, because the unperturbed problem is degenerate: $\det(\partial^2 H_0/\partial x^2) \equiv 0$. This difficulty is circumvented by observing that the canonical equations with the Hamiltonians H and $\exp H$ have the same trajectories (but not the same solutions). Consequently, either both or neither of these equations are integrable. It remains to note that $\exp H = \exp H_0 + \mu(\exp H_0) H_1 + \dots$ and $\det(\partial^2(\exp H_0)/\partial x^2) \neq 0$. We thus conclude that the equations of the restricted three-body problem in the form (6) do not possess a first integral $\Phi = \sum \Phi_s \mu^s$ independent of H and formally-analytic in the parameter μ, whose coefficients Φ_s are smooth functions on $D \times T^2\{y \bmod 2\pi\}$, where Φ is an arbitrary domain in the half-plane $x_1 > 0$.

b) "Let us proceed to another problem: that of the motion of a heavy body around a fixed point .... We can therefore ask whether the considerations presented in this section prevent the existence of a single-valued integral other than those of the vis viva and of area" (Poincaré [33], no. 86).

Corresponding to the symmetry group consisting of the rotations of the body around the vertical axis there is the linear integral $\langle k, \gamma \rangle$: the vertical projection of the angular momentum is constant. Fixing this constant, we reduce the number of degrees of freedom to two: on the four-dimensional invariant manifolds $M_c = \{\langle k, \gamma \rangle = c,\ \langle \gamma, \gamma \rangle = 1\}$ one obtains a Hamiltonian system with two degrees of freedom. Its Hamiltonian function, the total energy of the body for fixed value of the projection $\langle k, \gamma \rangle$, is $H_0 + \varepsilon H_1$, where H_0 is the kinetic energy (the Hamiltonian of the integrable Euler problem on the inertial motion of the body) and εH_1 is the potential energy of the body in a homogeneous gravitational field (ε is the product of the weight of the body and the distance from the center of mass to the fixed point). We shall assume that the parameter ε is small (cf. Ch. 5, §2.1, Example 2). This is equivalent to studying the rapid rotation of the body in a moderate force field. In the unperturbed integrable Euler problem we can introduce action-angle variables I, φ. The formulas for the transition from the special canonical variables L, G, l, g to the action-angle variables I, φ can be found, for example, in [20]. In the new variables, $H = H_0(I) + \varepsilon H_1(I, \varphi)$. The action variables I_1, I_2 vary in the domain $\Delta = \{|I_1| \leqq I_2,\ I_2 \geqq 0\}$. The Hamiltonian $H_0(I_1, I_2)$ is a homogeneous function of degree 2 and is analytic in each of the four connected subdomains into which Δ is partitioned by the three straight lines

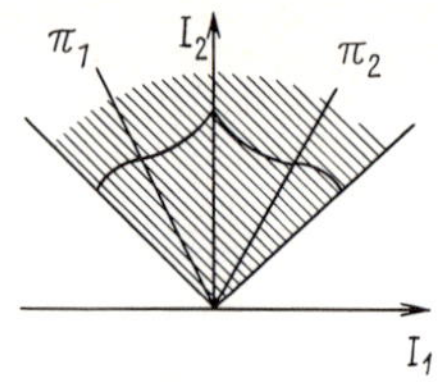

Fig. 57

π_1, π_2, and $I_1=0$. The lines π_1 and π_2 are given by the equation $2H_0/I_2^2=A_2^{-1}$. They are symmetric relative to the vertical axis and tend to the line $I_1=0$ as $A_2 \to A_1$, and to the pair of lines $|I_1|=I_2$ as $A_2 \to A_3$ (recall that A_1, A_2, A_3 are the principal moments of inertia of the body and $A_1 \geqq A_2 \geqq A_3$). The level lines of H_0 are shown in Fig. 57.

The expansion of the perturbing function H_1 in a multiple Fourier series in the angle variable φ_1 and φ_2 is in fact contained in one of the works of Jacobi and has the form

$$\sum_{m\in Z} h_{m,1}\, e^{i(m\varphi_1+\varphi_2)} + \sum_{m\in Z} h_{m,-1}\, e^{i(m\varphi_1-\varphi_2)} + \sum_{m\in Z} h_{m,0}\, e^{im\varphi_1}.$$

If the principal moments of inertia satisfy the inequalities $A_1 > A_2 > A_3$, the Poincaré set consists of infinitely many straight lines which pass through the point $I=0$ and accumulate at the pair of straight lines π_1 and π_2. One can show that the function H_0 is nondegenerate in the domain Δ. If H were analytic in I throughout Δ, then the results of §1.1 would be applicable: points I^0 lying on the lines π_1 and π_2 would satisfy the conditions of Theorem 1. The difficulty associated with the analytic singularities of the Hamiltonian in the action-angle variables can be circumvented by considering the problem of an additional integral analytic on the entire invariant level manifold M_c. Using Poincaré's method one can prove the following theorem:

Theorem 6. *The equations of motion of a dynamically asymmetric heavy rigid body do not have a formal first integral* $\sum F_s \varepsilon^s$, *independent of* $H_0+\varepsilon H_1$, *whose coefficients are analytic on the level manifold* M_c ([12]).

This result gives a negative answer to a question posed by Poincaré in [33] (no. 86).

§2. Splitting of Asymptotic Surfaces

Nondegenerate unstable periodic solutions have asymptotic manifolds filled out by trajectories that approach asymptotically the periodic trajectories as $t \to \pm\infty$. In integrable Hamiltonian systems these surfaces, as a rule, are pairwise identical. In nonintegrable systems the situation is different: asymptotic surfaces may intersect forming a rather tangled net in the intersection (see Fig. 44). In this section we describe a method for proving nonintegrability which goes back to Poincaré and rests on the analysis of asymptotic surfaces of near-integrable Hamiltonian systems.

2.1. Conditions for Splitting. Let V be the smooth n-dimensional configuration manifold of a Hamiltonian system, T^*V its phase space, and H: $T^*V\times R\{t\}\to R$ its Hamiltonian function. In the extended phase space $M=T^*V\times R^2\{E, t\}$ the equations of motion are again Hamiltonian:

$$\dot{x}=\frac{\partial K}{\partial y},\quad \dot{y}=-\frac{\partial K}{\partial x},\quad \dot{E}=\frac{\partial K}{\partial t},\quad \dot{t}=-\frac{\partial K}{\partial E},\tag{7}$$

where $K=H(y, x, t)-E$, $x\in V$, $y\in T_x^* V$.

A smooth surface $\Lambda^{n+1}\subset M$ is called a *Lagrangian surface* if for any closed contour γ contractible to a point, the integral

$$\oint_\gamma y\,dx-E\,dt$$

($H(y, x, t)=E$ on Λ^{n+1}) is equal to zero. Lagrangian surfaces are invariant under the action of the phase flow of system (7). In the autonomous case Lagrangian surfaces $\Lambda^n\subset T^*V$ are defined by the condition

$$\oint_\gamma y\,dx=0 \qquad (\text{where } \gamma\subset\Lambda^n, \partial\gamma=0).$$

If the Lagrangian surface Λ^{n+1} is mapped by the projection $M\to V\times R\{t\}$ in one-to-one manner onto $D\times R\{t\}$, where $D\subset V$, then it can be represented as a graph

$$y=\frac{\partial S(x,t)}{\partial x},\qquad H(y,x,t)=-\frac{\partial S(x,t)}{\partial t},$$

with S: $D\times R\to R$ a smooth function. In the autonomous case Λ^n is given by a graph

$$y=\frac{\partial S(x)}{\partial x},\qquad x\in D.$$

The function $S(x, t)$ satisfies the Hamiltonian-Jacobi equation

$$\frac{\partial S}{\partial t}+H\left(\frac{\partial S}{\partial x}, x, t\right)=0.$$

In this section we shall be concerned with Lagrangian surfaces formed by asymptotic trajectories. Such surfaces will be naturally called *asymptotic.*

We assume that the Hamiltonian is 2π-periodic in t and depends also on a parameter ε: $H=H(y, x, t, \varepsilon)$. Suppose that for $\varepsilon=0$ the function $H(y, x, t, 0)=H_0(y, x)$ does not depend on time and satisfies the following conditions:

1) There exist two critical points y_-, x_- and y_+, x_+ of H_0, at which the eigenvalues of the linearized Hamiltonian system

$$y = -\frac{\partial H_0}{\partial x}, \qquad x = \frac{\partial H_0}{\partial y},$$

are real and different from zero. In particular, the 2π-periodic solutions $x_\pm(t) = x_\pm$, $y_\pm(t) = y_\pm$ are of hyperbolic type.

2) If Λ^+ (Λ^-) designates the stable (respectively, unstable) asymptotic manifold in T^*V passing through x_+, y_+ (respectively, x_-, y_-), then $\Lambda^+ = \Lambda^-$. Hence, in particular, $H_0(y_+, x_+) = H_0(y_-, x_-)$.

3) There is a domain $D \subset V$ which contains the points $x_\pm$, such that in $T^*D \subset T^*V$ the equation of the surface $\Lambda^+ = \Lambda^-$ can be written in the form $y = \partial S_0/\partial x$, where S_0 is an analytic function in D. It is useful to consider the differential equation

$$\dot{x} = \left.\frac{\partial H_0}{\partial y}\right|_{y(x)}, \qquad y = \frac{\partial S_0}{\partial x}. \tag{8}$$

In a small neighborhood of the point $x_\pm$ its solutions tend to $x_\pm$ as $t \to \pm\infty$.

4) Equation (8) has a doubly-asymptotic solution in D: $x_0(t) \to x_\pm$ as $t \to \pm\infty$ (Fig. 58).

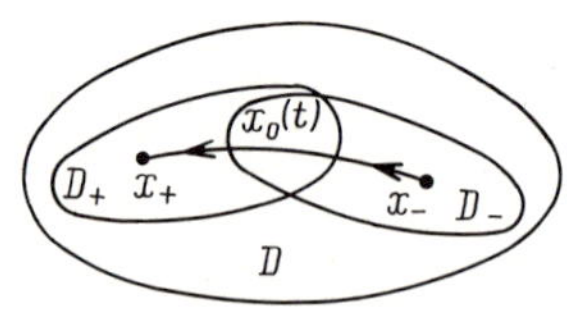

Fig. 58

The Hamiltonian system with Hamiltonian function H_0 should be regarded as the unperturbed system. In applications it is most frequently completely integrable. Let D_+ (D_-) be a subdomain of D which contains the point x_+ (respectively, x_-) but not x_- (respectively, x_+). For small values of ε, the asymptotic surfaces Λ^+ and Λ^- do not disappear, but go over into "perturbed" surfaces Λ_ε^+ and Λ_ε^-. More precisely, in the domain $D_\pm \times R\{t\}$ the equation of the asymptotic surface can be written in the form

$$y = \frac{\partial S^\pm}{\partial x},$$

where $S^\pm(x, t, \varepsilon)$ is 2π-periodic in t, and is defined and analytic for $x \in D$ and small ε (Poincaré [33]). The function $S^\pm$ must, of course, satisfy the Hamilton-Jacobi equation

$$\frac{\partial S^\pm}{\partial t} + H\left(\frac{\partial S^\pm}{\partial x}, t, \varepsilon\right) = 0. \tag{9}$$

By hypothesis, for $\varepsilon=0$ the surfaces Λ_0^+ and Λ_0^- coincide. However, as Poincaré first noted [33], in the general case, for small values of $\varepsilon \neq 0$ the surfaces Λ_ε^+ and Λ_ε^-, viewed as subsets of $T^*(D_+ \cap D_-) \times R$, no longer coincide. This phenomenon is called *splitting* (or *bifurcation*) *of asymptotic surfaces.* Obviously, Λ_ε^+ coincides with Λ_ε^- if and only if equation (9) admits a solution $S(x, t, \varepsilon)$ which is analytic in x throughout D.

Theorem 7 (Poincaré). *If $H_1(y_+, x_+, t) = H_1(y_-, x_-, t)$ and*

$$\int_{-\infty}^{\infty} \{H_0, H_1\}(y(x_0(t)), x_0(t), t)\, dt \neq 0, \tag{10}$$

then for small values of the parameter $\varepsilon \neq 0$ the perturbed asymptotic surfaces Λ_ε^+ and Λ_ε^- do not coincide.

◁ Suppose that equation (9) has an analytic solution $S(x, t, \varepsilon)$ that can be expressed for small ε as a convergent power series

$$S = S_0(x, t) + \varepsilon S_1(x, t) + \dots .$$

The function $S_0(x, t)$ must satisfy the equation

$$\frac{\partial S_0}{\partial t} + H_0\left(\frac{\partial S_0}{\partial x}, x\right) = 0.$$

Hence, $S_0 = -ht + W(x)$, where $h = H_0(y_\pm, x_\pm)$, and W is a solution of the equation

$$H_0\left(\frac{\partial W}{\partial x}, x\right) = h.$$

Clearly, W coincides with the function $S_0(x)$.

Let $H = H_0(y, x) + \varepsilon H_1(y, x, t) + \dots$. Then we obtain from (9) the following quasi-linear differential equation for S_1:

$$\frac{\partial S_1}{\partial t} + \left.\frac{\partial H_0}{\partial x}\right|_{y(x)} \cdot \frac{\partial S_1}{\partial x} + H_1(y(x), x, t) = 0. \tag{11}$$

Since equation (8) is autonomous, together with the solution $x_0(t)$, it has the family of solutions $x_0/(t+\alpha)$, $\alpha \in R$. It follows from (11) that along these solutions

$$S_1(x_0(t+\alpha), t) = S_1(x_0(\alpha), 0) - \int_0^t H_1(y(x_0(t+\alpha)), x_0(t+\alpha), t)\, dt. \tag{12}$$

With no loss of generality we may assume that $H_1(y_\pm, x_\pm, t) = 0$ for all t. If this is not the case, then instead of H_1 we must take $H_1 - H_1(y_\pm, x_\pm, t)$ for the perturbing function; this does not affect the Poisson bracket $\{H_0, H_1\}$.

Since the Taylor expansion of H_1 in the neighborhood of the points $y_\pm$, $x_\pm$ begins with linear terms in $y - y_\pm$, $x - x_\pm$, and since the functions $x_0(t) - x_\pm$,

$y(x_0(t)) - y_\pm$ tend to zero exponentially fast as $t \to \pm\infty$, the integral

$$J(\alpha) = \int_{-\infty}^{\infty} H_1(y_0(t+\alpha), x_0(t+\alpha), t)\, dt \tag{13}$$

converges. From equation (12) it also follows that the values of $S_1(x, t)$ at the points $x_\pm$ do not depend on t. Furthermore, by (12), the integral $J(\alpha)$ equals $S_1(x_+) - S_1(x_-)$ and hence it does not depend on α. To complete the proof it remains to compute the derivative

$$\left.\frac{dJ}{d\alpha}\right|_{\alpha=0} = \int_{-\infty}^{\infty} \sum \left(\frac{\partial H_1}{\partial x_s} \dot{x}_s + \frac{\partial H_1}{\partial y_s} \dot{y}_s\right) dt = \int_{-\infty}^{\infty} \{H_0, H_1\}\, dt = 0. \ \triangleright$$

A different proof of Poincaré's theorem can be found in [50].

In the autonomous case, the condition for the splitting of asymptotic surfaces lying on some fixed energy level can be written in the form

$$\int_{-\infty}^{\infty} \{F_0, H_1\}\, dt \neq 0, \tag{14}$$

where F_0 is a first integral of the unperturbed system. If at the points of unstable periodic trajectories $dF_0 = 0$, then the integral (14) manifestly converges.

2.2. Splitting of Asymptotic Surfaces is an Obstruction to Integrability. Consider a Hamiltonian system with Hamiltonian $H(z, t, \varepsilon) = H_0(z) + \varepsilon H_1(z, t) + O(\varepsilon^2)$ under the assumptions of §2.1. In particular, the unperturbed system has two hyperbolic equilibrium positions $z_\pm$, connected by a doubly asymptotic solution $t \mapsto z_0(t)$, $t \in R$.

Theorem 8 (S.V. Bolotin). *Suppose the following conditions are fulfilled:*

1) $\int_{-\infty}^{\infty} \{H_0, \{H_0, H_1\}\}\,(z_0(t), t)\, dt \neq 0,$

2) *for small ε the perturbed system has a doubly-asymptotic solution $t \mapsto z_\varepsilon(t)$, $t \in R$, close to $t \mapsto z_0(t)$.*

Then for small fixed values $\varepsilon \neq 0$, in any neighborhood of the closure of the trajectory $z_\varepsilon(t)$ the Hamiltonian equation $\dot{z} = \mathfrak{J}\, dH$ does not have a complete set of independent first integrals in involution.

Remark. Condition 1) may be replaced by the following one: *for some* $m \geqq 2$

$$\int_{-\infty}^{\infty} \underbrace{\{H_0, \dots, \{H_0}_{m}, H_1\} \dots\}\,(z_0(t), t)\, dt \neq 0.$$

If 1) holds, then the asymptotic surfaces are manifestly distinct. Needless to say, condition 2) is not always satisfied. We give a sufficient condition for the existence of doubly-asymptotic trajectories.

Let $H_0 = F_1, \ldots, F_n$ be commuting integrals of the unperturbed problem, which are independent on $\Lambda_0^+ = \Lambda_0^-$. If

$$\int_{-\infty}^{\infty} \{F_i, H_1\}(z_0(t), t)\, dt = 0$$

and

$$\det\left(\int_{-\infty}^{\infty} \{F_i, \{F_j, H_1\}\}(z_0(t), t)\, dt \right) \neq 0,$$

then there exists a family of asymptotic solutions $t \mapsto z_\varepsilon(t)$ which is analytic in ε. This result is readily deduced from the implicit function theorem. △

If we investigate the problem of the existence of independent first integrals $F_i(z, t, \varepsilon)$, $1 \leqq i \leqq n$, in involution which are analytic (or formally analytic) in the parameter ε, then 2) can be omitted. In particular, if 1) holds, then the series of perturbation theory diverge in the neighborhood of the split asymptotic surfaces.

Using the method of Birkhoff normal forms, one can find in the neighborhood of the unstable periodic solutions $z_\pm + O(\varepsilon)$ a 2π-periodic in t, formal canonical change of variables $z \mapsto u$, which transforms the Hamiltonian $H(z, t, \varepsilon)$ into a function $H^\pm(u, \varepsilon)$ that does not depend on t. This Birkhoff transformation may diverge due to commensurability of the characteristic exponents. However, in the case of one degree of freedom ($n = 1$) the formal series that give of the change of variables $z \mapsto u$ always converge and depend analytically on the parameter ε (see [145]).

Theorem 9. *Suppose that the Birkhoff transformation converges and depends analytically on ε. If condition 1) of Theorem 8 is satisfied, then for small $\varepsilon \neq 0$ Hamilton's equations do not have a complete set of independent analytic first integrals in involution.*

In particular, for $n = 1$, 1) is a sufficient condition for nonintegrability (S.L. Ziglin [195]).

Proof of Theorem 9. We define functions $R^\pm$ on the surfaces $\Lambda_0^\pm$ by the formulas

$$R^+(z) = -\int_0^{+\infty} \{H_0, \{H_0, H_1\}\}(z(t), t)\, dt$$

and

$$R^-(z) = \int_{-\infty}^{0} \{H_0, \{H_0, H_1\}\}(z(t), t)\, dt,$$

where $t \mapsto z(t)$ is the asymptotic motion of the unperturbed system with initial condition $z(0) = z$.

Lemma 2. *The functions $R^\pm$ are uniquely determined by H_0, the family of surfaces $\Lambda_\varepsilon^\pm$, and the symplectic structure.*

◁ In fact, according to the results of §2.1,

$$S^+(z) = -\varepsilon \int_0^{+\infty} (H_1(z(t),t) - H_1(z_+,t))\,dt,$$

and

$$S^-(z) = \varepsilon \int_{-\infty}^{0} (H_1(z(t),t) - H_1(z_-,t))\,dt$$

are generating functions of the Lagrangian surfaces $\Lambda_\varepsilon^\pm$ modulo $O(\varepsilon^2)$. But $\varepsilon R^\pm = \{H_0, \{H_0, S^\pm\}\}$. ▷

Composing Birkhoff's transformation and iterations of the flow map at a period we can extend $H^\pm$ from neighborhoods of the critical points $u_\pm(\varepsilon)$ to certain neighborhoods $W_\pm$ of the asymptotic surfaces $\Lambda_\varepsilon^\pm$. Since a possible splitting of the surfaces Λ_ε^+ and Λ_ε^- is of order ε, it follows that for small ε the neighborhoods W_+ and W_- do intersect.

Lemma 4. $\{H^+, H^-\} \not\equiv 0$ *for* $\varepsilon \neq 0$.

◁ Write $H^\pm(u,\varepsilon) = H_0^\pm(u) + \varepsilon H_1^\pm(u) + O(\varepsilon^2)$. Since $H_0^\pm(u) = H_0(u)$,

$$\{H^+, H^-\} = \varepsilon\{H_0, H_1^- - H_1^+\} + O(\varepsilon^2).$$

Further, since Λ_0^- is an invariant asymptotic manifold of the Hamiltonian system $\dot{u} = \mathfrak{J}\,dH_0$ it follows, by Lemma 3, that

$$\{H_0, H_1^-\}(u) = \int_{-\infty}^{0} \{H_0, \{H_0, H_1^-\}\}(u_0(t))\,dt = R^-(u), \qquad u \in \Lambda_0^-.$$

Similarly,

$$\{H_0, H_1^+\}(u) = \int_{0}^{+\infty} \{H_0, \{H_0, H_1^+\}\}(u_0(t))\,dt = R^+(u), \qquad u \in \Lambda_0^+.$$

Consequently,

$$\{H^+, H^-\} = \varepsilon \int_{-\infty}^{\infty} \{H_0, \{H_0, H_1\}\}(z_0(t),t)\,dt + O(\varepsilon^2)$$

by condition 1), for small $\varepsilon \neq 0$ the Poisson bracket $\{H^+, H^-\} \neq 0$. ▷

In the new variables u, the first integrals $F_1, \ldots, F_n$ do not depend on t. Suppose that for $\varepsilon \neq 0$, $F_1, \ldots, F_n$ are independent at a point of $W_+ \cap W_-$. Since $\{H^\pm, F_i\} = 0$, the vector $\mathfrak{J}\,dH^\pm$ is a linear combination of the $\mathfrak{J}\,dF_i$. Moreover, since $\{F_i, F_j\} = 0$, then obviously at this point $\{H^+, H^-\} = 0$. To complete the proof it remains to remark that the analytic function $\{H^+, H^-\}$ is different from zero on a everywhere dense set. ▷

Theorem 10. *Let* $n = 1$. *Suppose that:*

1) $\int_{-\infty}^{\infty} \{H_0, H_1\}(z_0(t),t)\,dt \neq 0,$

2) *for small* ε *the perturbed system has a doubly-asymptotic solution* $t \mapsto z_\varepsilon(t)$ *close to* $t \mapsto z_0(t)$.

Then for small $\varepsilon \neq 0$ *the Hamiltonian system* $\dot{z} = \mathfrak{J} dH$ *does not have an additional analytic first integral.*

◁ Consider the flow map g at a period from the section $t = t_0$ onto itself. For small ε, g has two fixed hyperbolic points, z_1 and z_2, with invariant separatrices $W_1^\pm$ and $W_2^\pm$. By the assumptions of the theorem, for $\varepsilon \neq 0$ the separatrices W_1^+ and W_2^- intersect and do not coincide. Let V be a small neighborhood of the point z_1 and Δ a small segment of the separatrix W_2^- such that $\Delta \cap W_1^+ \neq \emptyset$. For sufficiently large n the segment $g^n(\Delta)$ lies wholly in V, and again $g^n(\Delta) \cap W_1^+ \neq \emptyset$. By the *Grobman-Hartman theorem* [32], in the domain V the map g is topologically conjugate to a linear hyperbolic rotation. Consequently, for $n \to \infty$ the segment $g^n(\Delta)$ "stretches" along the separatrix W_1^-, approaching it without limit. Obviously, the union

$$\bigcup_{n=1}^{\infty} g^n(\Delta) \tag{15}$$

is a key set for the class of functions analytic in the section $t = t_0$.

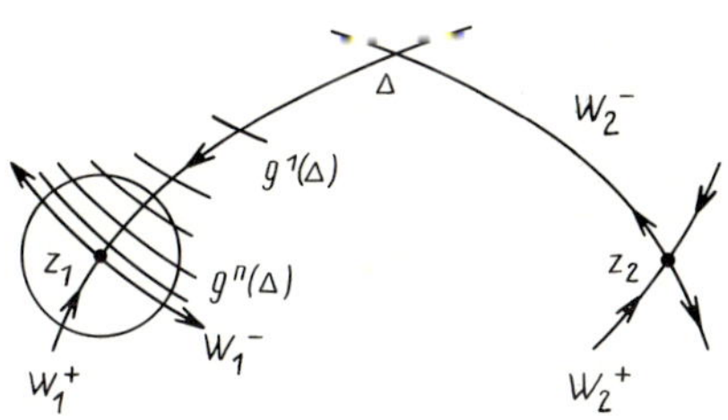

Fig. 59

Suppose now that Hamilton's equation possesses an analytic first integral $f(z, t)$. The function $f(z, t_0)$ is invariant under g and constant on W_2^- (since the sequence $g^n(z)$, $z \in W_2^-$, converges to z_2 as $n \to -\infty$). Consequently, the analytic function $f(z, t_0)$ is constant on the set (15), and is therefore constant for any t_0. ▷

Remark. Poincaré classified doubly-asymptotic solutions into two types: *homoclinic* (when $z_+ = z_-$) and *heteroclinic* (when $z_+ \neq z_-$). If $n = 1$, then for small ε the perturbed problem always has homoclinic solutions (provided, of course, that it has them for $\varepsilon = 0$).

2.3. Applications

a) We consider first the simplest problem of the oscillations of a pendulum with a vibrating point of suspension (see Ch. 5, §1.4, Example 6). The Hamiltonian H is $H_0+\varepsilon H_1$, where

$$H_0=\frac{y^2}{2}-\omega^2\cos x, \qquad H_1=-\omega^2 f(t)\cos x,$$

and f is a 2π-periodic function of time. For $\varepsilon=0$ the topmost position of the pendulum is an unstable equilibrium. The unperturbed problem has two families of homoclinic solutions:

$$\cos x_0=\frac{2e^{\pm\omega(t-t_0)}}{e^{\pm 2\omega(t-t_0)}+1}, \qquad x_0\to\pm\pi \quad \text{as} \quad t\to\pm\infty. \tag{16}$$

Since $\{H_0, H_1\}=-\omega^2 f(t)\dot{x}\sin x$, the integral (10) is equal to

$$\int_{-\infty}^{\infty} \dot{f}(t)\cos x_0\, dt$$

to within a constant factor. Let $f(t)=\sum f_n e^{int}$. Then the integral (10) can be expressed as a series

$$\sum_{n\in Z} 2inf_n J_n e^{int_0}, \qquad J_n=\int_{-\infty}^{\infty}\frac{e^{\pm\omega t}e^{int}}{e^{\pm 2\omega t}+1}dt.$$

The integrals J_n are readily calculated by residues:

$$J_n=\frac{-ie^{-n\pi/2\omega}}{2\omega(1+e^{\pm n/\pi\omega})}\neq 0.$$

Therefore, if $f(t)\neq\text{const}$ (i.e., $f_n\neq 0$ for some $n\neq 0$), then the integral (10) is different from zero on at least one of the doubly-asymptotic solutions of family (16). Thus, if $f(t)\neq\text{const}$, then according to the results of §2.2, for sufficiently small (but fixed) values of $\varepsilon\neq 0$ the problem under consideration has no analytic first integral $F(y, x, t)$ which is 2π-periodic in x and t. One can show that the equations governing the oscillations of the pendulum may be completely integrable only for a finite set of values of the parameter ε in the interval $[-a, a]$, where $a=1/\max|f(t)|$ (see [21]).

Remark. An example is given in [21] of a Hamiltonian system, with analytic Hamiltonian depending analytically on a parameter, which is completely integrable for a dense set of values of the parameter, and is nonintegrable for another such set. Hence, integrable cases are not always isolated.

b) In the problem of rapid rotation of a heavy asymmetric rigid body, the Hamiltonian is $H=H_0+\varepsilon H_1$, where

$$H_0=\langle AM, M\rangle/2, \qquad H_1=\langle x, e\rangle,$$

$$A=\operatorname{diag}(a_1, a_2, a_3), \qquad x=(x_1, x_2, x_3).$$

The numbers a_1, a_2, a_3 are the reciprocals of the principal moments of inertia of the body, and x_1, x_2, x_3 are the coordinates of the center of mass relative to the principal axes. For $\varepsilon=0$ we have Euler's integrable case. In this unperturbed problem, on every noncritical three-dimensional level $M_{h,c}=\{M, e: H_0=h, \langle M, e\rangle=c, \langle e, e\rangle=1\}$ there exist two unstable periodic solutions: if $a_1<a_2<a_3$, take

$$\begin{aligned} &M_1=M_3=0, && M_2=M_2^0=\pm\sqrt{2h/a_2}, && e_2=e_2^0=\pm c/M_2^0, \\ &e_1=\alpha\cos(a_2 M_2^0)t, && e_3=\alpha\sin(a_2 M_2^0)t; && \alpha^2=1-(c/M_2^0)^2. \end{aligned} \tag{17}$$

It follows from the inequality $\langle M, e\rangle^2 \leqq \langle M, M\rangle\langle e, e\rangle$ and the independence of the first integrals on $M_{h,c}$ that $\alpha^2>0$. The stable and unstable asymptotic surfaces of the periodic solutions (17) can be represented as the intersections of $M_{h,c}$ and the hyperplanes $M_1\sqrt{a_2-a_1}\pm M_3\sqrt{a_3-a_2}=0$. In Euler's problem the asymptotic surfaces are "doubled": they are filled out by doubly-asymptotic trajectories which approach unboundedly the periodic trajectories (17) as $t\to\pm\infty$. The splitting of these surfaces was studied in the works of V.V. Kozlov (1976) and S.L. Ziglin (1980). It turned out that under perturbations the asymptotic surfaces always split, except in the "Hess-Appel'rot case", in which

$$x_2=0, \quad x_1\sqrt{a_3-a_2}\pm x_3\sqrt{a_2-a_1}=0. \tag{18}$$

In this case one pair of separatrices splits and the other does not. The reason for absence of splitting is that under conditions (18) the perturbed problem has, for all values of ε, the "partial" first integral $F=M_1\sqrt{a_2-a_1} \pm M_3\sqrt{a_3-a_2}$ (i.e., $\dot{F}=0$ if $F=0$). One can show that for small ε the closed invariant surfaces $M_{h,c}\cap\{F=0\}$ are exactly one pair of doubled separatrices (see [20]).

In the problem of rapid rotation of a heavy asymmetric top, the split separatrices apparently do not always intersect. However, in this situation Theorem 9 applies and with its help it can be established that there is no additional analytic first integral of the perturbed problem for small, but fixed $\varepsilon\neq 0$ (S.L. Ziglin (1980)).

The behavior of the solutions of the perturbed problem has been investigated numerically (L. Galgani, A. Giorgilli, J.-M. Strelcyn (1980)). Figure 60 shows the computational results for various values of the perturbation parameter ε. It is clearly seen that the destruction of the portrait of invariant curves of the unperturbed problem begins exactly near the separatrices.

c) Consider now Kirchhoff's equations

$$\begin{aligned} &\dot{M}=M\times\omega+e\times u, \quad \dot{e}=e\times\omega; \quad \omega=\frac{\partial H}{\partial M}, \quad u=\frac{\partial H}{\partial e}, \\ &H=\tfrac{1}{2}\langle AM, M\rangle+\langle BM, e\rangle+\tfrac{1}{2}\langle Ce, e\rangle, \end{aligned} \tag{19}$$

which describe the rotation of a rigid body in an ideal fluid. The matrix $A=\operatorname{diag}(a_1, a_2, a_3)$ is diagonal, and the matrices B and C are symmetric.

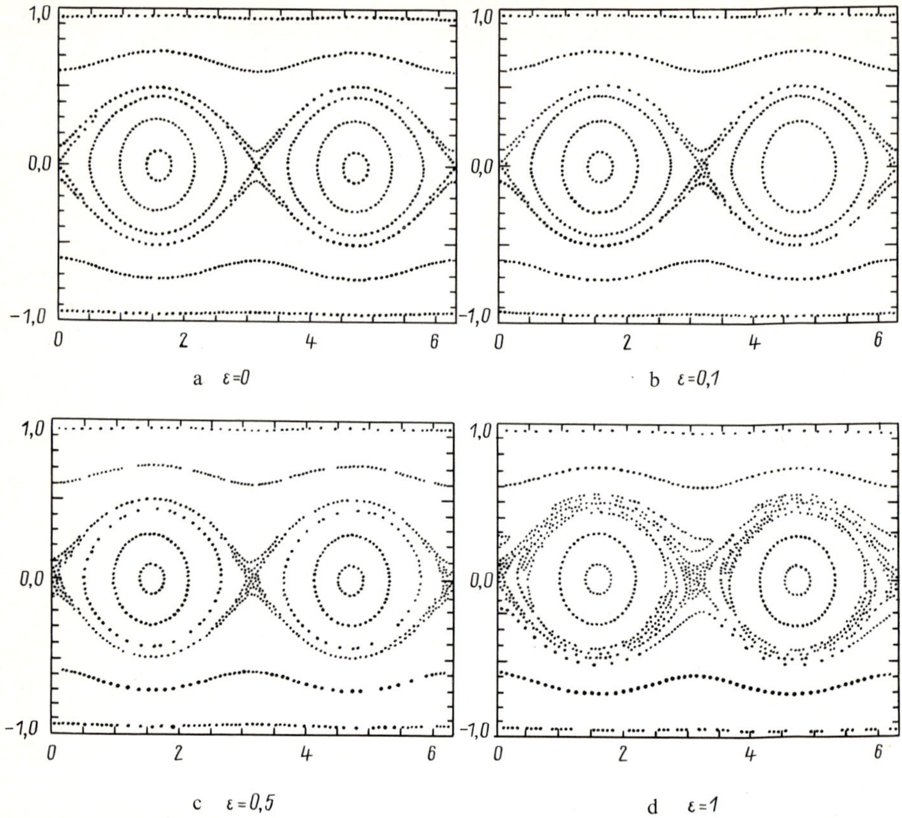

Fig. 60

Theorem 11. *Suppose that the numbers a_1, a_2, a_3 are all distinct. If Kirchhoff's equations have an additional first integral, independent of the functions $F_1=H$, $F_2=\langle M, e\rangle$, $F_3=\langle e, e\rangle$, and analytic in $R^6\{M, e\}$, then $B=\operatorname{diag}(b_1, b_2, b_3)$ and*

$$a_1^{-1}(b_2-b_3)+a_2^{-1}(b_3-b_1)+a_3^{-1}(b_1-b_2)=0. \tag{20}$$

If $B=0$, then an independent additional analytic first integral exists only when $\operatorname{diag}(c_1, c_2, c_3)$ and

$$a_1^{-1}(c_2-c_3)+a_2^{-1}(c_3-c_1)+a_3^{-1}(c_1-c_2)=0. \tag{21}$$

The matrix B in Steklov's integrable case is defined precisely by condition (20). Condition (21) gives Clebsch's integrable case. It is interesting to note that conditions (20) and (21) are identical in form.

Corollary. *In the general case Kirchhoff's equations are not integrable.*

The proof of Theorem 11, given by V.V. Kozlov and D.A. Onishchenko (1982), is also based on the phenomenon of splitting of separatrices: In equations (19) one introduces a small parameter ε upon replacing e by εe; for $\varepsilon=0$ one again has Euler's integrable problem whose doubled separatrices split upon adding the perturbation unless conditions (20) or (21) are satisfied. For details see [21].

d) Using the method of splitting of asymptotic surfaces one can establish the nonintegrability of the problem of the motion of four point-vortices (Ziglin [194]). More precisely, we consider this problem in a restricted setting: a vortex of zero intensity (i.e., simply a particle of ideal fluid) moves in the "field" of three vortices of equal intensity. It turns out that the equations of motion of the zero-intensity vortex can be put in Hamiltonian form with a time-periodic Hamiltonian, and that these equations have hyperbolic periodic motions with intersecting separatrices. Therefore, the restricted problem of four vortices is not completely integrable, although it has four independent noncommuting first integrals (as in the unrestricted setting).

§3. Quasi-Random Oscillations

Most methods for proving nonintegrability rest on the observation that a sufficiently complex topological behavior of phase curves is an obstruction to the existence of first integrals. One of the cases in which one succeeds in explicitly exhibiting such a complex topological behavior, and hence also nonintegrability, is the theory of *quasi-random oscillations*, considered here in the simplest model situation.

We consider, following V.M. Alekseev, a nonautonomous system with one degree of freedom whose motion is governed by the equation

$$\ddot{x}=-Q(x,t), \qquad x\in R. \tag{22}$$

We shall assume that the following conditions are satisfied:

a) Q is a smooth function and is 2π-periodic in t;

b) $Q(-x,t)=-Q(x,t)$; in particular, $Q(0,t)=0$, so that $x=0$ is an equilibrium position,

and

c) $Q>0$ for $x>0$, and

$$\int_0^\infty \int_0^{2\pi} Q(x,t)\,dx\wedge dt<\infty;$$

if the system is autonomous the last condition expresses the boundedness of the potential energy for $|x|\to\infty$.

d) $Q'_x\leqq 0$ for $x\geqq x_*>0$.

This means that the graph of the potential energy $U(x, t)$ (defined by $U_x' = -Q$) is convex for $|x| > x_*$.

The next two are technical conditions:

e) $|Q_t'| \leqq \Psi(x)$, $\int_0^\infty \Psi(x)\,dx < \infty$;

f) $\Psi(x)/Q^2(x, t) = O(1)$ for $x \geqq x_0$.

An important example of system (22) is furnished by a variant of the restricted three body problem, in which two points of equal masses move in the (x, y)-plane along elliptic orbits symmetric with respect to the z-axis, while a third point of zero mass lies always on this axis (see Fig. 61). The motion of the third point is governed by the differential equation

$$\ddot{z} = -z/[z^2 + r^2(t)]^{3/2}, \tag{23}$$

where

$$r(t) = \frac{1}{1 + e\cos\varphi(t)}, \qquad \dot{\varphi} = (1 + e\cos\varphi)^2, \ \varphi(0) = 0;$$

here e is the eccentricity of the elliptic orbit of the massive bodies. It is clear that in this example conditions a) through f) are satisfied.

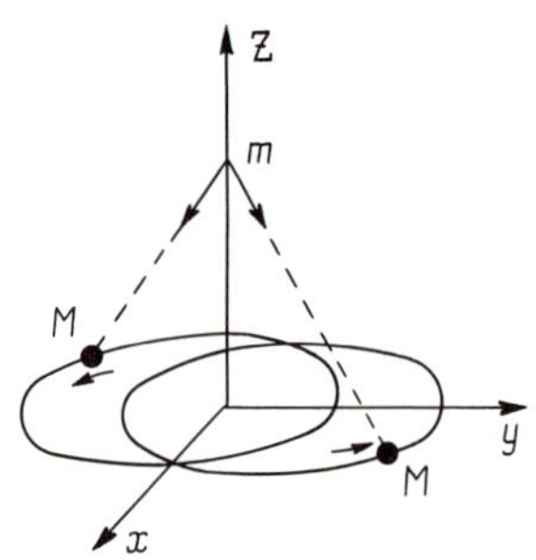

Fig. 61

3.1. The Poincaré Map

Definition. The solution $x(t)$ of equation (24) is called *future-* (or *forward-*) *hyperbolic* if the limit $\dot{x}(+\infty) = \lim_{t \to +\infty} \dot{x}(t) > 0$ exists, *parabolic* if $\dot{x}(+\infty) = 0$, and *oscillatory* if the function $x(\cdot)$ has infinitely many zeros for $t \to +\infty$. In analogous manner one defines three kinds of motions in the past (for $t \to -\infty$) (see Ch. 2, §4.1).

In the example of the restricted three-body problem indicated above the motions of the first (second) kind are called *hyperbolic-elliptic* (respectively, *parabolic-elliptic*) after Chazy. The final type of motions of the third kind has not been yet defined.

Proposition 2. *Every solution of equation* (22) *belongs to one of the three classes (for* $t \to +\infty$ *as well as for* $t \to -\infty$*) defined above.*

The proof is easy and uses properties a)–c) of the function Q.

Let $x(t)$ be a solution of equation (22) with initial conditions: $x(\tau)=0$, $\dot{x}(\tau)=v>0$. Here two cases are possible. The first is that x increases monotonically as $t \to +\infty$; this solution is either hyperbolic or parabolic. In the second case x attains a maximum $X^+(v, \tau)$ and then decreases to the value $x=0$. We introduce the function

$$h^+(v, \tau) = \begin{cases} \dfrac{\dot{x}^2(+\infty)}{2} + \int\limits_0^\infty Q_0(x)\,dx, & \text{in the first case}, \\ \int\limits_0^{X^+} Q_0(x)\,dx, & \text{in the second case}. \end{cases}$$

Here

$$Q_0 = \frac{1}{2\pi} \int\limits_0^\infty Q(x, t)\,dt.$$

The function $h^-(v, \tau)$ is defined in the same manner for $t \to -\infty$. In the stationary case $h^+(v) \equiv h^-(v)$. In the ensuing analysis an important role is played by the constant

$$J = \int\limits_0^\infty Q_0(x)\,dx,$$

whose existence is guaranteed by property c). If Q does not depend on time, then J is the total energy of the parabolic motion. One can show that $h^\pm$ are differentiable functions.

Now consider the two curves $\Pi^\pm = \{h^\pm = J\}$ in the plane Σ with polar coordinates $v, \tau \bmod 2\pi$. They are closed, differentiable, and bound on Σ some open domains $R^\pm$ which contain the point $v=0$.

Proposition 3. *If the point* (v, τ) *lies outside (on)* $\Pi^\pm$*, then the function* $x(t)$ *is monotonic and the motion is hyperbolic (respectively, parabolic) for* $t \to \pm\infty$. *If* (v,τ) *lies inside* $R^\pm$*, then* $x(t)$ *has at least one zero for* $t \gtrless \tau$.

◁ The inequality $h^+ \geqq J$ is equivalent to the condition $X^\pm(v, \tau) = \infty$, which in turn implies that $h^+ - J = \dot{x}^2(+\infty)/2$. If $\dot{x}(+\infty) > 0$ ($=0$), then the motion is hyperbolic (respectively parabolic). ▷

By Proposition 3, for the points $(v, \tau) \in R^+$ the natural map $S\colon (v, \tau) \mapsto (v', \tau')$ is well defined, where τ' is the zero of the function $x(t)$ closest to τ, and v' is the velocity of the point at time τ'; in view of the symmetry under the transformation $x \mapsto -x$, we may assume that $v'>0$ (see Fig. 62). Clearly, $SR^+ = R^-$ and $h^+ \circ S = h^-$.

Lemma 5. *The map* $S\colon R^+ \to R^-$ *preserves area on* Σ.

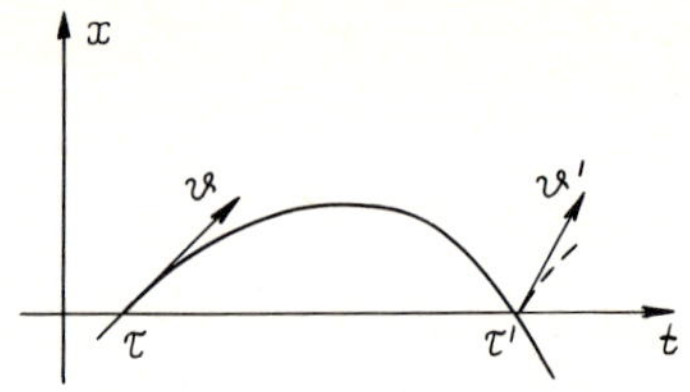

Fig. 62

◁ Equation (24) is, needless to say, Hamiltonian with the Hamiltonian function $H(y, x, t) = y^2/2 + U(x, t)$, where $y = \dot{x}$. Let Γ be a closed contour in the domain in R^+ and let $\Gamma' = S(\Gamma)$. By the theorem on the integral invariant

$$\oint_{\Gamma} y\,dx - H\,dt = \oint_{\Gamma'} y\,dx - H\,dt \quad \Leftrightarrow \quad \oint_{\Gamma} v^2\,d\tau = \oint_{\Gamma'} v^2\,d\tau. \;\triangleright$$

From this statement we deduce the following result.

Proposition 4. *Almost all solutions which are oscillatory in the past are also oscillatory in the future, and conversely.*

◁ We let A_m denote the set of points $(v, \tau) \in \Sigma$ for which the solution $x(t)$ has infinitely many zeros for $t \geqq \tau$ and exactly m zeros for $t < \tau$. Obviously, $S(A_m) = A_{m+1}$, $\operatorname{mes}(A_m) = \operatorname{mes}(A_{m+1})$ (Lemma 5), and $A_k \cap A_l = \emptyset$ for $k \neq l$. We are interested in the measure of the set $A = \bigcup_{m \geqq 0} A_m$. If $\operatorname{mes}(A_m) \neq 0$, then $\operatorname{mes}(A) = \infty$. But the measure of A is finite, since this set is contained in the disc of radius

$$2\left(\pi \int_0^\infty \Psi(x)\,dx\right)^{1/2}.$$

In fact, multiplying equation (24) by $\dot{x}$ and integrating the result from τ to t, we get

$$\frac{v^2}{2} - \frac{\dot{x}^2}{2} = -\int_t^\tau \dot{x} Q\,dt = \int_0^{x(t)} Q(x, t)\,dx.$$

It remains to use the inequality $Q < 2\pi \Psi(x)$ for $x > 0$, which follows from condition e). ▷

Since the domains R^+ and R^- have nonempty intersection and also equal measures, their boundaries Π^+ and Π^- must intersect in at least two points. We shall assume here that Π^+ and Π^- intersect transversally. For example, in the case of equation (25), for zero eccentricity $\Pi^+ = \Pi^- = \{v = \sqrt{2}\}$. In view of the symmetry of the problem relative to the moment of junction $\tau = 0$, in the general case Π^+ and Π^- have common points on the ray $\tau = 0$. One can show that at these points Π^+ and Π^- intersect transversally, at least for small values $e > 0$.

In a neighborhood of a point where Π^+ and Π^- intersect one can take the functions $\xi = h^+ - J$ and $\eta = h^- - J$ as local coordinates on Σ. Consider a small square $B = \{|\xi| \leqq \varepsilon, |\eta| \leqq \varepsilon\}$. It can be verified that for small values of ε the set $S(B \cap R^+)$ is a "spiral" that winds on the curve Π^-, whereas $S(B \cap R_-)$ is a similar spiral that winds on Π^+ (see Fig. 63). This is a consequence of the hyperbolicity of the map S (respectively, S^{-1}) in the neighborhood of the point $\xi = \eta = 0$: S is contracting in the direction of the η-axis and expanding in the direction of the ξ-axis (for details see [2]). The set $S(B \cap R^+) \cap B \cap R^+$ already has infinitely many (connected) components. Each such component is taken by S into a narrow spiral which lies inside the spiral $S(B \cap R^+)$. Iterating the maps S and S^{-1} we obtain in the square B strips of decreasing width which intersect one another transversally. In the limit we obtain a Cantor (i.e., a perfect and nowhere dense) set $\Lambda \subset B$ which invariant under all integer iterates of S. The orbit of any point $(v, \tau) \in \Lambda$ (i.e., the set $S^n(v, \tau)$, $n \in Z$) has a very intricate shape, typical for random wandering through the set Λ. Proofs of these statements can be found in the works of Alekseev [2]. We next illustrate the foregoing discussion on a (model) example.

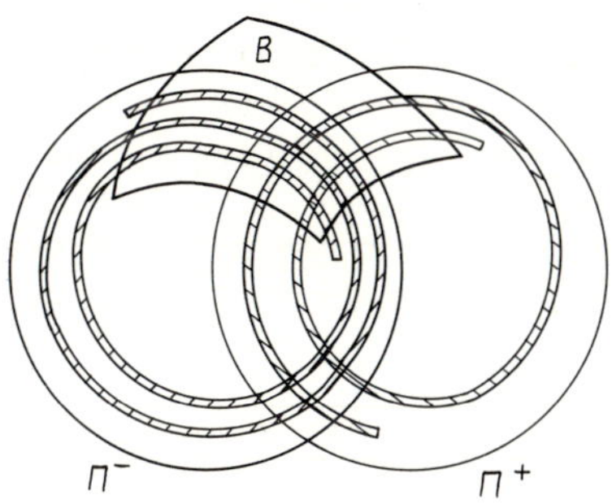

Fig. 63

3.2. Symbolic Dynamics. Consider the unit square $B = \{(x, y) \in R^2\colon 0 \leqq x, y \leqq 1\}$ and define a map S of B into itself by the rule:

$$\begin{aligned} &\text{if } 0 \leqq x \leqq 1/3, \quad \text{then } x \mapsto 3x, \qquad y \mapsto y/3, \\ &\text{if } 2/3 \leqq x \leqq 1, \quad \text{then } x \mapsto 3x - 2, \quad y \mapsto y/3 + 2/3. \end{aligned} \tag{24}$$

In the strip $1/3 < x < 2/3$, $0 < y < 1$, the map S is not defined. The geometric action of the transformation $S\colon B \to B$ is clearly displayed in Fig. 64. Let us clarify the structure of the sets $S^n B \cap B$ for $n > 1$. To obtain SB we must remove from the square B the horizontal strip $[0, 1] \times (1/3, 2/3)$. If from the remaining two strips we next remove the thinner strips $[0, 1] \times [1/9, 2/9]$ and $[0, 1] \times [7/9, 8/9]$, we obtain the set $S^2 B$, and so on (see Fig. 65). Continuing this process indefinitely, we obtain the set $[0, 1] \times K_{[0,1]} \subset B$ (here $K_{[0,1]}$ is the Cantor set on the segment $[0, 1]$) on which all the negative

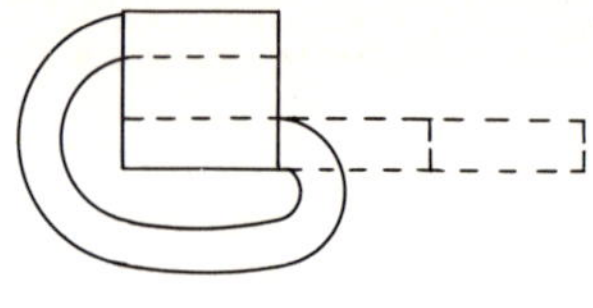

Fig. 64

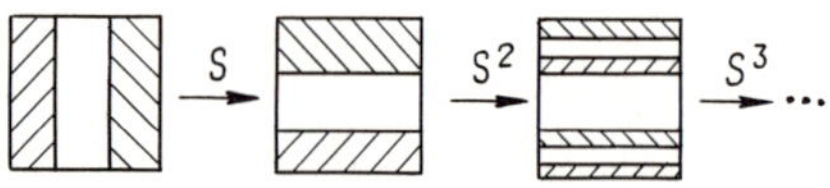

Fig. 65

iterates of S are defined. Arguing in exactly the same manner we conclude that on the set $K_{[0,1]} \times [0, 1]$ all positive iterates of mapping S are defined. Hence, on the direct product of Cantor sets $\Lambda = K_{[0,1]} \times K_{[0,1]}$ all integer iterates of S are defined.

What is the structure of the map $S: \Lambda \to \Lambda$? To answer this question we introduce the space Ω of doubly infinite sequences $\omega = \{\omega_n\}$, $n \in Z$, of zeros and ones. We endow Ω with a topology $\mathscr{T}$, defining convergence as follows: the sequence $\omega^{(k)} \in \Omega$ converges to $\omega \in \Omega$ if $\omega_n^{(k)} \to \omega_n$ for all n.

Lemma 6. *The space* $(\Omega, \mathscr{T})$ *is homeomorphic to* Λ.

$\triangleleft$ In fact, we can correspond to the sequence $\{\omega_n\}$ the two numbers

$$x = 2 \sum_{s \geqq 0} \omega_s / 3^{s+1}, \qquad y = 2 \sum_{s \geqq 0} \omega_{-s} / 3^s, \tag{25}$$

which obviously belong to $K_{[0,1]}$. It is readily verified that the mapping $\{\omega_n\} \mapsto (x, y)$ is a homeomorphism. $\triangleright$

Let T be the map of Ω into itself which sends $\omega = \{\omega_n\}$ into $\omega' = \{\omega_{n+1}\}$ (the shift of all indices by one).

Theorem 12. *There exists a homeomorphism* $f: \Lambda \to \Omega$ *such that the following diagram commutes:*

$$\begin{array}{ccc} \Lambda & \xrightarrow{S} & \Lambda \\ \downarrow f & & \downarrow f \\ \Omega & \xrightarrow{T} & \Omega \end{array}$$

The proof rests on simple juxtaposition of formulas (24) and (25).

Thus, to each trajectory $S^n(a)$, $a \in B$, $n \in Z$, contained in the square B we assigned a sequence of symbols $\omega = \{\omega_n\}$ in such a way that the action of the map S becomes the shift of all symbols to the left by one unit. This *method of coding trajectories*, which goes back to the works of Birkhoff,

M. Morse, and G. Hedlund, constitutes the basis of *"symbolic dynamics"*. For a more detailed presentation we refer the reader to [2, 34].

Theorem 12 has a number of important consequences.

Proposition 5. *The map $S: \Lambda \to \Lambda$ enjoys the following properties:*

1) *any two periodic trajectories can be connected by a doubly-asymptotic trajectory;*

2) *the periodic points are dense in Λ;*

3) *there exist trajectories that fill Λ densely.*

◁ In fact, to a periodic trajectory there corresponds a point $(a) = (\ldots a, a, a, \ldots) \in \Omega$, where a is a finite string of zeros and ones. Suppose that to the points $(a), (b) \in \Omega$ there correspond two periodic trajectories. Then to the sequence $(\ldots a, a, b, b, \ldots)$ there obviously corresponds the required doubly-asymptotic trajectory. Further, for every element $\omega \in \Omega$ we can consider the sequence $\omega^{(n)} = (a_n) \in \Omega$, where $a_n = \{\omega_{-n}, \ldots, \omega_n\}$. Evidently, $\omega^{(n)} \to \omega$. Finally, consider a point $\omega^* \in \Omega$ such that in the sequence ω_n^*, starting with some position, all possible finite strings of zeros and ones are written in succession. It can be readily verified that the closure of the trajectory $\bigcup T^n \omega^*$ $(n \in Z)$ is equal to Ω. ▷

3.3. Nonexistence of Analytic First Integrals

Theorem 13. *Under the assumptions of* §3.1, *the differential equation* (24) *does not have a first integral that is analytic in $\dot{x}$, x, t, and 2π-periodic in t.*

◁ If we assume that such an integral exists, then the map $S: R^+ \to R^-$ considered in §3.1 will have a nonconstant analytic invariant function $f(v, \tau)$. One can show that the restriction of S to the invariant Cantor set Λ has the properties enumerated in Proposition 5 (see [2]). In particular, by continuity, f is constant on Λ. It follows from the construction of the perfect set Λ that for every point $(v_0, \tau_0) \in \Lambda$ there are two sequences of points of Λ which converge to (v_0, τ_0) from two distinct directions. Consequently, the derivatives of any order of f with respect to v and τ vanish at the point (v_0, τ_0). To complete the proof it remains to use the analyticity of f. ▷

We conclude with a number of remarks.

1. Since Λ is nowhere dense, the above argument does not imply the nonexistence of smooth first integrals.

2. The symbolic dynamics in the restricted (and even in the unrestricted) three-body problem considered in §3.1 has been constructed in the works of V.M. Alekseev [2]. With its help Alekssev has found all logically possible combinations of final types of motions according to Chazy's classification.

3. A statement analogous to Theorem 12 holds true in the neighborhood of homoclinic trajectories with transverse asymptotic surfaces. The rigorous proof of this statement, attributed to Birkhoff (1935), is due to S. Smale (1965)

and L.P. Shil'nikov (1967) (see [32]). We remark that the proof of nonexistence of analytic integrals (Theorem 10) does not depend on the transversality property. However, the presence of nontransverse asymptotic surfaces may drastically affect the qualitative behavior of trajectories (see [32]).

4. One can show that the periodic trajectories lying in Λ are hyperbolic, and hence nondegenerate. On the other hand, they are dense in Λ, and Λ is a key set in B. It follows that the nonexistence of analytic first integrals can be established by Poincaré's method (see § 1.2).

§4. Nonintegrability in the Neighborhood of an Equilibrium Position (Siegel's Method)

There exists another method of proving nonintegrability, based on estimates from below of the coefficients of the power series for formal integrals. As it turns out, here divergence is due to anomalously *small denominators*, that is, ultimately, to the effect of the resonances close to the equilibrium position under study.

We consider the canonical system of differential equations

$$\dot{x}_k = \frac{\partial H}{\partial y_k}, \qquad \dot{y}_k = -\frac{\partial H}{\partial x_k} \quad (1 \leqq k \leqq n), \tag{26}$$

and assume that H is an analytic function in a neighborhood of the point $x=y=0$, and that $H(0)=0$, $dH(0)=0$. Let $H=\sum_{s\geqq 2} H_s$, where each H_s is a homogeneous polynomial of degree s in x and y.

Let $\lambda_1, \ldots, \lambda_{2n}$ be the eigenvalues of the linearized canonical system with the Hamiltonian H_2. We may assume that $\lambda_{n+k} = -\lambda_k$ $(1 \leqq k \leqq n)$. We consider the case where the numbers $\lambda_1, \ldots, \lambda_n$ are purely imaginary and independent over the field of rationals.

In this section we investigate the complete integrability of the equations (26) in the neighborhood of the equilibrium position $x=y=0$, and the convergence of the Birkhoff normalizing transformation.

We consider the set $\mathscr{H}$ of all power series

$$H = \sum h_{ks} x^k y^s, \qquad k=(k_1, \ldots, k_n), \; s=(s_1, \ldots, s_n),$$

that converge in some neighborhood of $x=y=0$. We endow $\mathscr{H}$ with the following topology $\mathscr{T}$: a neighborhood of the power series H^* with coefficients h^*_{ks} is the set of all power series H with coefficients h_{ks} such that $|h_{ks} - h^*_{ks}| < \varepsilon_{ks}$, where ε_{ks} is an arbitrary sequence of positive numbers.

Theorem 14. *In every neighborhood of an arbitrary point $H^* \in \mathscr{H}$ there is a Hamiltonian H such that the corresponding canonical system* (26) *does not admit a first integral independent of H and analytic in a neighborhood of the point $x=y=0$.*

Thus, nonintegrable systems are dense in $\mathscr{H}$. In particular, the set of Hamiltonians for which the Birkhoff transformation diverges is dense in $\mathscr{H}$. The idea of the proof of Theorem 14 is as follows: let

$$F=\sum f_{ks}\, x^k\, y^s \tag{27}$$

be a formal first integral of equations (26) which is independent of H. Its existence follows from Birkhoff's theorem (Ch. IV, §1.3). It can be shown that in every neighborhood of the point $H^*\in\mathscr{H}$ there is a Hamiltonian H, to which there corresponds a formal series (27), such that infinitely many coefficients f_{ks} satisfy the estimate $|f_{ks}|\geqq m^{m^2}$, where $m=|k|+|s|$. This is achieved by choosing eigenvalues $\lambda_1, \ldots, \lambda_n$ that are approximated sufficiently fast by rational numbers. On the other hand, if equations (26) admit an analytic first integral independent of H, then the estimate $|f_{ks}|<c\,m^m$, $c=\text{const}$, holds. The details of this proof can be found in [34].

Concerning the divergence of Birkhoff's transformation, the following stronger result holds:

Theorem 15 ([35]). *The Hamiltonians for which Birkhoff's transformation converges form a set of first Baire category (i.e., representable as a finite or countable union of nowhere dense sets) in the topology $\mathscr{T}$ on $\mathscr{H}$.*

More precisely, Siegel proved the existence of an infinite countable set of independent analytic power series $\Phi_1, \Phi_2, \ldots$ in the infinitely many variables h_{ks}, which are absolutely convergent for $|h_{ks}|<\varepsilon$ (for all k, s), and such that if the Hamiltonian $H\in\mathscr{H}$ is reduced to normal form by a convergent Birkhoff transformation, then at the point H almost all (i.e., all except, possibly, finitely many) of the Φ_s vanish. Since the functions Φ_s are analytic, their zeros are nowhere dense in $\mathscr{H}$. Consequently, the set of points of $\mathscr{H}$ which satisfy at least one of the equations $\Phi_s=0$ is of the first Baire category. If we attempt to investigate the convergence of Birkhoff's transformation for a concrete Hamiltonian system, then we are led to checking infinitely many conditions. No finite method for doing this is known, although all the coefficients of the series Φ_s can be calculated explicitly. The proof of the theorem rests on a careful analysis of isolated long-periodic solutions in the neighborhood of the equilibrium position; thus, as far as its idea is concerned it goes back, too, to earlier investigations of Poincaré (see §1.2).

Remark. We endow $\mathscr{H}$ with a new topology $\mathscr{T}'$, in which the neighborhoods of a power series with coefficients h^*_{ks} are the sets consisting of all power series whose coefficients h_{ks} satisfy the inequalities $|h_{ks}-h^*_{ks}|<\varepsilon$ for $|k|+|s|\leqq N$, for some $\varepsilon>0$ and $N\geqq 3$. It can be shown that in the topology $\mathscr{T}'$ the set of Hamiltonians for which the Birkhoff transformation converges is dense in $\mathscr{H}$. In fact, if in the formal power series that give the Birkhoff transformation we omit the terms of order higher than N and subsequently correct the coefficients of the higher order terms in the series of the given Hamiltonian, we obtain a convergent canonical transformation which reduces the Hamiltonian modified in the indicated manner to normal form. Note that topology $\mathcal{T}'$ is considerably weaker than $\mathcal{T}$.

Using Siegel's method one can prove that nonintegrable systems are dense in certain subspaces of $\mathcal{H}$. As an example, we consider the equation

$$\ddot{x} = -\frac{\partial U}{\partial x}, \qquad x \in R^n, \tag{28}$$

which describes the motion of a point mass in a force field with potential $U(x)$. This equation can be, of course, written in Hamiltonian form:

$$\dot{x} = H'_y, \quad \dot{y} = -H'_x; \qquad H = \frac{y^2}{2} + U(x).$$

Suppose that $U(0)=0$ and $dU(0)=0$. Then the point $x=0$ is an equilibrium position. We write $U = \sum_{s \geqq 2} U_s$ and assume that $U_2 = \sum \omega_k^2 x_k^2/2$. The frequencies $\omega_1, \ldots, \omega_n$ of small oscillations are assumed to be rationally independent.

We introduce the space $\mathcal{U}$ of formal power series

$$\sum_{|k| \geqq 2} u_k x^k$$

that converge in some neighborhood of the point $x=0$. We equip $\mathcal{U}$ with the topology $\mathcal{T}$ used above for the space $\mathcal{H}$.

Theorem 16. *In the space $\mathcal{U}$ with the topology $\mathcal{T}$, the points for which equations* (28) *have no first integral $F(\dot{x}, x)$ that is analytic in the neighborhood of the point $\dot{x}=x=0$ and independent of the energy integral $E=\dot{x}^2/2+U(x)$ form a dense set.*

It seems that the points $U \in \mathcal{U}$ for which Birkhoff's transformation to normal form converges form a subset of the first category in $\mathcal{U}$. The proof of Theorem 16 can be found in [21].

In connection with the analysis of normal forms, it is useful to keep in mind the following important fact: a divergent Birkhoff transformation may converge on some analytic invariant manifold Λ containing the equilibrium position. The dynamical system arising in this way on Λ is integrable. A classical example of such a situation is presented in the following theorem:

Theorem 17 (A.M. Lyapunov). *If for all $s>1$ the ratio λ_s/λ_1 is not an integer, then there exists an invertible analytic canonical transformation $x, y \mapsto \xi, \eta$ which reduces the Hamiltonian $H(x, y)$ to the form*

$$\Phi(\rho) + O(|\zeta|^2),$$

where Φ is a function of the single variable $\rho = \xi_1^2 + \eta_1^2$, and $\zeta = (\xi_2, \ldots, \xi_n, \eta_2, \ldots, \eta_n)$.

Thus, on the invariant manifold $\Lambda=\{\zeta=0\}$ the Hamiltonian system (28) reduces to the system with one degree of freedom

$$\dot{\xi}_1=2\Phi'_\rho\eta_1, \qquad \dot{\eta}_1=-2\Phi'_\rho\xi_1.$$

Therefore, $\rho=\text{const}$ and $\xi_1+i\eta_1=c\exp(-2i\Phi'_\rho t)$. The phase plane $R_2=\{\xi_1,\eta_1\}$ is foliated into invariant concentric circles $\xi_1^2+\eta_1^2=\rho$ on which the motion is uniform with frequency Φ'_ρ, which depends on ρ; moreover, $\Phi'_\rho(0)=\lambda_1/2$.

An analogous statement is valid also in the case where the characteristic numbers λ_s include a real pair $\lambda_1, -\lambda_1$. In this case $\rho=\xi_1\eta_1$ (for details see [36]).

Remark. As Siegel showed, the condition $\lambda_s/\lambda_1\notin Z$ $(s>1)$ in Lyapunov's theorem cannot be dropped. Generalizations of the theorem to the case where this condition is not fulfilled are given in Roels' works (1971).

The foregoing discussion can be carried over, with the necessary modifications, to the case of normal forms of Hamiltonian systems in the neighborhood of periodic trajectories. A thorough analysis of the convergence of normalizing transformations (and not only for Hamiltonian equations) can be found in Bryuno's book [70].

§5. Branching of Solutions and Nonexistence of Single-Valued First Integrals

In the majority of problems of Hamiltonian mechanics that have been integrated the known first integrals admit continuation to holomorphic or meromorphic functions in the complex domain of variation of the canonical variables. In this section we show that, in the general case, branching of solutions of Hamiltonian systems in the complex time plane is an obstruction to the existence of new single-valued first integrals.

5.1. Branching of Solutions is an Obstruction to Integrability. Let

$$D_{C,\delta}=\{I\in C^n\colon\ \operatorname{Re} I\in D\subset R^n,\ |\operatorname{Im} I|<\delta\},$$

let $T^n_C=C^n/2\pi Z^n$ be a complex torus (over R it is $T^n\times R^n$) with complex angular coordinates $\varphi_1, \ldots, \varphi_n \bmod 2\pi$, and let E be a neighborhood of zero in C. Let $H\colon D_{C,\delta}\times T^n_C\times E\to C$ be a holomorphic function which for real values of I, φ, and ε assumes real values, and such that $H(I,\varphi,0)=H_0(I)$ depends only on I.

The direct product $D_{C,\delta}\times T^n_C$ is endowed with the obvious symplectic structure, relative to which Hamilton's equations with Hamiltonian H have canonical form:

$$\frac{dI}{dt} = -\frac{\partial H}{\partial \varphi}, \quad \frac{d\varphi}{dt} = \frac{\partial H}{\partial I}; \qquad H = H_0 + \varepsilon H_1 + \dots. \tag{29}$$

All solutions of the system with Hamiltonian H_0 are single-valued in the complex time plane $t \in C$: indeed, they are given by

$$I = I^0, \qquad \varphi = \varphi^0 + \omega(I^0)\, t.$$

For $\varepsilon \neq 0$ the solutions of the "perturbed" equations (29), generally speaking, are no longer single valued. Let γ be a contour in the complex time plane. According to a well-known *theorem of Poincaré*, the solutions of equations (29) can be expanded in power series

$$\begin{gathered} I = I^0 + \varepsilon I^1(t) + \dots, \qquad \varphi = \varphi^0 + \omega t + \varepsilon \varphi^1(t) + \dots, \\ I^1(0) = \dots = \varphi^2(0) = \dots = 0, \end{gathered} \tag{30}$$

which converge for sufficiently small values of the parameter ε, if $t \in \gamma$.

We say that the analytic vector-valued function $f(t)$, $t \in C$, is not single-valued along the contour γ if it undergoes a jump $\Delta f = \xi \neq 0$ on a circuit of γ. If, for example, the function $I^1(t, I^0, \varphi^0)$ is not single-valued along γ, then for small values of ε the perturbed solution (30) is also not single-valued along γ. The jump ΔI^1 is obviously equal to

$$\xi = \int_\gamma \Phi(t)\, dt, \quad \text{where} \quad \Phi(t) = -\left.\frac{\partial H_1}{\partial \varphi}\right|_{I^0,\, \varphi^0 + \omega(I^0)t}.$$

If for fixed I the function H_1 is holomorphic on T_C^n, then, of course, $\xi = 0$. However, in important concrete cases this function does have singularities (for example, poles). For this reason we shall only assume that H_1 is holomorphic in $D_{C,\delta} \times \Omega \times E$, where Ω is a connected domain in T_C^n which contains the real torus T_R^n and the closed contour Γ – the image of γ under the map $\varphi = \varphi^0 + \omega(I^0)t$, $t \in \gamma$.

We fix initial data I^0, φ^0, and deform continuously γ so that Γ will not meet any singular point of H. Then, by Cauchy's theorem, the function $I^1(t)$ will change, on going around the deformed contour once, by the same quantity $\xi \neq 0$. On the other hand, since the solutions (30) are continuous in the initial data, function $I^1(t, I^0, \varphi^0)$ is not single-valued along γ, for all initial values close to I^0, φ^0.

Theorem 18. *Suppose that the following conditions are satisfied:*

1) $\det(\partial^2 H_0/\partial I^2) \not\equiv 0$ *in* $D_{C,\delta}$,

2) *for some initial data* I^0, φ^0 *the function* I^1 *is not single-valued along the closed contour* $\gamma \subset C\{t\}$.

Then equations (29) *do not have a complete set of independent formal*[52] *first integrals*

$$F_s = \sum_{i=0}^{\infty} F_i^s(I, \varphi)\, \varepsilon^i \qquad (1 \leqq s \leqq n),$$

whose coefficients are single-valued holomorphic functions in the direct product $V \times \Omega \subset D_{C,\delta} \times T_C^n$, *where* V *is a neighborhood of the point* I^0 *in* $D_{C,\delta}$ ([111, 20]).

◁ We indicate the key steps of the proof. As always, we first show that the functions $F_0^s(I, \varphi)$ do not depend on φ. Let $(I, \varphi) \in D \times T_R^n$ and $F_0^s = \Phi_0^s + i\Phi_0^s$. Then Φ_0^s and Ψ_0^s are first integrals of the nondegenerate unperturbed system. By Lemma 1 of §1, they do not depend on $\varphi \in T_R^n$. When $\varphi \in \Omega$ the constancy of F_0^s follows from the connectedness of the domain Ω and the uniqueness of analytic continuation.

Next we prove that the function $F_0^1(I), \ldots, F_n^1(I)$ are dependent in a domain $V \subset D_{C,\delta}$. In fact, since $F_s(I, \varphi, \varepsilon)$ is a first integral of the canonical system (29), this function is constant on solutions (30). Consequently, its values at time $t \in \gamma$ and after a circuit of γ coincide:

$$\begin{aligned} &F_0^s(I^0 + \varepsilon I^1(\tau) + \ldots) + \varepsilon F_1^s(I^0 + \varepsilon I^1(\tau) + \ldots, \varphi^0 + \omega\tau + \varepsilon\varphi^1(\tau) + \ldots) + \ldots \equiv \\ &\equiv F_0^s(I^0 + \varepsilon(I^1(\tau) + \xi(I^0)) + \ldots) + \varepsilon F_1^s(I^0 + \ldots, \varphi^0 + \omega\tau + \ldots) + \ldots. \end{aligned}$$

Expanding this identity in power series in ε and equating the coefficients of ε $(=\varepsilon^1)$ we get

$$\left\langle \frac{\partial F_0^s}{\partial I}, \xi \right\rangle = 0, \qquad 1 \leqq s \leqq n.$$

Since the jump ξ is different from zero in the neighborhood of I^0, the Jacobian

$$\frac{\partial(F_0^1, \ldots, F_0^n)}{\partial(I_1, \ldots, I_n)} \equiv 0$$

in a whole domain V containing the point I^0.

On the other hand, applying Poincaré's method presented in §1 one can prove the existence of independent first integrals

$$\Phi_s(I, \varphi, \varepsilon) = \sum_{i \geqq 0} \Phi_i^s(I, \varphi)\, \varepsilon^i$$

with coefficients holomorphic in the domain $W \times \Omega$ (where W is a small subdomain of V), such that the functions $\Phi_0^1, \ldots, \Phi_0^n$ are independent. ▷

Example 1. Again we consider the problem of rapid rotation of an asymmetric heavy rigid body around a fixed point. The Hamiltonian of this prob-

[52] We again regard the formal series $F = \sum F_i \varepsilon^i$ as a first integral of the canonical equations (1.1) if formally $\{H, F\} \equiv 0$. It is readily verified that in this case the composition of the power series (1.2) and $\sum F_i \varepsilon^i$ is a power with constant coefficients.

lem is $H_0(I)+\varepsilon H_1(I, \varphi)$, $I\in\Delta\subset R^2\{I\}$, $\varphi\in T^2$ (see §1.3). The perturbing function H_1 can be expressed as a sum

$$h_1(I, \varphi_1)\, e^{i\varphi_2}+h_2(I, \varphi_1)\, e^{-i\varphi_2}+h_3(I, \varphi_1);$$

moreover, for fixed $I\in\Delta$, the functions $h_s(I, z)$ $(1\leqq s\leqq 3)$ are elliptic (double-periodic meromorphic functions of $z\in C$). It follows that the Hamiltonian can be continued to a single-valued meromorphic function on T_C^2.

Let $\varphi^0=0$ and let I^0 belong to the Poincaré set of the perturbed problem. In the complex plane $t\in C$ we consider the closed contour γ given by the boundary of the rectangle ABCD (see Fig. 66). Here T and iT' are respectively the real and the purely imaginary period of the elliptic functions $f_s(I^0, \omega_1 z)$, where $\omega_1=\partial H_0/\partial I_1$. The number τ is chosen so that these meromorphic functions will have no poles on γ. On can show that the function $I_2^1(t, I^0, 0)$ is not single-valued along the contour γ [111]. Consequently, the solutions of the perturbed problem branch in the complex-time plane, thereby preventing the appearance of a new single-valued first integral. △

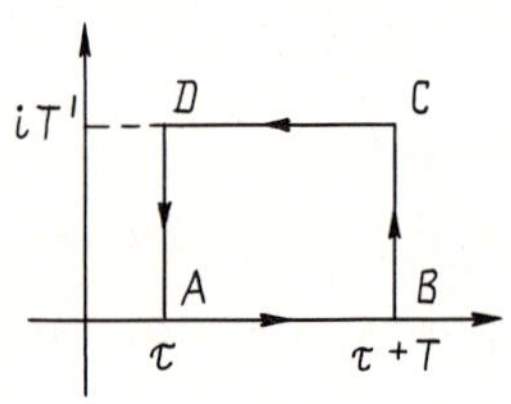

Fig. 66

Using the branching of solutions one can establish the nonexistence of single-valued analytic first integrals for small, but fixed values of $\varepsilon\neq 0$ (see [196]).

5.2. Monodromy Groups of Hamiltonian Systems with Single-Valued First Integrals. The presence of multi-valued solutions can be established not only by means of expansion in series of powers of a small parameter. An alternate method, based on the analysis of the variational equations for known single-valued solutions, was proposed by Lyapunov in 1894. We have already applied Lyapunov's method in our investigation of multiple collision singularities in the n-body problem (see Ch. 2, §2.4). In this section we will begin by investigating linear Hamiltonian equations with holomorphic coefficients.

Let $H=\langle z, A(t)\, z\rangle/2$ be a quadratic form in $z\in C^{2n}$, where $A(t)$ is a given $2n\times 2n$ matrix whose entries are holomorphic functions defined on a Riemann surface X. If, for example, the entries of $A(t)$ are functions meromorphic on C, then X is the complex plane with some points (poles) removed. The linear Hamiltonian equations with Hamiltonian function H have the form

$$\dot z=\Im dH=\Im A(t)\, z. \tag{31}$$

Locally, for a given initial condition $z(t_0)=z_0$ a uniquely determined holomorphic solution always exists. This can be continued along any curve in X, but in general the continuation will no longer be a single-valued function on X. The branching of solutions of the linear system (31) is described by its monodromy group G: to each element σ of the fundamental group $\pi_1(X)$ there corresponds a $2n \times 2n$ matrix T_σ, such that after going around any closed path of homotopy class σ, the value of $z(t)$ becomes $T_\sigma z(t)$. If τ is another element of $\pi_1(X)$, then $T_{\tau\sigma}=T_\tau T_\sigma$. Thus, the mapping $\sigma \mapsto T_\sigma$ is a group homomorphism $\pi_1(X) \to G$.

We are interested in determining whether equations (31) possess holomorphic first integrals $F: C^{2n} \times X \to C$. Since every first integral $F(z, t)$ is constant along the solutions of (31), for each $t_0 \in X$ the function $F(z, t_0)$ is invariant under the action of the monodromy group G. This property imposes severe constraints on the form of the first integral: if G is sufficiently "rich", then the only invariant functions (integrals) are the constants.

Since system (31) is Hamiltonian, the transformations of the monodromy group are symplectic. The problem of integrals of groups of symplectic transformations has been studied by Ziglin [197]. We briefly discuss his results.

According to Proposition 1, the eigenvalues $\lambda_1, \ldots, \lambda_{2n}$ of any symplectic transformation $g: C^{2n} \to C^{2n}$ are divided into pairs $\lambda_1=\lambda_{n+1}^{-1}, \ldots, \lambda_n=\lambda_{2n}^{-1}$. We say that a transformation $g \in G$ is nonresonant if from any equality $\lambda_1^{m_1} \ldots \lambda_n^{m_n}=1$ with integers $m_1, \ldots, m_n$ it follows that all $m_k=0$. For $n=1$ this condition means that λ is not a root of unity. Let T be the matrix of a nonresonant symplectic transformation g. Since none of the eigenvalues of T equals 1, the equation $Tz=z$ has only the trivial solution $z=0$.

It is convenient to pass to a *symplectic basis* for the map g: if $z=(x, y)$, $x=(x_1, \ldots, x_n)$, $y=(y_1, \ldots, y_n)$ are the coordinates in this basis, then $g: (x, y) \mapsto (\lambda x, \lambda^{-1} y)$. A symplectic basis exists if all $\lambda_k \neq 1$ $(1 \leqq k \leqq n)$ (this is proved, for example, in [36]).

Let $F(z)=\sum_{s \geqq 1} F_s(z)$ be a first integral of g. Then all the homogeneous forms F_s are also first integrals. Let $F_s(x, y)=\sum_{k+l=s} f_{kl} x^k y^l$. Then, obviously,

$$\sum f_{kl} x^k y^l=\sum \lambda^{k-l} f_{kl} x^k y^l.$$

If g is nonresonant, then s is even and $f_{kl}=0$ for $k \neq l$.

Theorem 19 ([197]). *Let $g \in G$ be nonresonant. If the given Hamiltonian system has n independent holomorphic first integrals $F: C^{2n} \times X \to C$, then every transformation $g' \in G$ has the same fixed point as g, and takes the eigendirections of g again into eigendirections. If, in addition, there are no k eigenvalues $(k \geqq 2)$ of g' which form a regular polygon centered at zero in the complex plane, then g' commutes with g.*

The last condition is evidently satisfied if g' is also nonresonant.

We now prove Theorem 19 in the simple, yet practically important case $n=1$. Suppose that the eigenvalues of g are not roots of unity and let $(x, y)=z$ be a symplectic basis for g. The eigendirections of g are the two straight lines $x=0$ and $y=0$. We have already seen that every homogeneous integral of g has the form $c(xy)^s$, $s \in N$. Let g' be another element of group G. Since the function $(xy)^s$ is invariant under the action of g', the set $xy=0$ is fixed by g'. Also, since g' is a nonsingular linear map, the point $x=y=0$ is fixed by g', and g' either preserves the eigendirections of g or permutes them. In the first case g' clearly commutes with g, whereas in the second it has the form

$$x \mapsto \alpha y, \qquad y \mapsto \beta x.$$

But g' is a symplectic map, so its matrix

$$S=\begin{pmatrix} 0 & \alpha \\ \beta & 0 \end{pmatrix}$$

satisfies the condition

$$S^* \mathfrak{J} S = \mathfrak{J},$$

whence $\alpha\beta=-1$. However, since in this case the eigenvalues of S are $\pm i$, they form precisely that exceptional regular polygon mentioned in the conclusion of the theorem. This completes the proof.

We now consider the case where the entries of $A(t)$ are single-valued doubly-periodic meromorphic functions of time $t \in C$, which have only one pole inside the parallelogram of periods. We may assume that $A(t)$ is a meromorphic function on the complex torus X obtained by factoring the complex plane C by the lattice of periods. Consider the two symplectic maps g and g' at the periods of $A(t)$. Suppose that their eigenvalues satisfy the conditions of Theorem 18. Then for equation (31) to have n independent analytic integrals it is necessary that g and g' commute. Consequently, to a circuit around the singular point (i.e., to the element $g g' g^{-1} g'^{-1} \in G$), there corresponds the identity map of C^{2n}.

Suppose that the nonlinear Hamiltonian system

$$\dot{z} = \mathfrak{J}\, dH, \qquad z \in C^{2n} \tag{32}$$

has a particular solution $z_0(t)$ which is single-valued on its Riemann surface X. Set $u=z-z_0(t)$. Then we can rewrite equation (32) in the form

$$\dot{u} = \mathfrak{J} H''(z_0(t))u + \dots . \tag{33}$$

The linear nonautonomous system

$$\dot{u} = \mathfrak{J} H''(t) u$$

is the variational equation for the solution $z_0(t)$, and is clearly Hamiltonian, with Hamiltonian function $\frac{1}{2}\langle u, H''(t)u\rangle$. To the first integral $H(z)$ of the autonomous system (32) there corresponds the linear first integral

$\langle H'(z_0(t)), u\rangle$ of the variational equation. It may serve, for example, to reduce the number of degrees of freedom of system (33) by one.

Suppose that the nonlinear system (32) has several independent holomorphic first integrals $F_s(z)$ $(1 \leqq s \leqq m)$. Then equation (33) will also have first integrals, namely, the homogeneous terms of the expansion of the functions F_s in power series in u:

$$\langle F_s'(z_0(t)), u\rangle + \dots .$$

These forms are holomorphic functions on the direct product $C^{2n} \times X$. The following lemma is valid:

Lemma 7 ([197]). *If system* (32) *has m independent first integrals, then the variational equation has m independent polynomial first integrals.*

Thus, the problem of the complete integrability of Hamiltonian systems in the complex domain reduces to an investigation of the integrability of linear canonical systems.

Following Lyapunov, Ziglin applied this result to the problem of the rotation of a heavy rigid body around a fixed point. It turned out that an additional holomorphic (or even meromorphic) first integral exists only in the three classical integrable cases of Euler, Lagrange, and Kovalevskaya. If the area integral constant is fixed to be zero, we must add the Goryachev-Chaplygin case to this list.

Using this method one can prove the nonintegrability of the Hénon-Heiles Hamiltonian system (Ch. 5, §1.3, Example 2) not only in the complex, but also in the real domain. An analogous result is valid for the homogeneous two-component model of the *Yang-Mills equations*, described by the Hamiltonian system with Hamiltonian

$$H = \tfrac{1}{2}(p_1^2 + p_2^2) + q_1^2 q_2^2.$$

The more difficult question of the existence of an additional real analytic first integral for an arbitrary mass distribution of the rigid body is still open.

Remark. Recently researchers became again interested in the integration of differential equations of mechanics in terms of ϑ-functions (the so-called "algebraic integrability"). The search for necessary conditions of algebraic integrability follows the method of S. Kovalevskaya that she applied in 1888 to the dynamics of the rigid body. The reader may familiarize himself with the current state of this problem by referring to [45, 46, 87] (translator's note: see also M. Adler and P. van Moerbeke, *A Systematic Approach Towards Solving Integrable Systems*, Academic Press, to be published).

§6. Topological and Geometrical Obstructions to Complete Integrability of Natural Systems with Two Degrees of Freedom

According to results of variational calculus, every closed one-dimensional cycle on the configuration manifold can be realized as the trajectory of a periodic solution of sufficiently high fixed energy. On the other hand, almost all periodic solutions of a completely integrable system with n degrees of freedom lie on n-dimensional tori that form smooth families. Hence, a sufficiently complex topological structure of the configuration manifold of a natural system is an obstruction to its complete integrability. This idea can be fully implemented for systems with two degrees of freedom.

6.1. Topology of the Configuration Space of Integrable Systems. Let M be a connected, compact, orientable, analytic surface which serves as the configuration manifold of a natural mechanical system. The topological structure of such surfaces is well known: they are spheres with a certain number κ of handles attached. The number κ is a topological invariant of the surface called its *genus.*

The state space (or velocity phase space), i.e., the tangent bundle TM, has a natural structure of a four-dimensional analytic manifold. We shall assume that the Lagrangian $L=T+V$ is a real analytic function on TM. The total energy $H=T-V$ is constant on the trajectories of the equations of motion $[L]=0$.

Theorem 19. *If the genus of M is greater than one (i.e., M is not homeomorphic to the sphere S^2 or the torus T^2), then the equation of motion does not have a first integral which is analytic on TM and independent on the energy integral.*

There are numerous well-known examples of integrable systems with configuration space S^2 or T^2. In the infinitely differentiable case Theorem 19 is false: for any smooth surface M one can indicate a "natural" Lagrangian L such that Lagrange's equation $[L]=0$ has a smooth first integral independent of (more precisely, not everywhere dependent on) H (see [21]).

Theorem 19 is a consequence of a stronger result which establishes the nonintegrability of the equations of motion for fixed sufficiently large values of the total energy. For all values $h>h^*=\max_M(-V)$, the level set of the total energy, $M_h=\{H=h\}$, is a three-dimensional analytic invariant manifold on which there naturally arises an analytic differential equation. We shall refer to the latter as the *reduced* equation.

Theorem 20. *If the genus of M is greater than one, then for all $h>h^*$ the reduced equation on M_h does not have a first integral analytic on the entire M_h (see [21]).*

Remark. Theorems 19 and 20 are also valid in the nonorientable case, if M is not the projective plane RP^2 or the Klein bottle K. In fact, the standard regular double covering $N \to M$, where N is an orientable surface, induces a natural mechanical system on N, which has an additional first integral if the given system on M has one. It remains to observe that the genus of surface N is greater than one if M is not homeomorphic to RP and K. △

Let k be the Gaussian curvature of the Maupertuis Riemannian metric $(ds)^2 = 2(h+V)\,T(dt)^2$ on M. By the Gauss-Bonnet formula

$$\int_M k\,d\sigma = 2\pi\,\chi(M),$$

where $\chi(M)$ is the Euler characteristic of the compact surface M. If the genus of M is greater than one, then $\chi(M) < 0$, and hence the Gaussian curvature is negative in the mean. If the curvature is negative everywhere, then the reduced dynamical system on M_h is an *Anosov system* and, in particular, is ergodic on M_h (see [4]). These conclusions are valid also in the higher-dimensional case (we need only require that the curvature be negative in all two-dimensional directions). Here the reduced equation on M_h does not admit even a continuous first integral, since almost all trajectories are dense in M_h. Needless to say, from the fact that the curvature is negative in the mean it does not necessarily follow that it is negative everywhere.

Let us outline the key steps in the proof of Theorem 20. According to the principle of least action, the trajectories on M with total energy h are geodesics of the Maupertuis Riemannian metric ds^2. We fix a point $x \in M$ and consider the tangent vectors $v \in T_x M$ satisfying the equality $H(v, x) = h$. We call the vector v *critical* if the value of the first integral $f\colon M_h \to R$ at the point (x, v) is critical. We first show that there are infinitely many critical velocities. Assume the contrary. Then the circle $S_x = \{v \in T_x M : H(v, x) = h\}$ breaks into a finite number of arcs Δ_i such that all the points $v \in \Delta_i$ are noncritical. By Theorem 3 of Chapter 5, to each vector $v \in \Delta_i$ there corresponds a unique torus T_V^2 on which the motion $z(t)$ with initial data $z(0) = x$, $\dot z(0) = v$ lies. The union $D_i = \bigcup_{v \in \Delta_i} T_v^2$ is obviously diffeomorphic to $\Delta_i \times T^2$. Let $\pi\colon TM \to M$ be the canonical projection; we set $X_i = \pi(D_i)$. We claim that the homology groups $H_1(X_i) \subset H_1(M)$ cover "almost the entire" group $H_1(M)$, i.e., all $H_1(M)$ except possibly for the elements that belong to some finite set of rank-one subgroups. This results can be derived from a *theorem of E.V. Gaidukov* (1966): for every nontrivial class of freely homotopic paths on M there exists a geodesic semitrajectory $\gamma(t)$, emanating from the point x, which asymptotically approaches some closed geodesic from the given homotopy class. If the velocity $\dot\gamma(0)$ is not critical, then $\gamma(t)$ is closed. The exceptional rank-one subgroups of $H_1(M)$ mentioned above are generated precisely by the closed geodesics on which asymptotic semitrajectories that are distinct from them "wind". Since the continuous mapping $D_i \to X_i$ induces a homomorphism of homology groups $H_1(D_i) \to H_1(X_i)$ and since $H_1(D_i) \simeq Z^2$, the group $H_1(M)$ is the union of a finite number of groups of rank not exceeding 2. It is known that $H_1(M) \simeq Z^{2\kappa}$, where κ is the genus of M. Since $\kappa > 1$, we have reached a contradiction.

Thus, there are infinitely many distinct critical velocities. Since every analytic function on a compact analytic manifold has only a finite number of critical values, the first integral $f(v, x)$ must be constant on the circles S_x. Consequently, f is a function on M. Since M is connected and compact, any two of its points can be joined by geodesic of shortest length. Hence, $f \equiv \text{const}$.

Remark. The inequality rank $H_1(M) \leqq \dim M$ holds for all known multidimensional completely integrable natural mechanical systems with compact analytic configuration manifold M. △

6.2. Geometrical Obstructions to Integrability. Let N be a closed submanifold with boundary on an analytic (not necessarily compact) surface M. We let N_h denote the set of all points of the level set M_h which are taken by the projection $\pi: TM \to M$ into points of N. We shall say that N is geodesically convex if for any two close points of the boundary ∂N, the shortest-length geodesic of the Maupertuis metric joining these points is entirely contained in N.

Theorem 21. *If in the analytic surface M there is a compact geodesically convex subdomain N of negative Euler characteristic, then the reduced system on M_h does not have an analytic first integral. Moreover, an analytic first integral does not even exist in a neighborhood of the set N_h.*

The proof of Theorem 21 rests on the arguments given §6.1. A slight difference is that instead of the homology group $H_1(M)$ one uses the free homotopy classes of closed paths on M.

Theorem 21 was successfully used by S.V. Bolotin to prove that the *problem* of the motion of a point in the gravitational field of *n fixed centers* is not integrable for $n > 2$ (see [64]). We recall that to the values $n = 1$ and $n = 2$ there correspond the integrable cases of Kepler and Euler, respectively.

Chapter 7
Theory of Small Oscillations

The study of the oscillations of a system in the neighborhood of an equilibrium position or a periodic motion usual begins with linearization. The linearized system can be integrated. After this is done, the main properties of the oscillations in the original system can frequently be determined using the theory of normal forms of Poincaré-Birkhoff. This theory is an analog of perturbation theory (Ch. 5, §2). Here the linearized system plays the role of the unperturbed system with respect to the original one. In this chapter we describe the basic elements of such an approach.

Central to the theory of small oscillations is the investigation of the stability of the equilibrium position or periodic motion in question. A considerable amount of literature is devoted to the theory of stability (see the survey [19] and: V.I. Arnold, Yu.S. Il'yashenko, Ordinary Differential Equations, Vol. 1 of this Encyclopaedia). Here we discuss briefly only some results of this theory by which we can draw conclusions on stability based on the study of normal forms. We also present results related to the search for a converse to Lagrange's theorem on the stability of equilibria in a potential field.

§ 1. Linearization

We consider a natural Lagrangian system

$$\frac{d}{dt}\left(\frac{\partial L}{\partial \dot{q}}\right)-\frac{\partial L}{\partial q}=0, \qquad L=T-U(q),\ \ T=(A(q)\dot{q},\dot{q})/2. \tag{1}$$

The equilibrium positions of system (1) are precisely the critical points of the potential energy U. In order to linearize system (1) near the equilibrium position $q=0$, it suffices to replace the kinetic energy T by its value T_2 at $q=0$, and the potential energy U by its quadratic part U_2 in the neighborhood of zero.

Example 1. For a one-dimensional system, $L=a(q)\dot{q}^2/2-U(q)$ and $L_2=T_2-U_2=(a\dot{q}^2-bq^2)/2$, with $a=a(0)$, $b=\partial^2 U/\partial q^2|_{q=0}$. The linearized equation of motion is $a\ddot{q}+bq=0$. △

We consider now a Hamiltonian system. Its equilibrium positions are the critical points of the Hamiltonian. In order to linearize a Hamiltonian system near an equilibrium position it suffices to replace the Hamiltonian by its quadratic part in the neighborhood of this point.

The linearization of Hamiltonian systems near periodic trajectories is considered in §3.2.

§2. Normal Forms of Linear Oscillations

2.1. Normal Form of Linear Natural Lagrangian Systems. Consider a dynamical system with a quadratic Lagrangian $L_2 = T_2 - U_2$, $T_2 \geqq 0$. Its oscillations take a particularly simple form in special coordinates known as principal or normal.

Theorem 1. *There exists a linear change of coordinates $Q = Cq$ which reduces the quadratic Lagrangian L_2 to the diagonal form*

$$L_2 = (\dot{Q}_1^2 + \dots + \dot{Q}_n^2)/2 - (\lambda_1 Q_1^2 + \dots + \lambda_n Q_n^2)/2, \tag{2}$$

and reduces the equations of motion to the corresponding form

$$\ddot{Q}_i = -\lambda_i Q_i, \qquad i = 1, \dots, n. \tag{3}$$

The eigenvalues λ_i are the roots of the characteristic equation

$$\det(B - \lambda A) = 0;$$

here $T_2 = (A\dot{q}, \dot{q})/2$ and $U_2 = (Bq, q)/2$.

◁ According to a theorem of Linear Algebra, the pair of quadratic forms T_2 and U_2, with T_2 positive definite, can be simultaneously reduced to principal axes by a linear change of coordinates. The new coordinates can be chosen so that the form T_2 becomes a sum of squares. ▷

Corollary 1. *A system performing small oscillations is the direct product of n linear one-dimensional systems.*

For each one-dimensional system (3) there are three possible cases:

1) $\lambda_i = \omega^2 > 0$; the solution is $Q = c_1 \cos \omega t + c_2 \sin \omega t$ (oscillation),
2) $\lambda_i = 0$; the solution is $Q = c_1 + c_2 t$ (neutral equilibrium),
3) $\lambda_i = -k^2 < 0$; the solution is $Q = c_1 \cosh kt + c_2 \sinh kt$ (instability).

Corollary 2. *Suppose that one of the eigenvalues is positive: $\lambda = \omega^2 > 0$. Then the system can perform a periodic oscillation of the form $q(t) = (c_1 \cos \omega t + c_2 \sin \omega t)\xi$, where ξ is an eigenvector corresponding to ω: $B\xi = \lambda A\xi$.*

Such a periodic motion is called a *characteristic* or *principal oscillation* (or a *normal mode*), and the number ω is called the *characteristic* (*principal* or *normal*) *frequency*.

These results remain valid also in the case where there are multiple eigenvalues: in a natural Lagrangian system, as opposed to a general system of linear equations (and even to a general Hamiltonian system), resonant terms of the form $t \sin \omega t$, etc., do not arise even in the case of multiple eigenvalues (except for $\lambda = 0$, when Jordan blocks of order two may arise).

2.2. The Rayleigh-Fischer-Courant Theorems on the Behavior of Characteristic Frequencies under an Increase in Rigidity and under Imposition of Constraints. Given two linear Lagrangian systems with the same kinetic energy, the *more rigid* (or *stiff*) system is, by definition, the one for which the potential energy is the largest.

Theorem 2. *Under an increase in the rigidity of a system which performs small oscillations all the characteristic frequencies are increased.*

We say that a natural Lagrangian system with $n-1$ degrees of freedom is obtained from a system with n degrees of freedom which performs small oscillations by *imposition of a linear constraint* if its kinetic and potential energies are the restrictions of the kinetic and potential energies of the original system to an $(n-1)$-dimensional subspace.

Theorem 3. *The characteristic frequencies ω_i', $i = 1, \ldots, n-1$, of the system with constraint separate the characteristic frequencies ω_i of the original system* (Fig. 67).

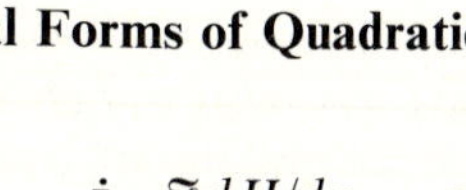

Fig. 67

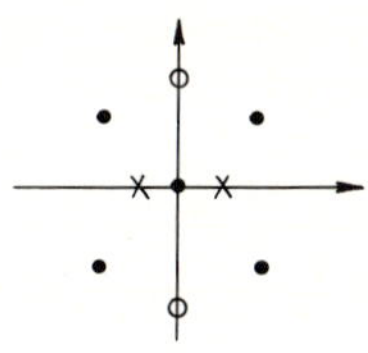

Fig. 68

2.3. Normal Forms of Quadratic Hamiltonians. Consider the Hamiltonian system

$$\dot{z} = \mathfrak{I}\, dH/dz, \qquad z \in R^{2n}, \qquad \mathfrak{I} = \begin{bmatrix} 0 & -E_n \\ E_n & 0 \end{bmatrix},$$

with quadratic Hamiltonian: $H = \frac{1}{2}(\Omega z, z)$. Consider the characteristic equation

$$\det(\mathfrak{I}\Omega - \lambda E_{2n}) = 0.$$

Its roots are called the *eigenvalues* of the Hamiltonian H.

Theorem 4. *The eigenvalues of the Hamiltonian in the complex λ-plane lie symmetrically with respect to the "cross" of coordinate axes: if λ is an eigenvalue, then so are $\bar{\lambda}$, $-\lambda$,* and $-\bar{\lambda}$ (Fig. 68).

◁ If λ is an eigenvalue, then

$$\det(\Im\Omega-\bar{\lambda}E_{2n})=\overline{\det(\Im\Omega-\lambda E_{2n})}=0$$

and

$$\det(\Im\Omega+\lambda E_{2n})=\det(\Omega\Im+\lambda E_{2n}=\det((-\Im\Omega+\lambda E_{2n})^T)=0,$$

where the superscript T denotes transposition. ▷

Corollary 1. *Stability in Hamiltonian systems is always neutral: if the equilibrium is stable, then all eigenvalues have zero real part.*

Corollary 2. *Under a small perturbation of the Hamiltonian, purely imaginary (real simple) eigenvalues remain on the imaginary (respectively, real) axis.*

Corollary 3. *If $\lambda=0$ is an eigenvalue, then its multiplicity is necessarily even.*

According to Theorem 4, the eigenvalues are of four types: real pairs $(a, -a)$, purely imaginary pairs $(ib, -ib)$, quadruplets $(\pm a\pm ib)$, and zero eigenvalues.

The next result replaces, for Hamiltonian systems, the theorem on reduction of the matrix of a linear differential equation to Jordan form.

Theorem 5 (J. Williamson [40]). *The Hamiltonian can be reduced by a symplectic change of variables to a sum of partial Hamiltonians (functions of distinct subsets of canonically conjugate variables). The matrix of the corresponding Hamiltonian system is thereby reduced to block form. Each partial Hamiltonian corresponds to either a real pair, a purely imaginary pair, a quadruplet of eigenvalues, or to a zero eigenvalue. The partial Hamiltonians are uniquely determined, up to a sign, by the Jordan blocks of the operator $\Im\Omega$.*

The list of partial Hamiltonians is given in [8, 92].

A Hamiltonian in general position has only simple eigenvalues. For a simple real pair $(a, -a)$, the partial Hamiltonian is $\mathscr{H}=-ap_1q_1$. For a simple purely imaginary pair $(ib, -ib)$, the Hamiltonian is $\mathscr{H}=\pm b(p_1^2+q_1^2)/2$ (the Hamiltonians with the + and − signs are not equivalent, i.e., cannot be transformed into one another by a symplectic change of coordinates). For a quadruplet $(\pm a\pm ib)$, the Hamiltonian is $\mathscr{H}=-a(p_1q_1+p_2q_2)+b(p_1q_2-p_2q_1)$. For a purely imaginary pair, one often uses *symplectic polar coordinates* ρ, φ: $p=\sqrt{2\rho}\cos\varphi$, $q=\sqrt{2\rho}\sin\varphi$, in which the Hamiltonian takes the form $\mathscr{H}=\pm b\rho$.

Corollary 1. *Suppose there exists a simple purely imaginary eigenvalue $\lambda=i\omega$. Then the system can perform a periodic oscillation of the form* $z=\mathrm{Re}(\xi\exp(i\omega(t+t_0)))$, *where ξ is an eigenvector corresponding to λ:* $(\Im\Omega-i\omega E_{2n})\xi=0$. *This motion is called a characteristic oscillation, and the number ω is called as characteristic frequency.*

Corollary 2. *If the eigenvalues are all distinct and purely imaginary, then the Hamiltonian can be reduced to the normal form*

$$H=\tfrac{1}{2}\omega_1(p_1^2+q_1^2)+\ldots+\tfrac{1}{2}\omega_n(p_n^2+q_n^2) \tag{4}$$

or, in symplectic polar coordinates,

$$H=\omega_1\rho_1+\ldots+\omega_n\rho_n.$$

Every motion in this system is a sum of characteristic oscillations.

Remark. If the Hamiltonian is of form (4), then the equilibrium is stable whether or not the Hamiltonian is positive definite (in contrast to linear natural Lagrangian systems, in which the equilibrium is stable only if the total energy is positive definite). △

One is often led to considering not an individual Hamiltonian, but rather a whole family depending on parameters. In such a family, for some exceptional values of the parameter singularities (multiple eigenvalues and, correspondingly, Jordan blocks of order greater than one in the matrix of the system) can arise which cannot be removed by a small change of the family. For every finite l, the nonremovable singularities arising in l-parameter families of quadratic Hamiltonians are indicated in [92]. Also calculated therein are the versal deformations of these singularities, i.e., the normal forms to which any family of quadratic Hamiltonians that depends smoothly on the parameters can be reduced in the neighborhood of the singular values of the parameters by means of a linear symplectic change of coordinates that depends smoothly on the parameters. In particular, in one-parameter families of Hamiltonians one can encounter, generally speaking, only the following three types of singularities: real pair of multiplicity two, $(\pm a)^2$, with two Jordan blocks of order 2; purely imaginary pair of multiplicity two, $(\pm ib)^2$, again with two Jordan blocks of order 2; and a zero eigenvalue of multiplicity two, $(0)^2$, with a single Jordan block of order 2. The versal deformations of these singularities are given below (δ_1, δ_2 are deformation parameters):

$$\begin{aligned}
&(\pm a)^2: && \mathscr{H}=-(a+\delta_2)(p_1q_1+p_2q_2)+p_1q_2+\delta_1 p_2q_1,\\
&(\pm ib)^2: && \mathscr{H}=\pm(p_1^2+p_2^2)/2+(b+\delta_2)(p_2q_1-p_1q_2)+\delta_1(q_1^2+q_2^2)/2,\\
&(0)^2: && \mathscr{H}=\pm p_1^2/2+\delta_1 q_1^2/2.
\end{aligned} \tag{5}$$

§3. Normal Forms of Hamiltonian Systems Near Equilibria

3.1. Reduction to Normal Form. Let the origin of coordinates be an equilibrium position of a Hamiltonian system with n degrees of freedom. Suppose that all the eigenvalues of the quadratic part of the Hamiltonian in the neighborhood of the equilibrium position are distinct and purely imaginary.

According to §1 and §2.2, the Hamiltonian can be expressed as

$$H=\tfrac{1}{2}\omega_1(p_1^2+q_1^2)+\ldots+\tfrac{1}{2}\omega_n(p_n^2+q_n^2)+H_3+H_4+\ldots, \tag{6}$$

where H_m is a form of degree m in the phase variables p, q (some of the frequencies ω_i may be negative).

Definition. We say that the characteristic frequencies $\omega_1, \ldots, \omega_n$ *satisfy a resonance relation of order* $l>0$ if there exist integers k_i such that $k_1\omega_1+\ldots+k_n\omega_n=0$ and $|k_1|+\ldots+|k_n|=l$.

Definition. A *Birkhoff normal form* of degree L for Hamiltonians is a polynomial of degree L in the symplectic phase variables P, Q, which is actually a polynomial of degree $[L/2]$ in the variables $\rho_i=(P_i^2+Q_i^2)/2$.

Example 2. For systems with two degrees of freedom the Birkhoff normal form of degree 4

$$H=\omega_1\rho_1+\omega_2\rho_2+\omega_{11}\rho_1^2+2\omega_{12}\rho_1\rho_2+\omega_{22}\rho_2^2. \tag{7}$$

The terms quadratic in ρ describe the dependence of the frequencies of oscillations on amplitudes. △

Theorem 6 (Birkhoff [10]). *Suppose that the characteristic frequencies do not satisfy any resonance relation of order less than or equal to L. Then there exists a symplectic change of coordinates $p, q \mapsto P, Q$ in a neighborhood of the equilibrium position such that in the new coordinates the Hamiltonian is reduced to a Birkhoff normal form of degree L up to terms of order L+1:*

$$H(p,q)=\mathscr{H}_L(\rho)+R, \qquad R=O(|P|+|Q|)^{L+1}. \tag{8}$$

If we discard the remainder R in (8), we obtain a completely integrable system for which the symplectic polar coordinates ρ_i, φ_i, given by

$$P_i=\sqrt{2\rho_i}\cos\varphi_i, \qquad Q_i=\sqrt{2\rho_i}\sin\varphi_i, \tag{9}$$

are action-angle variables. The trajectories of this system are conditionally-periodic windings on the tori $\rho=\text{const}$ with frequencies $\partial\mathscr{H}_L/\partial\rho$. Most of these tori exist in the original system too (this follows the KAM theory; Ch. 5, § 3.5.B).

Birkhoff's normalization reduces to Lindstedt's procedure for eliminating fast variables upon rescaling the deviation from the equilibrium position by a small quantity ε ($p=\varepsilon\hat{p}$, $q=\varepsilon\hat{q}$, $\hat{H}=H/\varepsilon^2$) and passing to symplectic polar coordinates.

The normalization procedure is described below in a more general situation (see the proof of Theorem 7). The normalizing transformations are constructed as polynomials of degree $L-1$ in the phase variables. The modification of the terms of degree l in the original Hamiltonian does not affect the terms of degree less than l in the normal form and those of degree less than $l-1$ in the normalizing transformation.

On considering the normalization for $L\to\infty$ we are led to the notion of formal normal form that was discussed in Ch. 4, §1.3.

For the cases where the characteristic frequencies satisfy resonance relations, the definition of the normal form must be modified. The same modification is advisable for near-resonant frequencies. Let K be a subgroup of the lattice of integers Z^n.

Definition. A *resonant normal form* of degree L for Hamiltonians under resonances from K (or a *K-resonant normal form* of degree L) is a polynomial of degree L in the symplectic variables P_i, Q_i, which, when expressed in the polar coordinates (9), depends on the phases only through combinations (k,φ) with $k\in K$.

Theorem 7 ([96]). *Suppose that the eigenvalues do not satisfy any resonance relation of order less than or equal to L, except, possibly, for relations $(k,\omega)=0$ with $k\in K$. Then there exists a symplectic change of coordinates $p,q\mapsto P,Q$ in the neighborhood of the equilibrium position, such that in the new coordinates the Hamiltonian is reduced to a K-resonant normal form of degree L up to terms of order $L+1$.*

◁ We subject the system with Hamiltonian (6) to a change of coordinates with generating function $Pq+S(P,q)$, $S=S_3+\ldots+S_L$. The new Hamiltonian has the form

$$\mathscr{H}=\tfrac{1}{2}\omega_1(P_1^2+Q_1^2)+\ldots+\tfrac{1}{2}\omega_n(P_n^2+Q_n^2)+\mathscr{H}_3+\mathscr{H}_4+\ldots.$$

Here S_l and $\mathscr{H}_l$ are forms of degree l in (P,q) and (P,Q), respectively. The new and old Hamiltonians are connected by the relation

$$H(P+\partial S/\partial q,\ q)=\mathscr{H}\left(P,q+\frac{\partial S}{\partial P}\right).$$

Equating here the forms of the same degree in P, q we get

$$\sum_{j=1}^{n}\omega_j\left(P_j\frac{\partial S_l}{\partial q_j}-q_j\frac{\partial S_l}{\partial P_j}\right)=\mathscr{H}_l-F_l,\qquad l=3,\ldots,L.$$

The form F_l is uniquely determined if S_ν, $\mathscr{H}_\nu$ with $\nu\leqq l-1$ are known. In symplectic polar coordinates ρ, φ, the last equation becomes

$$\omega\frac{\partial S_l}{\partial\varphi}=\mathscr{H}_l-F_l.$$

We choose

$$S_l=\sum\frac{if_k(\rho)}{(k,\omega)}\exp(i(k,\varphi)),\qquad k\notin K,$$

where f_k are the coefficients of the Fourier series of F_l. Then $\mathscr{H}_l$ has the desired normal form. In this way we can successively determine all the forms S_l and $\mathscr{H}_l$. To complete the proof, we return to Cartesian coordinates. ▷

Suppose the Hamiltonian is in resonant normal form. If r denotes the rank of the subgroup $K \subset Z^n$ that defines the possible resonances, then the system has $n-r$ independent first integrals in involution, which are linear combinations of the quantities $\rho_i = (P_i^2 + Q_i^2)/2$ (cf. Theorem 12, Ch. 5). In particular, if only one resonance is possible, then the system in the normal form is integrable.

The resonant normalization reduces to von Ziepel's procedure for eliminating fast variables (Ch. 5, §2.2) upon rescaling the deviation by a small quantity ε and passing to symplectic polar coordinates.

3.2. Phase Portraits of Systems with Two Degrees of Freedom in the Neighborhood of an Equilibrium Position under Resonance. Any system with two degrees of freedom whose Hamiltonian is a resonant normal form is integrable. One can reduce it to a system with one degree of freedom depending on the constant value of the first integral as a parameter, and then draw phase portraits. If the coefficients of the lowest-order terms in the normal form are in general position, then for the given resonance there are only finitely many types of phase portraits, and these types can be distinguished through the lowest-order terms in the normal form. The phase portraits are qualitatively distinct only for finitely many resonances. In the general-position case, enumerating the portraits provides complete information on the motion near resonance for systems in normal form. Correspondingly, one obtains considerable information on the motion in systems for which the lowest-order terms in the Hamiltonian can be reduced to the given normal form. We give below the phase portraits and their bifurcations. For lack of space we confine ourselves to the case where the frequencies ω_1 and ω_2 have opposite signs, which is the most interesting from the viewpoint of the theory of stability (if $\omega_1 \omega_2 > 0$, then the energy level $H = h \ll 1$ is a sphere, and the equilibrium is stable). The information necessary for building these phase portraits is contained in works of K.T. Alfriend, A.H.P. van der Burgh, J. Henrard, A.P. Markeev, J. Roels, J. Sanders, D.S. Schmidt, and others. Portraits for resonances of order $\geqq 4$ can be found in [176].

Let k_1 and k_2 be coprime positive coefficients of the resonance relation. There exist coprime integers l_1, l_2 such that $k_1 l_2 - k_2 l_1 = 1$. In the neighborhood of the equilibrium we pass to canonical polar coordinates (9) and then perform the change of variables $\rho_1, \rho_2, \varphi_1, \varphi_2 \mapsto G, I, \psi, \chi$ with generating function $S = (k_1\varphi_1 + k_2\varphi_2)G + (l_1\varphi_1 + l_2\varphi_2)I$:

$$\psi = k_1\varphi_1 + k_2\varphi_2, \qquad G = l_2\rho_1 - l_1\rho_2,$$

$$\chi = l_1\varphi_1 + l_2\varphi_2, \qquad I = -k_2\rho_1 + k_1\rho_2.$$

Since, by assumption, the Hamiltonian is in normal form, it does not depend on χ; accordingly, I is a first integral of the system. Next, we subject the system to isoenergetic reduction on the energy level $H = h$ (see [8]) and take the phase χ as the new time variable. We thus obtain the reduced system with one degree of freedom, whose Hamiltonian depends on the parameter h; it is its phase portrait that we must investigate. In the general-position case the portrait depends essentially on another parameter, namely, the resonance detuning $\delta = k_1\omega_1 + k_2\omega_2$.

The neighborhood of the origin of coordinates in the plane h, δ is partitioned into domains corresponding to different types of phase portraits. These partitions are shown, for various resonances, in Figs. 69a–74a; the corresponding bifurcations of the phase portraits as one moves around the origin of coordinates in clockwise direction are shown in Figs. 69b–74b. Portraits are labelled according to the labelling of domains in the parameter plane. Unlabelled portraits correspond to the curves which separate the domains, and they are shown only in Figs. 69–71.

The normal forms for which we give the bifurcations are

$$H_{k_1,k_2} = \omega_1\rho_1 + \omega_2\rho_2 + F(\rho_1, \rho_2) + B\rho_1^{k_1/2}\rho_2^{k_2/2}\cos(k_1\varphi_1 + k_2\varphi_2 + \psi_0).$$

Here F is a polynomial in ρ_1 and ρ_2, which starts with a quadratic form $F_2(\rho_1, \rho_2)$ (in the Hamiltonian $H_{2,1}$ the term F must be omitted), and B and ψ_0 are constants. The required general position conditions are $B \neq 0$, $A = F_2(k_1, k_2) \neq 0$, and, for the Hamiltonian $H_{3,1}$, $|A| \neq 3\sqrt{3}|B|$. The pictures correspond to the case $\omega_1 > 0$, $A > 0$, $B > 0$ (which does not result in a loss of generality). They are given for the following resonant vectors (k_1, k_2): (2, 1) in Fig. 69; (3, 1) in Fig. 70 if $A < 3\sqrt{3}B$, and in Fig. 71 if $A > 3\sqrt{3}B$; (4, 1) in Fig. 72; (2, 3) in Fig. 73; and (4, 3) in Fig. 74. For the resonant vectors $(n, 1)$ with $n \geqq 5$, the bifurcations are the same as for the case (4, 1); for $(n, 2)$ with $n \geqq 5$, the same as for (3, 2), except that the domain (5) in Fig. 73 does not appear; for $(n, 3)$ with $n \geqq 5$, the same as for (4, 3); and for (n, m) with $m \geqq 4$, $n \geqq 5$, the same as for (4, 3), except that the domain (2) in Fig. 74 does not appear. (Needless to say, the number of singular points of each

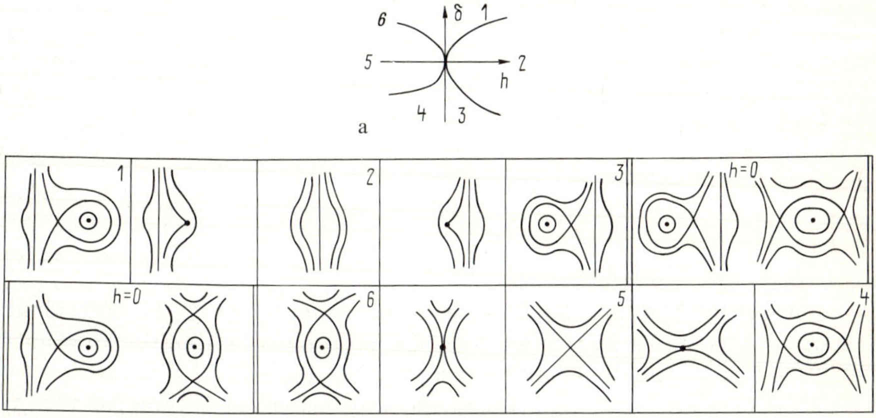

b

Fig. 69

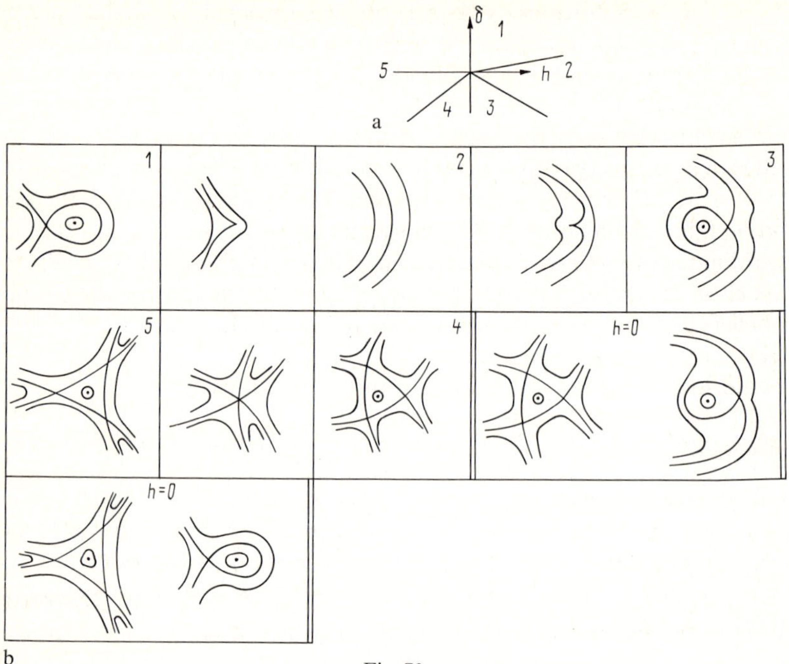

Fig. 70

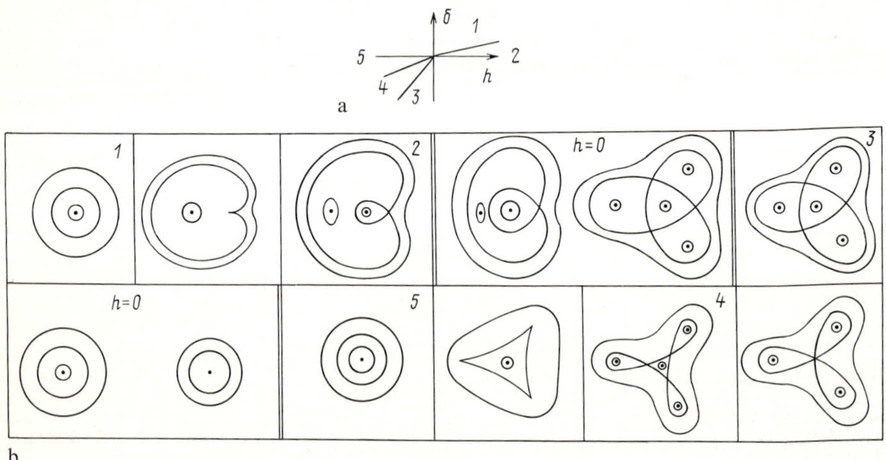

Fig. 71

type must be changed in accordance with the symmetry of the Hamiltonian). The axis $\delta = 0$ is not a bifurcation line. The positions of the bifurcation lines relative to it may actually be different from those shown in the pictures.

We make a few more remarks on the way in which the information is presented. In order that the phase portraits display no singularities, in the

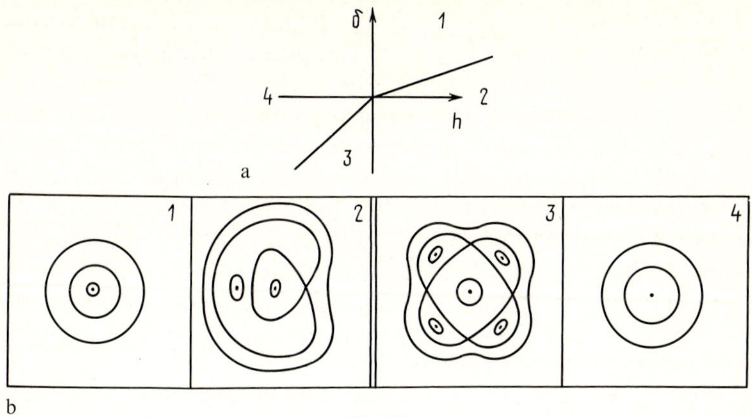

Fig. 72

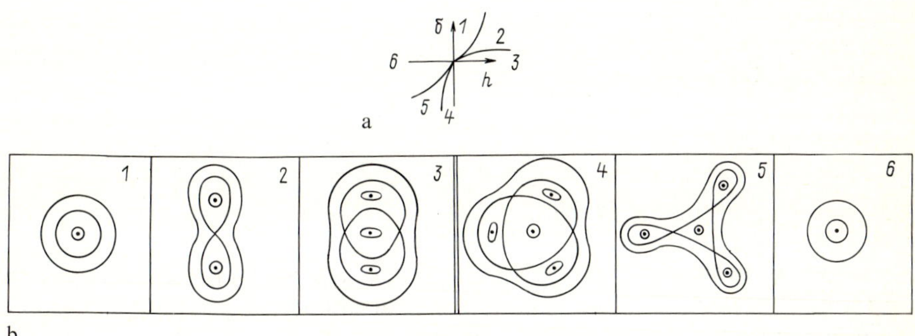

Fig. 73

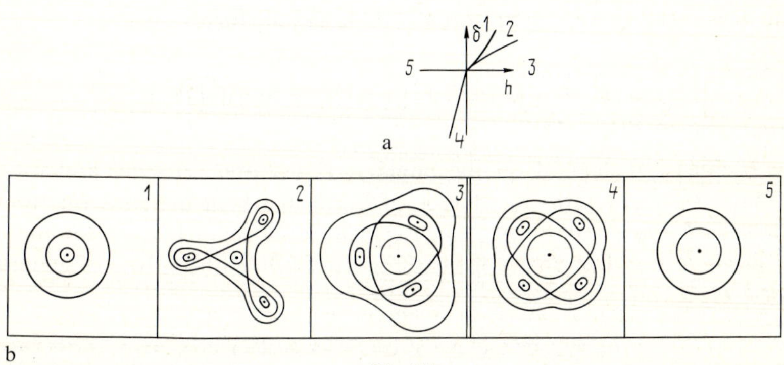

Fig. 74

case $h>0$ they are shown in polar coordinates $\rho_2, \psi/k_2$, and for $h<0$ in polar coordinates $\rho_1, \psi/k_1$. For $h=0$ the phase portrait in Figs. 69–71 is shown in both sets of coordinates. For $h>0$ ($h<0$) the portrait may be thought of as the section of the three-dimensional energy level by the plane $\varphi_1=0$ (respectively, $\varphi_2=0$). Corresponding to the equilibrium positions in the phase portrait are periodic solutions of the original system with two degrees of freedom[53], and corresponding to the closed curves are two-dimensional invariant tori. Moreover, to equilibrium positions that are obtained from one another by rotation through the angle $2\pi/k_2$ ($2\pi/k_1$) in the region $h>0$ (respectively, $h<0$) there corresponds exactly one periodic solution which crosses the section surface k_2 (respectively, k_1) times. A similar statement is valid for the two-dimensional tori.

To complete the analysis of resonances in systems with two degrees of freedom, it remains to consider those resonances which have an essential role already in the quadratic terms of the Hamiltonian, i.e., the case of multiple eigenvalues and the case of a zero eigenvalue.

For multiple eigenvalues the matrix of the linear Hamiltonian systems has, in the typical case, two Jordan blocks of second order (see §2.3). For near-multiple eigenvalues the quadratic part of the Hamiltonian reduces to the form $(\pm ib)^2$ given in the list (5). According to [179], in this case the terms of order $\leqq 4$ in the Hamiltonian can be reduced to the following form, which is also called a normal form:

$$H=a(p_1^2+p_2^2)/2+\omega(p_2q_1-p_1q_2)+\delta(q_1^2+q_2^2)/2+ \\ +(q_1^2+q_2^2)[D(q_1^2+q_2^2)+B(p_2q_1-p_1q_2)+C(p_1^2+p_2^2)], \qquad a=\pm 1. \tag{10}$$

The formal normal form is a series in $q_1^2+q_2^2$, $p_1^2+p_2^2$, and $p_2q_1-p_1q_2$. Following [109] and [181], we pass to polar coordinates r, χ in the plane q_1, q_2 and introduce the corresponding momenta P, I:

$$\begin{aligned} q_1&=r\cos\chi, & p_1&=P\cos\chi-I\sin\chi/r, \\ q_2&=r\sin\chi, & p_2&=P\sin\chi+I\cos\chi/r. \end{aligned} \tag{11}$$

In the new variables the Hamiltonian (10) takes the form

$$H=\frac{1}{2}a\left(P^2+\frac{I^2}{r^2}\right)+\omega I+r^2\left(\frac{\delta}{2}+Dr^2+BI+C\left(P^2+\frac{I^2}{r^2}\right)\right). \tag{12}$$

Since H does not depend on the angle χ, I is a first integral, and for P, r we obtain a system with one degree of freedom which depends on the two parameters I and δ. Its bifurcation diagram is given in Fig. 75 for the case $a=1$, $D>0$, and in Fig. 76 for the case $a=1$, $D<0$. Without loss of generality we took $I\geqq 0$.

[53] For $h=0$, to the equilibrium position at the center of the portrait there corresponds the equilibrium position of the original system.

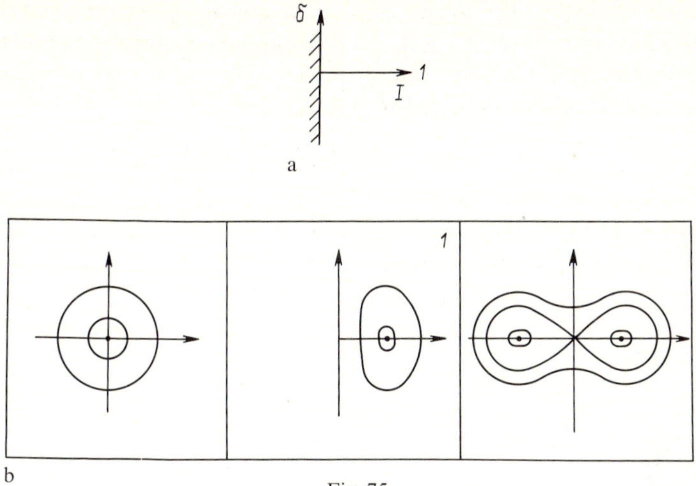

Fig. 75

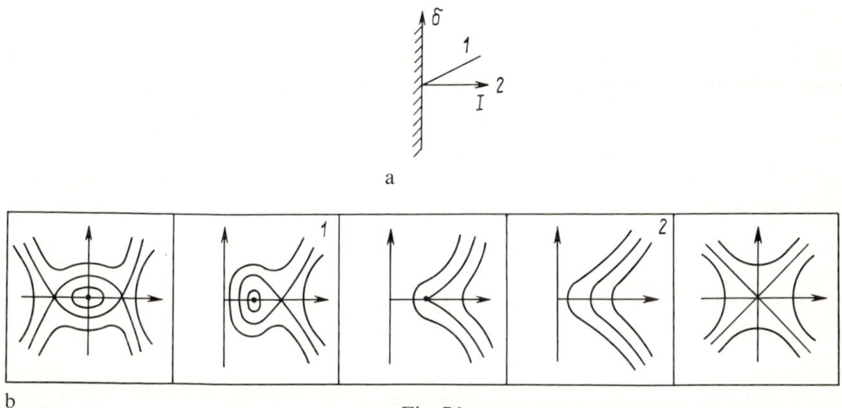

Fig. 76

The left and right extreme portraits in Figs. 75 and 76 correspond to $I=0$; to ensure that they will have no singularities, we must consider that r assumes values of both signs. The curves in the phase portrait which are symmetric with respect to the axis $r=0$ correspond to the same invariant surfaces in the phase space of the system with two degrees of freedom.

Finally, we examine the case of a zero eigenvalue (degenerate equilibrium). It arises even in systems with one degree of freedom, and it is precisely such a system that we will be concerned with here[54]. We assume that in the linearized system to the zero eigenvalue there corresponds a Jordan block of order two (§2.3). If the equilibrium position is nearly degenerate, then

[54] In the case of two degrees of freedom the order can be reduced to one using the first integral corresponding to the nonzero eigenvalue.

it cannot be shifted to the origin of coordinates by means of a coordinate transformation depending smoothly on the parameters of the problem. Consequently, the linear part in the Hamiltonian is preserved. The terms of order $\leqq 3$ in the Hamiltonian can be reduced to the form

$$H=\delta q+a p^2/2+b q^3, \quad a=\pm 1. \tag{13}$$

We shall assume that $a=1$ and $b>0$. The bifurcation of the phase portrait which occurs when passing from negative to positive values of δ is shown in Fig. 77. Two equilibrium positions merge and disappear.

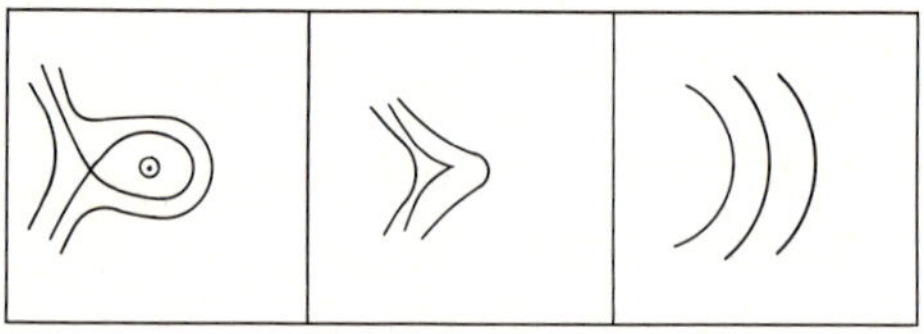

Fig. 77

The diagrams given above exhaust all bifurcations connected with resonances which occur in general-position one-parameter families of Hamiltonians with two degrees of freedom, and can be calculated from the normal form.

These diagrams are also useful in the case of a higher number of degrees of freedom. Indeed, suppose that in a system with n degrees of freedom a single resonance condition holds approximately between two frequencies. Then the normal form of this system has $n-2$ first integrals $\rho_i=\text{const}$, and reduces to a system with two degrees of freedom. This results in one of the normal forms considered above, the coefficients of which depend on the parameters $\rho_i \ll 1$.

The investigation of multiple resonances in systems with many degrees of freedom is presently in its early stage. In [43] the case of frequency ratios 1:2:1 is studied, and its periodic solutions and first integrals, arising for special values of the parameters, are found. In [44] it is shown that for the resonance 1:2:2 the normal form of degree three enjoys an additional symmetry and the corresponding system is completely integrable. In [88] it is shown that for the resonance 1:1:2 the normal form of degree three generates a nonintegrable system [55].

3.3. Stability of Equilibria in Systems with Two Degrees of Freedom under Resonance. The investigation of a normal form furnishes a considerable

[55] In these works it is assumed that, corresponding to the multiple eigenvalue, in the matrix of the linear system there are four Jordan blocks of order one, rather than two blocks of order two, i.e., an additional degeneracy is present [92]: to obtain such a case in a general-position system one needs four parameters.

amount of information on the motion in the original system, if in the latter the lowest-order terms can be reduced to this form. For example, if the normal form has a nondegenerate periodic solution, then the original system has close to it a periodic solution. This results from the implicit function theorem. Most invariant tori existing for the normal form exist also, in the general case, for the original system. This follows from the results of the KAM theory (one uses Theorem 14 of Ch. 5, §3). As is always the case with systems with two degrees of freedom, the presence of invariant tori allows us to draw conclusions on stability.

If among the characteristic frequencies of a system with two degrees of freedom there are no resonance relations of order $\leqq 4$, then the equilibrium is stable (under the additional assumption of isoenergetic nondegeneracy); we have already discussed this result in Ch. 5, §3.5.B. For the remaining finite number of resonance cases, the following result holds:

Theorem 8 ([81, 109, 131, 179–181]). *If the eigenvalues satisfy a resonance relation of order* $\leqq 4$ *and the general-position conditions of* §3.2 *are fulfilled, then the equilibrium of the original system is stable (unstable) if and only if the equilibrium of the normal form is stable (respectively, unstable).*

◁ Stability is proved using the KAM theory. Instability is proved either by comparing the rates of the drift away from the equilibrium in the original system and the normal form, or by constructing a Chetaev function. ▷

In the notations of §2.3 we have the following results:

Corollary 1 ([131]). *For the resonance* (2, 1) *the equilibrium is unstable if* $B \neq 0$ (Fig. 69).

Corollary 2 ([131]). *For the resonance* (3, 1) *the equilibrium is stable if* $|A| > 3\sqrt{3}|B| > 0$ (Fig. 71) *and unstable if* $0 < |A| < 3\sqrt{3}|B|$ (Fig. 70).

Corollary 3 ([109, 179, 181]). *If the linearized system has a pair of Jordan blocks of order* 2 *corresponding to a multiple nonnull characteristic frequency, then the equilibrium of the full system is stable for* $aD > 0$ (Fig. 75), *and unstable for* $aD < 0$ (Fig. 76).

Corollary 4 ([81, 180]). *If the linearized system has a Jordan block of order* 2 *corresponding to a null characteristic frequency, then the equilibrium of the full system is unstable for* $b \neq 0$ (Fig. 77).

For the case in which some of the general-position conditions formulated above are violated the stability is analysed in [81, 131, 179, 180].

Generally speaking, the separatrices arising in the phase portraits of the normal form split on passing to the exact system, as described in Ch. 5, §3.3.B.

§4. Normal Forms of Hamiltonian Systems Near Closed Trajectories

4.1. Reduction to the Equilibrium of a System with Periodic Coefficients. Suppose given a Hamiltonian system with $n+1$ degrees of freedom having a closed trajectory which is not an equilibrium position. Such trajectories are not isolated, occuring, as a rule, in one-parameter families. Let us reduce the problem of oscillations in the neighborhood of such a family to a convenient form.

Proposition 1. *In a neighborhood of the given closed trajectory one can choose new symplectic coordinates* φ mod 2π, J, *and* $z \in R^{2n}$, *such that on the trajectory considered* $J=0$, $z=0$, *while the coordinate* φ *varies on a full circuit by* 2π; *on the trajectory itself* $\dot{\varphi}=\text{const}$. *In the new coordinates, the Hamiltonian takes the form* $H=f(J)+\mathscr{H}(z, \varphi, J)$, $f_J' \neq 0$; *the expansion of* $\mathscr{H}$ *in powers of* z *and* J *begins with terms of order two.*

Next, we carry out isoenergetic reduction (see [8]), choosing on the energy level $H=h$ the phase φ as the new time (and denoting it now by t). The Hamiltonian of the problem takes the form $F=F(z, t, h)$. For $h=0$, the origin of coordinates is an equilibrium position of the system. We assume that it is nondegenerate (i.e., all the multipliers are different from one; the degenerate case is considered in §4.3). Then for small values of h the system also has a nondegenerate equilibrium position. By a coordinate change which is smooth in the parameter, one can shift this equilibrium to the origin of coordinates. The Hamiltonian takes the form

$$F=\tfrac{1}{2}(\Xi(t, h)z, z)+G(z, t, h), \tag{14}$$

where the expansion of G in z starts with terms of order three; F has period 2π in time t.

We now consider the linearized system.

Theorem 9 (see, for example [189]). *A linear,* 2π*-periodic in time Hamiltonian system can be reduced to autonomous form by a linear sympletic change of coordinates. If the system has no (has) real negative multipliers, then the coordinate change effecting the reduction can be chosen to be* 2π*-periodic (respectively,* 4π*-periodic) in time. If the system depends smoothly on a parameter, then the coordinate change is also chosen to be smooth in this parameter.*

Suppose that all the multipliers of the linearized system lie on the unit circle and are distinct. Then, according to Theorem 9 and §2.2, the Hamiltonian (14) can be reduced by a linear, 2π-periodic change of coordinates to the form

$$\Phi=\omega_1(p_1^2+q_1^2)/2+\ldots+\omega_n(p_n^2+q_n^2)/2+\Psi(p, q, t, h), \tag{15}$$

where the expansion of Ψ in the phase variables starts with terms of third order, and Ψ has period 2π in time t.

4.2. Reduction of Systems with Periodic Coefficients to Normal Form

Definition. For 2π-periodic systems, we say that the characteristic frequencies $\omega_1, \ldots, \omega_n$ *satisfy a resonance relation of order* $l>0$ if there exist integers $k_0, k_1, \ldots, k_n$, such that $k_1\omega_1+\ldots+k_n\omega_n+k_0=0$ and $|k_1|+\ldots+|k_n|=l$.

Theorem 10 (Birkhoff [10]). *Suppose that the characteristic frequencies ω_i of the 2π-periodic system (15) do not satisfy any resonance relation of order $\leqq L$. Then there is a symplectic change of coordinates, 2π-periodic in time, reducing the Hamiltonian to the same Birkhoff normal form of degree L as if the system were autonomous, with the only difference being that the remainder terms of order $\geqq L+1$ will depend 2π-periodically on time.*

The normalization procedure is analogous to that described in §3.1. If the system depends smoothly on a parameter, the normalizing transformation is also chosen to be smooth in the parameter.

For resonance cases one uses the resonant normal form. Let K be a subgroup of the integer lattice Z^{n+1}.

Definition. A *nonautonomous K-resonant* (i.e., relative to the resonances from K) *normal form* of degree L for Hamiltonians is a polynomial of degree L in the symplectic variables P_i, Q_i, which in the polar coordinates (9) depends on phases φ_i and time t only through combinations $k_1\varphi_1+\ldots+k_n\varphi_n+k_0 t$ with $(k_1, \ldots, k_n, k_0)\in K$.

Theorem 11. *Suppose that the characteristic frequencies do not satisfy any resonance relation of order $\leqq L$, except, possibly, resonances from K. Then there exists a symplectic change of coordinates, 2π-periodic in time, which reduces the Hamiltonian to a nonautonomous K-resonant normal form of degree L to within terms of degree $L+1$.*

If $r=\operatorname{rank} K$, then a system in a K-resonant normal form has $n-r$ indepedent first integrals in involution, which are linear combinations of the variables $\rho_i=(P_i^2+Q_i^2)/2$. In particular, if there is only one resonance relation, then the system is integrable.

4.3. Phase Portraits of Systems with Two Degrees of Freedom Near a Closed Trajectory under Resonance. The oscillations about a closed trajectory in a system with two degrees of freedom are described by a time-periodic system with one degree of freedom which depends on a parameter (§4.1). The system having the resonant normal form for this problem can be reduced to one with a single degree of freedom and its phase portraits can then be drawn. If the coefficients of the lowest-order terms in the normal form are in general position, then for each given resonance there are only finitely many types of phase portraits, and these types are distinguished by the lowest-order terms in the normal form. The phase portraits differ qualitatively only for a finite number of resonances. The list of these phase portraits and the description of the bifurcations that they undergo when the parameters of the system pass through an exact resonance are given in [68, 69], and are reproduced below.

In the variables ρ, $\psi=\varphi+k_0 t/k+\psi_0$, the normal forms H_{k,k_0} for the resonances (k, k_0) are

$$H_{3,k_0}=\delta\rho+B\rho^{3/2}\cos 3\psi,$$

and

$$H_{k,k_0}=\delta\rho+\rho^2 A(\rho)+B\rho^{k/2}\cos k\psi, \qquad k\geqq 4.$$

Here ρ and ψ are canonically conjugate variables, $\delta=\omega+k_0/k$ is the resonance detuning, A is a polynomial in ρ, and B, ψ_0 are constants. The required

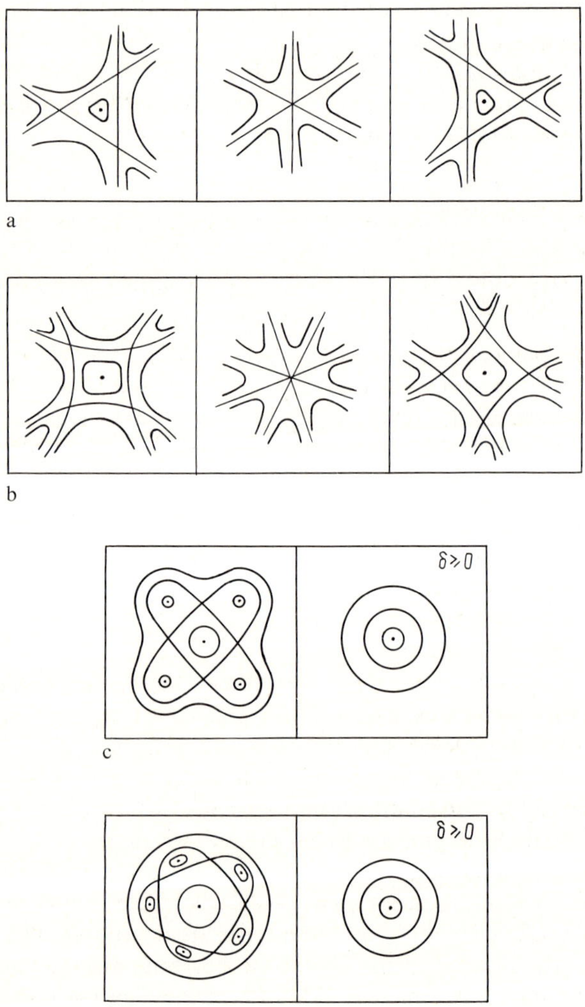

Fig. 78

general-position conditions are written: $B \neq 0$, $A(0) \neq 0$ for $k \geqq 4$, and $|A(0)| \neq |B|$ for $k=4$. All the coefficients depend also on a parameter h. We shall assume that $d\delta/dh \neq 0$, so that instead of h we can use δ. Under the indicated assumptions, a small change in B and in the coefficients of A does not lead to bifurcations; consequently, the dependence of A and B on the parameter can be ignored. We shall assume, with no loss of generality, that $B>0$ and $A(0)>0$.

The metamorphoses of the phase portrait as δ grows and passes through zero are shown in Fig. 78a for $k=3$, in Fig. 78b for $k=4$ and $A(0)<B$, in Fig. 78c for $k=4$ and $A(0)>B$, and in Fig. 78d for $k=5$.

For $k \geqq 6$ the metamorphose is the same as for $k=5$, except that around the origin of coordinates there are $2k$ singular points, and not 10. For $k \geqq 5$ these points lie at a distance of order $\sqrt{\delta}$ from the origin of coordinates. The "oscillation islands" surrounding the stable singular points have a width of order $\delta^{(k-2)/4}$. Hence, for $k \geqq 5$ these islands cover only a small portion of the neighborhood of the origin of coordinates considered; the remaining phase curves are close to circles.

There are two more resonance cases that appear already in the quadratic terms of the Hamiltonian, namely, that the multipliers of the closed trajectory are equal to -1 or 1.

If the multipliers are equal to -1 (the $(2, k_0)$ resonance), then the lowest-order terms of the Hamiltonian can be reduced, in the typical case, to the normal form

$$H=\delta q^2 + a p^2/2 + D q^4, \qquad a = \pm 1,$$

by means of a 4π-periodic change of coordinates. The metamorphose is shown in Fig. 79a for $a=1$, $D>0$, and in Fig. 79b for $a=1$, $D<0$.

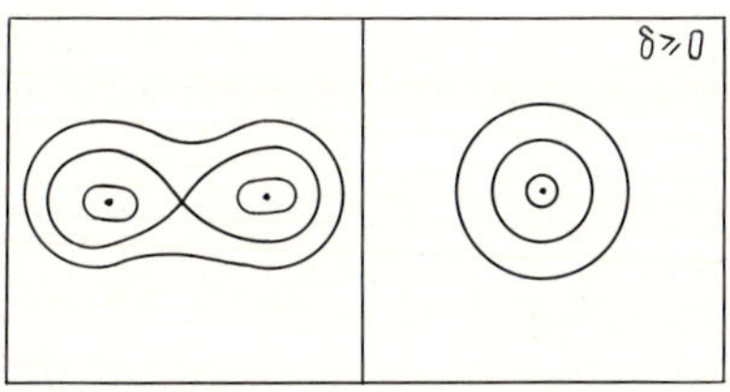

a

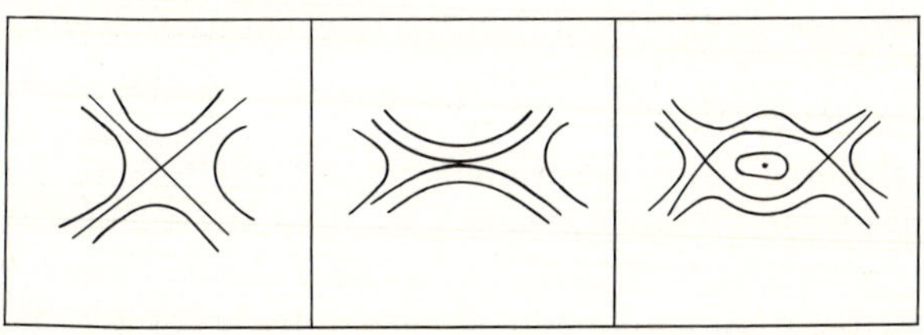

b

Fig. 79

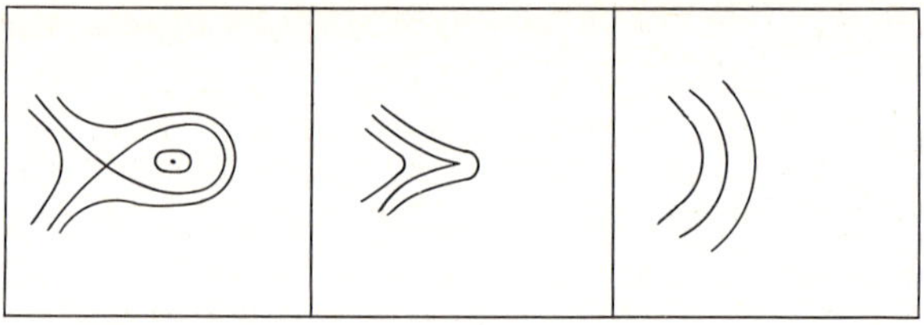

Fig. 80

If the multipliers are equal to 1 (the $(1, k_0)$ resonance), then the lowest-order term of the Hamiltonian can be reduced to the normal form

$$H=\delta q+ap^2/2+bq^3, \quad a=\pm 1.$$

The metamorphose is shown in Fig. 80 (under the assumption that $a=1$ and $b>0$).

The phase portraits drawn above reveal many properties of the original system in case that its lowest-order terms can be reduced to the corresponding normal form. For instance, corresponding to nondegenerate equilibrium positions on the phase portraits there are periodic trajectories of the original system which make k circuits around the given periodic trajectory. For a resonance of order three there is only one such trajectory; it is unstable and coalesces with the given trajectory at the moment of exact resonance ($\delta=0$). For a resonance of order $k\geqq 5$ there are two such trajectories, one stable and one unstable, which branch from the given trajectory on passing through resonance along the δ-axis from a fixed direction. For a resonance of order four the picture is, depending on the value of the parameters, the same as for $k=3$, or as for $k\geqq 5$. At passage through a resonance of order two (multipliers equal to -1), the given trajectory loses or acquires stability, and then a periodic trajectory making two circuits around it branches out. Finally, at a resonance of order one (multipliers equal to 1), the given trajectory vanishes, coalescing with another trajectory of the same period (or, if one moves in opposite direction with respect to the parameter, two periodic trajectories are created).

To most invariant curves on the phase portraits there correspond (according to the KAM theory) two-dimensional invariant tori of the original system carrying conditionally-periodic motions.

Under the general-position conditions formulated above, the stability or instability of the given closed trajectory is determined by the normal form (cf. Theorem 8). For $k=3$, the trajectory is unstable if $B\neq 0$, for $k=4$ it is stable if $|A(0)|>|B|>0$, and unstable if $0<|A(0)|<B$, and for $k\geqq 5$ it is stable if $A(0)B\neq 0$. If the multipliers are equal to -1, the trajectory is stable if $aD>0$, and unstable if $aD<0$. Finally, if the multipliers are equal to 1, it is unstable if $ab\neq 0$.

On passing from the normal form to the exact system the separatrices present in the phase portrait undergo, generally speaking, a splitting, in a manner analogous to that described in Ch. 5, §3.3.B.

§5. Stability of Equilibria in Conservative Fields

Theorem 12 (Lagrange-Dirichlet). *If at the given equilibrium position the potential has a strict local minimum, then the corresponding equilibrium state is stable.*

◁ One can take as a Lyapunov function the total mechanical energy. ▷

Example (Painlevé-Wintner). Consider the infinitely differentiable potential

$$U(q)=\exp(-q^{-2})\cos q^{-1}, \quad q\neq 0, \; U(0)=0.$$

The equilibrium position $q=0$ is stable although the point $q=0$ is clearly not a local minimum of function U (Fig. 81). △

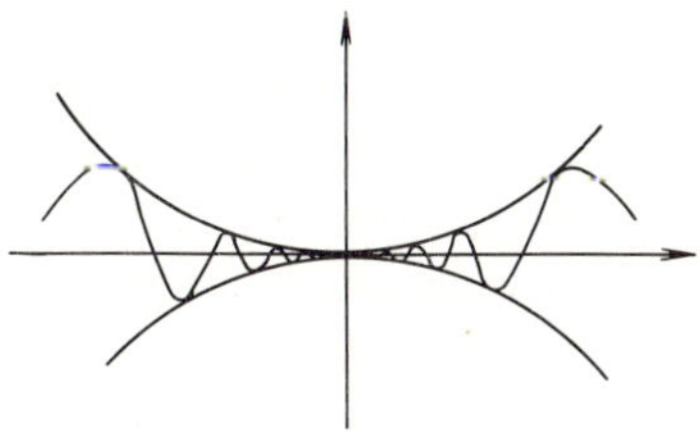

Fig. 81

In 1892 Lyapunov posed the problem of proving the converse of Lagrange's theorem for the case in which the coefficients of the quadratic form $T=\sum a_{ij}(q)\dot{q}_i\dot{q}_j$ and the potential U are analytic in the neighborhood of the equilibrium position. This problem is presently settled only in particular cases.

Theorem 13. *Suppose that the equilibrium position $q=0$ is not a strict local minimum of the analytic potential U. The equilibrium state $(q, \dot{q})=(0, 0)$ is unstable if one of the following supplementary conditions is satisfied:*

1) *the number of degrees of freedom $n\leqq 2$;*

2) *U is a quasi-homogeneous or semi-quasihomogeneous function;*

3) *the first nontrivial homogeneous term in the Maclaurin expansion of the function U does not have a local minimum at zero.*

The case $n=1$ is trivial. The instability of the equilibrium in the case $n=2$ was established by V.P. Palamodov [162]. His proof rests on the following assertion, which goes back to N.G. Chetaev [78].

Lemma 1 ([162]). *Suppose that the point $q=0$ is not a local minimum of the function U. Suppose also that in the domain $\Sigma_\varepsilon=\{q\colon U(q)<0, |q|<\varepsilon\}$ there is a smooth vector field v such that*

1) *$\langle v, U'\rangle\leqq 0$ in Σ_ε,*

2) *$\langle v'\xi, \xi\rangle\geqq c\langle\xi, \xi\rangle$ for all $\xi\in R^n$ and $q\in\Sigma_\varepsilon$ (where $c>0$),*

and

3) $|v(q)|=O(|q|)$ *for* $|q|\to 0$.

Then there is an $\varepsilon_0>0$ *such that if* $q(\cdot)$ *is any motion of the given mechanical system with negative energy and* $q(0)\in\Sigma_{\varepsilon_0}$, *then* $|q(t)|>\varepsilon_0$ *for some* $t>0$.

Remark. Let $q(\cdot)$ be a motion with zero total energy. If the equilibrium position $q=0$ is isolated then (under the assumptions of Lemma 1) the point $q(t)$ either leaves some domain $|q|\leqq\varepsilon_0$ in finite time, or tends to zero as $t\to\infty$. △

The main difficulty in proving Palamodov's theorem is precisely the construction of the required vector field v.

If U is a quasi-homogeneous function with exponents $\alpha_1, \ldots, \alpha_n$, then one can take for v the vector field Ax, where $A=\operatorname{diag}(\alpha_1, \ldots, \alpha_n)$. In the semi-quasihomogeneous case v is a certain perturbation of the field Ax (see [112] for details).

Remark. In applications the potential U most often depends also on a parameter. A function of general form depending on at most 6 parameters can be reduced in the neighborhood of a critical point to quasi-homogeneous form by means of a suitable change of coordinates (see, for example, [52]). Consequently, the converse of Lagrange's theorem is valid for such potentials. For all practical purposes this is more than enough, but the general case does not, of course, reduce to this one. △

The proof of Theorem 12 in case 3) is based on a different idea: if the equations of motion have a solution $q(t)$ which tends asymptotically to the point $q=0$ as $t\to+\infty$, then the equilibrium state $(q, \dot{q})=(0, 0)$ is unstable. Here we have to distinguish between three cases, namely, that the expansion of the potential starts with quadratic terms, with a form of odd degree and, finally, with an even form of degree greater than two. The first case has already been treated by Lyapunov: the equation of motion has the asymptotic solution $w(t)=\sum x_k e^{-k\lambda t}$, $k\geqq 1$, $x_k\in R^n$, $\lambda>0$. If $U=U_{2m+1}+\ldots$ $(m\geqq 1)$ and $U_{2m+1}\not\equiv 0$ (the second case), then the asymptotic solution can be represented as a series

$$\sum x_k/t^{k\mu}, \quad k\geqq 1, \ \mu=2/(2m-1), \tag{16}$$

which converges for sufficiently large values of t [114]. Finally, if $U=U_{2m}+\ldots$ $(m\geqq 2)$ and $U_{2m}\not\equiv 0$, series (16) is replaced by:

$$\sum \frac{x_{ij}(\ln t)^j}{t^{i/(m-1)}}, \quad x_{ij}\in R^n, \ i, j=0, 1, \ldots; \ j\leqq i/(m-1)$$

(see [118]).

From assertion 3) of Theorem 13 one can derive a number of corollaries:

a) Let $f\colon R^n\to R$ and $df(0)=0$. Then the equilibrium state $(q, \dot{q})=0$ is unstable for $U=f$ or for $U=-f$.

b) If the potential is analytic and has a local maximum at the equilibrium position, then the equilibrium state is unstable.

c) The equilibria of a mechanical system in a conservative force field with a harmonic potential U (i.e., U satisfies the Laplace equation $\Delta U=0$) are unstable. A particular case is the following theorem of S. Irnshaw: the equilibrium of a system of electrical charges in a stationary electric field is always unstable. Before the works [114, 118] appeared, Irnshaw's theorem was proved only under the assumption that the characteristic numbers of the first approximation were different from zero.

Suppose the given mechanical system is acted upon by a supplementary nonconservative force $F(q, \dot{q})$; the motion is then described by Lagrange's equation

$$\frac{d}{dt}\left(\frac{\partial L}{\partial \dot{q}}\right)-\frac{\partial L}{\partial q}=F, \qquad L=T-U. \tag{17}$$

Definition. We call F a *totally dissipative force of viscous friction* if $F(q, 0)=0$ and $(T+U)^{\cdot}=F\dot{q}<0$ for $\dot{q}\neq 0$.

The equilibrium positions will coincide with the critical points of the potential U even after introducing forces of viscous friction. Also, in this case those equilibrium states which are stable by virtue of Lagrange's theorem, remain stable when dissipation of energy occurs. Moreover, if the potential is an analytic function, then these equilibrium states become asymptotically stable.

Theorem 14 (see [113]). *Suppose that the point $q=0$ is not a local minimum of the function U, $U(0)=0$. The equilibrium state $(q, \dot{q})=(0, 0)$ of system* (17) *is unstable whenever one of the following conditions is satisfied:*

a) *U is analytic in the neighborhood of the point $q=0$;*

b) *U is smooth and for some $\varepsilon>0$ has no critical points in the domain $\Sigma_\varepsilon=\{q: U(q)<0, |q|<\varepsilon\}$.*

In the analytic case, condition b) is clearly fulfilled.

◁ Consider a motion $q(\cdot)$ with negative total energy and initial condition $q(0)\in\Sigma_\varepsilon$. We claim that the point $q(t)$ leaves Σ_ε within a finite time. In fact, along such a motion $\dot{q}(t)\neq 0$. Consequently, the total energy $E=T+U$ decreases. If $q(t)\in\Sigma_\varepsilon$ for all $t>0$ and $E(t)$ tends to a finite limit as $t\to+\infty$, then $\dot{q}(t)\to 0$. But for small values of the velocity the friction forces are small compared with the conservative forces, which impart a sufficiently large velocity to the system. ▷

Comments on the Bibliography

The fundamental principles of mechanics are discussed in sufficiently complete and detailed manner in books [1, 5, 8, 39]. In [24] one can become acquainted with the genesis of the fundamental concepts of mechanics. The book [17] contains an original construction of dynamics in which the notion of accelerating force does not appear: the bending of trajectories is caused only by the constraints imposed on the system. The collection of papers [38] gives a good picture of the development of variational methods of classical mechanics up to 1950. In [13] a systematic approach to Hamiltonian mechanics is developed, based on integral invariants. The work [16] contains a theory of Hamiltonian systems with constraints.

The works [33] and [10] have exerted a crucial influence upon the contemporary evolution of the theory of differential equations and classical mechanics. Therein new concepts and methods were introduced which have now become classical. Among the works devoted to qualitative aspects of the theory of dynamical systems we mention [2, 4, 9, 20, 23, 28, 32, 37].

The mathematical aspects of celestial mechanics are discussed in the books [15, 33, 36, 42]. In particular, [36] and [42] treat the problems of celestial mechanics as problems of the qualitative and analytic theory of differential equations; [15] contains a thorough introduction to perturbation theory. Paper [3] surveys results of the qualitative analysis of final motions in the three-body problem (see also [28]).

Problems pertaining to the reduction of order of Hamiltonian systems are discussed in detail in [13, 33] (see also [132]). Concerning the reduction of order in the sense of Routh for Lagrangian systems we refer the reader to [104, 186]. Paper [37] contains a detailed investigation of the energy-momentum mapping.

Various questions in the theory of integrable systems are discussed in [8, 14, 18, 21, 27, 29, 45, 46, 87]. In [14, 18, 27] the analytical aspects are stressed while [29, 45, 46, 87, 148] discuss the contemporary algebro-geometric methods for integrating Hamiltonian systems.

Perturbation theory for systems of differential equations of general form is treated in [9, 11, 12, 25, 26] and for Hamiltonian systems in [15, 31, 33]. The KAM theory was initiated by Kolmogorov's work [22]. Papers [6] and [7] contain the first detailed proofs of the theorems on conservation of invariant tori of Hamiltonian systems. The theory of perturbation of conditionally-periodic solutions of non-Hamiltonian systems of differential equations is constructed in [26].

Paper [21] is an exposition of various methods for proving nonintegrability of Hamiltonian systems. Papers [34] and [35] contain detailed proofs of the nonintegrability of Hamilton's equations in the neighborhood of an equilibrium position. Problems concerning the qualitative analysis of the behavior of trajectories in nonintegrable dynamical systems are considered in [2, 4, 28, 32].

A survey of results in the theory of stability of equilibrium positions and stationary motions of mechanical systems is contained in [19]. Elements of the theory of oscillations are given in [8, 9, 39]. The works [40, 41] contain a full description of normal forms for linear Hamiltonian systems.

For the convenience of the reader, references to reviews in *Zentralblatt für Mathematik* (Zbl.) and *Jahrbuch über die Fortschritte der Mathematik* (Jrb.) have, as far as possible, been included in this bibliography.

Recommended Reading

[1] Abraham, R.; Marsden, J.E.: Foundations of mechanics. 2nd ed. Reading, Mass.: The Benjamin/Cummings Publishing Company, Inc. m-XVI, XXII, 806 p. (1978). Zbl. 397.70001

[2] Alekseev, V.M.: Quasirandom dynamical systems. I, II, III Mat. Sb., Nov. Ser. *76*, No. 1, 72–134 (1968) (Russian); *77*, No. 4, 545–601 (1968) (Russian); *78*, No. 1, 3–50 (1968) (Russian); English transl.: Math. USSR, Sb. *5*, No. 1, 73–128 (1968); *6*, No. 4, 505–560 (1968); *7*, No. 1, 1–43 (1969). Zbl. 198, 569; Zbl. 198, 570; Zbl. 198, 570

[3] Alekseev, V.M.: Final motions in the three-body problem and symbolic dynamics Usp. Mat. Nauk *36*, No. 4, 161–176 (1981) (Russian); English transl.: Russ. Math. Surv. *36*, No. 4, 181–200 (1981). Zbl. 503.70006

[4] Anosov, D.V.: Geodesic flows on closed Riemannian manifolds of negative curvature Tr. Mat. Inst. Steklova *90*, 210 p. (1967) (Russian); English transl.: Proc. Steklov Inst. Math. *90* (1967). Zbl. 163, 436

[5] Appell, P.: Traité de mécanique rationnelle. Tomes 1–2. 4e éd. Paris: Gauthier-Villars (1919/1924)

[6] Arnol'd, V.I.: Proof of A.N. Kolmogorov's theorem on the preservation of quasi-periodic motions under small perturbations of the Hamiltonian. Usp. Mat. Nauk *18*, No. 5, 13–40 (1963) (Russian); English transl.: Russ. Math. Surv. *18*, No. 5, 9–36 (1963). Zbl. 129, 166

[7] Arnol'd, V.I.: Small denominators and problems of stability of motion in classical and celestial mechanics. Usp. Mat. Nauk *18*, No. 6, 91–192 (1963) (Russian); English transl.: Russ. Math. Surv. *18*, No. 6, 85–192 (1963). Zbl. 135, 427

[8] Arnol'd, V.I.: Mathematical methods of classical mechanics. Moskva: Nauka. 431 p. (1974). (Russian); English transl.: New York-Heidelberg-Berlin: Springer-Verlag. X, 462 p. (1978). Zbl. 386.70001

[9] Arnol'd, V.I.: Geometrical methods in the theory of ordinary differential equations. Moskva: Nauka. (1978) (Russian); English transl.: New York-Heidelberg-Berlin: Springer-Verlag. XI, 334 p. (1983). Zbl. 507.34003

[10] Birkhoff, G.D.: Dynamical systems. Am. Math. Soc. Colloq. Publ. IX. New York: American Mathematical Society. VIII, 295 p. (1927).

[11] Bogolyubov, N.N.: On some statistical methods in mathematical physics. L'vov: Akad. Nauk Ukr. SSR. 139 p. (1945). (Russian)

[12] Bogolyubov, N.N.; Mitropol'skij, Yu.A.: Asymptotic methods in the theory of nonlinear oscillations. 2nd ed. Moskva: Nauka. 408 p. (1958). (Russian); English transl.: Delhi: Hindustan Publ. Corp.; New York: Gordon and Breach Science Publ. V, 537 p. (1961). Zbl. 83, 81

[13] Cartan, E.: Leçons sur les invariants intégraux. Paris: Hermann. X, 210 p. (1922). Jrb. 48, 538

[14] Chaplygin, S.A.: Investigations in the dynamics of nonholonomic systems. Moskva-Leningrad. (1949). (Russian)

[15] Charlier, C.L.: Die Mechanik des Himmels. Bd. I, II. 2. Aufl. Berlin: Walter de Gruyter. VIII, 488 p.; VIII, 478 p. (1927). Jrb. 53, 892

[16] Dirac, P.A.M.: On generalized Hamiltonian dynamics. Can. J. Math. *2*, No. 2, 129–148 (1950). Zbl. 36, 141

[17] Hertz, H.: Die Prinzipien der Mechanik in neuem Zusammenhange dargestellt. Ges. Werke, Bd. 3. Leipzig: Barth. (1910). English transl.: New York: Dover Publications, Inc. 274 p. (1956). Zbl. 74, 388

[18] Jacobi, C.G.J.: Vorlesungen über Dynamik. Berlin: G. Reimer Verlag (1884)

[19] Karapetyan, A.V.; Rumyantsev, V.V.: Stability of conservative and dissipative systems. Itogi Nauki Tekh., Ser. Obshch. Mekh. *6*, 132 p. (1983). (Russian). Zbl. 596.70024

[20] Kozlov, V.V.: Methods of qualitative analysis in the dynamics of a rigid body. Moskva: Izdatel'stvo Moskovskogo Universiteta. 232 p. (1980). (Russian). Zbl. 557.70009

[21] Kozlov, V.V.: Integrability and non-integrability in Hamiltonian mechanics. Usp. Mat. Nauk *38*, No. 1, 3–67 (1983) (Russian); English transl.: Russ. Math. Surv. *38*, No. 1, 1–76 (1983). Zbl. 525.70023

[22] Kolmogorov, A.N.: On conservation of conditionally periodic motions under small perturbations of the Hamiltonian. Dokl. Akad. Nauk SSSR *98*, No. 4, 527–530 (1954). (Russian). Zbl. 56, 315

[23] Kolmogorov, A.N.: General theory of dynamical systems and classical mechanics. Proc. Int. Congr. Math, 1954, Amsterdam 1, 315–333 (1957). (Russian); English transl.: Appendix in [1]. Zbl. 95, 171

[24] Lagrange, J.L.: Mécanique analytique. Œuvres de Lagrange, Vols. 11–12 Paris: Gauthier-Villars (1888–1889)

[25] Moiseev, N.N.: Asymptotic methods of nonlinear mechanics. 2nd ed. Moskva: Nauka. 400 p. (1981). (Russian). Zbl. 527.70024

[26] Moser, J.: Convergent series expansions for quasi-periodic motions. Math. Ann. *169*, 136–176 (1967). Zbl. 149, 299

[27] Moser, J.: Lectures on Hamiltonian systems. Mem. Am. Math. Soc. 81, 60 p. (1968). Zbl. 172, 114

[28] Moser, J.: Stable and random motions in dynamical systems. Ann. Math. Stud. 77, VIII, 199 p. (1973). Zbl. 271.70009

[29] Moser, J.: Various aspects of integrable Hamiltonian systems. Dynamical systems, C.I.M.E. Lect., Bressanone 1978. Prog. Math. 8, 233–290 (1980). Zbl. 468.58011

[30] Nejmark, Yu.I.; Fufaev, N.A.: Dynamics of nonholonomic systems. Moskva: Nauka. 520 p. (1967). (Russian); English transl.: Transl. Math. Monogr., Am. Math. Soc. 33. IX, 518 p. (1972). Zbl. 171, 455

[31] Nekhoroshev, N.N.: An exponential estimate of the time of stability of nearly-integrable Hamiltonian systems. Usp. Mat. Nauk *32*, No. 6, 5–66 (1977) (Russian); English transl.: Russ. Math. Surv. *32*, No. 6, 1–65 (1977). Zbl. 383.70023

[32] Nitecki, Z.: Differentiable dynamics. An introduction to the orbit structure of diffeomorphisms. Cambridge, Mass.-London: The M.I.T. Press. XV, 282 p. (1971). Zbl. 246.58012

[33] Poincaré, H.: Les méthodes nouvelles de la mécanique céleste. Vols. 1–3. Paris: Gauthier-Villars. (1892/1893/1899); New York: Dover Publications, Inc. Vol. I: 382 p.; Vol. II: 479 p.; Vol. III: 414 p. (1957). Zbl. 79, 238

[34] Siegel, C.L.: On the integrals of canonical systems, Ann. Math., II. Ser. *42*, No. 3, 806–822 (1941). Zbl. 25, 265

[35] Siegel, C.L.: Über die Existenz einer Normalform analytischer Hamiltonscher Differentialgleichungen in der Nähe einer Gleichgewichtslösung. Math. Ann. *128*, 144–170 (1954). Zbl. 57, 320

[36] Siegel, C.L.; Moser, J.: Lectures on celestial mechanics. Berlin-Heidelberg-New York: Springer-Verlag. XII, 290 p. (1971). Zbl. 312.70017

[37] Smale, S.: Topology and mechanics. I, II. Invent. Math. *10*, No. 6, 305–331 and *11*, No. 1, 45–64 (1970). Zbl. 202, 232; Zbl. 203, 261

[38] Variational principles of mechanics. Collection of papers. Moskva: Fizmatgiz. 932 p. (1959). (Russian). Zbl. 87, 170

[39] Whittaker, E.T.: A treatise on the analytical dynamics of particles and rigid bodies. 4th ed. Cambridge: Cambridge University Press XIV, 456 p. (1960). Zbl. 91, 164

[40] Williamson, J.: On the algebraic problem concerning the normal forms of linear dynamical systems. Am. J. Math. *58*, No. 1, 141–163 (1936). Zbl. 13, 284

[41] Williamson, J.: The exponential representation of canonical matrices, Am. J. Math. *61*, No. 4, 897–911 (1939). Zbl. 22, 100

[42] Wintner, A.: The analytical foundations of celestial mechanics. Princeton: Princeton University Press. XII, 448 p. (1941). Zbl. 26, 23

Bibliography

[43] Aa, E. van der; Sanders, J.A.: On the 1:2:1-resonance, its periodic orbits and integrals. Asymptotic analysis, Lect. Notes Math. 711, 187–208 (1979). Zbl. 407.58027

[44] Aa, E. van der; Verhulst, F.: Asymptotic integrability and periodic solutions of a Hamiltonian system in 1:2:2 resonance. SIAM J. Math. Anal. *15*, 890–911 (1984). Zbl. 554.70017

[45] Adler, M.; van Moerbeke, P.: Completely integrable systems, Euclidean Lie algebras, and curves. Adv. Math. *38*, No. 3, 267–317 (1980). Zbl. 455.58017

[46] Adler, M.; van Moerbeke, P.: The algebraic integrability of the geodesic flow on $SO(4)$. Invent. Math. *67*, 297–331 (1982). Zbl. 539.58012

[47] Akulenko, L.D.; Leshchenko, D.D.; Chernous'ko, F.L.: Rapid motion of a rigid body around a fixed point in a resisting medium. Izv. Akad. Nauk SSSR, Mekh. Tverd. Tela, No. 3, 5–13 (1982). (Russian)

[48] Andoyer, M.H.: Cours de mécanique céleste, Tome I. Paris: Gauthier-Villars. 438 p. (1923). Jrb. 49, 665

[49] Anosov, D.V.: Averaging in systems of ordinary differential equations with rapidly oscillating solutions. Izv. Akad. Nauk SSSR, Ser. Mat. *24*, No. 5, 721–742 (1960). (Russian). Zbl. 102, 80

[50] Arnol'd, V.I.: Instability of dynamical systems with several degrees of freedom. Dokl. Akad. Nauk SSSR *156*, No. 1, 9–12 (1964) (Russian); English transl.: Sov. Math., Dokl. *5*, 581–585 (1964). Zbl. 135, 426

[51] Arnol'd, V.I.: Conditions for the applicability and estimate of the error of an averaging method for systems which pass through states of resonance in the course of their evolution. Dokl. Akad. Nauk SSSR *161*, No. 1, 9–12 (1965) (Russian); English transl.: Sov. Math., Dokl. *6*, 331–334 (1965). Zbl. 143, 120

[52] Arnol'd, V.I., Gusejn-Zade, S.M.; Varchenko, A.N.: Singularities of differentiable maps, Vol. I.: The classification of critical points, caustics, and wave fronts. Moskva: Nauka. 304 p. (1982). (Russian); English transl.: Monogr. Math., Basel, 82. Boston-Basel-Stuttgart: Birkhäuser. X, 382 p. (1985). Zbl. 554.58001

[53] Aubry, S.: The twist map, the extended Frenkel-Kontorova model, and the devil's staircase. Order in chaos, Proc. Int. Conf., Los Alamos/N.M. 1982, Physica *7D*, 240–258 (1983). Zbl. 559.58013

[54] Bakaj, A.S.: Resonance phenomena in nonlinear systems. Differ. Uravn. *2*, No. 4, 479–491 (1966) (Russian); English transl.: Differ. Equations *2*, No. 4, 242–249 (1969). Zbl. 178, 437

[55] Bakaj, A.S.; Stepanovskij, Yu.P.: Adiabatic invariants. Kiev: Naukova Dumka. 284 p. (1981). (Russian). Zbl. 486.70023

[56] Beletskij, V.V.: Essays on the motion of cosmic bodies, 2nd ed. Moskva: Nauka (1977). (Russian)

[57] Birkhoff, G.: Hydrodynamics. 2nd ed. Princeton: Princeton University Press. XI, 184 p. (1960). Zbl. 95, 203

[58] Bliss, G.A.: Lectures on the calculus of variations. Chicago: University of Chicago Press. IX, 296 p. (1946)

[59] Bogolyubov, N.N.: Perturbation theory in nonlinear mechanics. Sb. Stroit. Mekh. Akad. Nauk Ukr. SSR *14*, 9–34 (1950). (Russian)

[60] Bogolyubov, N.N.; Zubarev, D.N.: The method of asymptotic approximation for systems with rotating phase and its application to the motion of charged particles in a magnetic field. Ukr. Mat. Zh. *7*, No. 1, 5–17 (1955). (Russian). Zbl. 65, 427

[61] Bogolyubov, N.N.; Mitropolskij, Yu.A.; Samojlenko, A.M.: Methods of accelerated convergence in nonlinear mechanics. Kiev: Naukova Dumka. 248 p. (1969). (Russian); English transl.: Delhi, India: Hindustan Publishing Corporation; Berlin-Heidelberg-New York: Springer-Verlag. VIII, 291 p. (1976). Zbl. 187, 342

[62] Bohm, D.: Quantum theory. London: Constable & Co. Ltd. 646 p. (1951). Zbl. 48, 218

[63] Bolotin, S.V.: Libration motions of natural dynamical systems. Vestn. Mosk. Univ., Ser. I 1978, No. 6, 72–77 (1978). (Russian). Zbl. 403.34053

[64] Bolotin, S.V.: Nonintegrability of the n-center problem for $n>2$. Vestn. Mosk. Univ., Ser. I 1984, No. 3, 65–68 (1984). (Russian). Zbl. 551.70008

[65] Bolotin, S.V.: Comments on the Routh method and the Hertz hypothesis. Vestn. Mosk. Univ., Ser. I 1986, No. 5, 51–53 (1986). (Russian)

[66] Bolotin, S.V.; Kozlov, V.V.: Libration in systems with many degrees of freedom. Prikl. Mat. Mekh. *42*, No. 2, 245–250 (1978) (Russian); English transl.: J. Appl. Math. Mech. *42*, No. 2, 256–261 (1980). Zbl. 497.70033

[67] Brendelev, V.N.: On the realization of constraints in nonholonomic mechanics. Prikl. Mat. Mekh. *45*, No. 3, 481–487 (1981) (Russian); English transl.: J. Appl. Math. Mech. *45*, No. 3, 351–355 (1982). Zbl. 496.70026

[68] Bryuno, A.D.: Instability in a Hamiltonian system and the distribution of asteroids. Mat. Sb., Nov. Ser. *83 (125)*, No. 2, 273–312 (1970) (Russian); English transl.: Math. USSR, Sb. *12*, No. 2, 271–312 (1970). Zbl. 217, 124

[69] Bryuno, A.D.: Researches on the restricted three-body problem. I: Periodic solutions of a Hamiltonian system. Preprint No. 18. Moskva: Inst. Prikl. Mat. Akad. Nauk SSSR. 44 p. (1972). (Russian). Zbl. 389.70009

[70] Bryuno, A.D.: Local method of nonlinear analysis of differential equations. Moskva: Nauka. 253 p. (1979). (Russian). Zbl. 496.34002

[71] Burgh, A.H.P. van der: Studies in the asymptotic theory of nonlinear resonance. Diss. Delft: Technische Hogeschool. III, 85 p. (1974)

[72] Carathéodory, C.: Der Schlitten. Z. Angew. Math. Mech. *13*, 71–76 (1933). Zbl. 6, 373

[73] Cassels, J.W.S.: An introduction to diophantine approximation. Cambridge: Cambridge University Press. X, 166 p. (1957). Zbl. 77, 48

[74] Chebotarev, N.G.: The theory of Lie groups. Moskva-Leningrad: Gostekhizdat. (1940). (Russian)

[75] Chelomej, V.N.: On the possibility of raising the stability of elastic systems by means of vibrations, Dokl. Akad. Nauk SSSR *110*, No. 3, 345–347 (1956). (Russian). Zbl. 72, 198

[76] Chelomej, V.N.: Mechanical paradoxes caused by vibrations. Dokl. Akad. Nauk SSSR *270*, No. 1, 62–67 (1983) (Russian); English transl.: Soviet Phys. Dokl. *28*, No. 5, 387–390 (1983).

[77] Chernous'ko, F.L.: On resonance in an essentially non-linear system. Zh. Vychisl. Mat. Mat. Fiz. *3*, 131–144 (1963). (Russian); English transl.: U.S.S.R. Comput. Math. Math. Phys. 3, 168–185 (1965). Zbl. 237.70035

[78] Chetaev, N.G.: On instability of equilibria when the force function has not a minimum. Prikl. Mat. Mekh. *16*, No. 1, 89–93 (1952). Zbl. 47, 179

[79] Chirikov, B.V.: Passage of a nonlinear oscillatory system through resonance. Dokl. Akad. Nauk SSSR *125*, No. 5, 1015–1018 (1959). (Russian); English transl.: Sov. Phys., Dokl. 4, 390–394 (1959). Zbl. 102, 397

[80] Chirikov, B.V.: A universal instability of many-dimensional oscillator systems. Phys. Rep. *52*, No. 5, 263–379 (1979)

[81] Chudnenko, A.N.: To stability of equilibrium states of Hamiltonian systems with two degrees of freedom in the case of double zero root. Mekh. Tverd. Tela, 10, 54–60 (1978). (Russian). Zbl. 498.70031

[82] Cushman, R.: Examples of nonintegrable analytic Hamiltonian vector fields with no small divisors. Trans. Am. Math. Soc. *238*, 45–55 (1978). Zbl. 388.58008

[83] Deprit, A.; Deprit-Bartolome, A.: Stability of the triangular Lagrangian points. Astron. J. *75*, No. 2, 173–179 (1967)

[84] Deprit, A.; Henrard, J.; Rom, A.: Analytical lunar ephemeris. I. Definition of the main problem. Astron. Astrophys. 10, 257–269 (1971). Zbl. 216, 503

[85] Douady, R.: Une démonstration directe de l'équivalence des théorèmes de tores invariants pour difféomorphismes et champs des vecteurs C.R. Acad. Sci., Paris, Sér. I *295*, 201–204 (1982). Zbl. 502.58013

[86] Douady, R.; Le Calvez, P.: Example de point fixe elliptique non topologiquement stable en dimension 4. C.R. Acad. Sci., Paris, Sér. I *296*, 895–898 (1983). Zbl. 535.58015

[87] Dubrovin, B.A.; Matveev, V.B.; Novikov, S.P.: Non-linear equations of the Korteweg-de-Vries type, finite-zone linear operators, and Abelian varieties, Usp. Mat. Nauk *31*, No. 1, 55–136 (1976) (Russian); English transl.: Russ. Math. Surv. *31*, No. 1, 59–146 (1976). Zbl. 326.35011

[88] Duistermaat, J.J.: Non-integrability of the 1:1:2 resonance. Ergodic Theory Dyn. Syst. *4*, 553–568 (1984). Zbl. 552.58015

[89] Dykhne, A.M.: Quantum transitions in the adiabatic approximation. Zh. Ehksp. Teor. Fiz. *38*, No. 2, 570–578 (1960) (Russian); English transl.: Sov. Phys., JETP *11*, No. 2, 411–415 (1960). Zbl. 98, 197

[90] Evtushenko, Yu.G.: Influence of tangential acceleration on the motion of a satellite. Prikl. Mat. Mekh. *30*, No. 3, 594–598 (1966) (Russian); English transl.: J. Appl. Math. Mech. *30*, No. 3, 710–716 (1967). Zbl. 147, 432

[91] Fedoryuk, M.V.: An adiabatic invariant of a system of linear oscillators and scattering theory. Differ. Uravn. *12*, No. 6, 1012–1018 (1976) (Russian); English transl.: Differ. Equations *12*, No. 6, 713–718 (1977). Zbl. 338.34057

[92] Galin, D.M.: Versal deformations of linear Hamiltonian systems. Tr. Semin. Im. I.G. Petrovskogo 1, 63–74 (1975) (Russian); English transl.: Transl., II. Ser., Am. Math. Soc. *118*, 1–12 (1982). Zbl. 343.58006

[93] Gardner, C.S.: Adiabatic invariants of periodic classical systems. Phys. Rev. *115*, No. 4, 791–794 (1959)

[94] Giacaglia, G.E.O.: Perturbation methods in non-linear systems. Applied Mathematical Science, 8. New York-Heidelberg-Berlin: Springer-Verlag. VIII, 369 p. (1972). Zbl. 282.34001

[95] Graff, S.M.: On the conservation of hyperbolic invariant tori for Hamiltonian systems. J. Differ. Equations *15*, No. 1, 1–69 (1974). Zbl. 268.34051

[96] Gustavson, F.: On constructing formal integrals of a Hamiltonian system near an equilibrium point. Astron. J. *71*, No. 8, 670–686 (1966)

[97] Hamilton, R.S.: The inverse function theorem of Nash and Moser. Bull. Am. Math. Soc., New Ser. 7, No. 1, 65–222 (1982). Zbl. 499.58003

[98] Kampen, E.R. van; Wintner, A.: On a symmetrical canonical reduction of the problem of three bodies. Am. J. Math. *59*, No. 1, 153–166 (1937). Zbl. 15, 421

[99] Kaplansky, I.: An introduction to differential algebra. Paris: Hermann. 62 p. (1957). Zbl. 83, 33

[100] Kapitsa, P.L.: Dynamic stability of a pendulum with a vibrating point of suspension Zh. Ehksp. Teor. Fiz. *21*, No. 5, 588–598 (1951). (Russian)

[101] Karapetyan, A.V.: On realizing nonholonomic constraints by viscous friction forces and Celtic stones stability, Prikl. Mat. Mekh. *45*, No. 1, 42–51 (1981) (Russian); English transl.: J. Appl. Math. Mech. *45*, No. 1, 30–36 (1982). Zbl. 493.70008

[102] Kasuga, T.: On the adiabatic theorem for the Hamiltonian system of differential equations in classical mechanics. I, II, III. Proc. Japan Acad. *37*, 366–382 (1961). Zbl. 114, 149

[103] Khapaev, M.M.: On averaging in multifrequency systems. Dokl. Akad. Nauk SSSR *217*, No. 5, 1021–1024 (1974) (Russian); English transl.: Sov. Math., Dokl. *15*, No. 4, 1180–1184 (1975). Zbl. 316.34047

[104] Kharlamov, M.P.: Characteristic class of a bundle and the existence of a global Routh function. Funkts. Anal. Prilozh. *11*, No. 1, 89–90 (1977) (Russian); English transl.: Funct. Anal. Appl. *11*, No. 1, 80–81 (1977). Zbl. 355.58003

[105] Kharlamova-Zabelina, E.I.: Rapid rotation of a rigid body about a fixed point under the presence of a nonholonomic constraint. Vestn. Mosk. Univ., Ser. Mat. Mekh. Astron. Fiz. Khim. 12, No. 6, 25–34 (1958). (Russian). Zbl. 81, 179

[106] Klingenberg, W.: Lectures on closed geodesics. Berlin-Heidelberg-New York: Springer-Verlag. IX, 227 p. (1978). Zbl. 397.58018

[107] Kolmogorov, A.N.: On dynamical systems with integral invariant on the torus. Dokl. Akad. Nauk SSSR *93*, No. 5, 763–766 (1953). (Russian). Zbl. 52, 319

[108] Kovalev, A.M.; Savchenko, A.Ya.: Stability of uniform rotations of a rigid body about a principal axis. Prikl. Mat. Mekh. *39*, No. 4, 650–660 (1975) (Russian); English transl.: J. Appl. Math. Mech. *39*, No. 4, 623–633 (1975). Zbl. 339.70004

[109] Kovalev, A.M.; Chudnenko, A.N.; Chudnenko, O.M.: On the stability of the equilibrium of a two-dimensional Hamiltonian system in case of equal frequencies. Dopov. Akad. Nauk Ukr. RSR, Ser. A, 1977, 1010–1013 (1977). (Ukrainian). Zbl. 376.70020

[110] Kozlov, V.V.: On the geometry of regions of possible motions with boundary. Vestn. Mosk. Univ., Ser. I 1977, No. 5, 118–120 (1977). (Russian). Zbl. 381.70017

[111] Kozlov, V.V.: Nonexistence of univalued integrals and branching of solutions in solid body dynamics. Prikl. Mat. Mekh. *42*, No. 3, 400–406 (1978) (Russian); English transl.: J. Appl. Math. Mech. *42*, No. 3, 420–426 (1979). Zbl. 444.70019

[112] Kozlov, V.V.: On equilibrium instability in a potential field. Usp. Mat. Nauk *36*, No. 3, 215–216 (1981) (Russian); English transl.: Russ. Math. Surv. *36*, No. 3, 256–257 (1981). Zbl. 488.70003

[113] Kozlov, V.V.: Equilibrium instability in a potential field, taking account of viscous friction, Prikl. Mat. Mekh. *45*, No. 3, 570–572 (1981) (Russian); English transl.: J. Appl. Math. Mech. *45*, No. 3, 417–418 (1982). Zbl. 511.70026

[114] Kozlov, V.V.: Asymptotic solutions of equations of classical mechanics. Prikl. Mat. Mekh. *46*, No. 4, 573–577 (1982) (Russian); English transl.: J. Appl. Math. Mech. *46*, No. 2, 454–457 (1983). Zbl. 522.70020

[115] Kozlov, V.V.: Dynamics of systems with non-integrable restrictions. I: Vestn. Mosk. Univ., Ser. I 1982, No. 3, 92–100 (1982). (Russian); English transl.: Mosc. Univ. Mech. Bull. 37, No. 3–4, 27–34 (1982)
II: Vestn. Mosk. Univ., Ser. I 1982, No. 4, 70–76 (1982). (Russian); English transl.: Mosc. Univ. Mech. Bull. 37, No. 3–4, 74–80 (1982)
III: Vestn. Mosk. Univ., Ser. I 1983, No. 3, 102–111 (1983). (Russian); English transl.: Mosc. Univ. Mech. Bull. 38, No. 3, 40–51 (1983). Zbl. 501. 70016; Zbl. 508. 70012; Zbl. 516.70017

[116] Kozlov, V.V.: Realization of nonintegrable constraints in classical mechanics. Dokl. Akad. Nauk SSSR *272*, No. 3, 550–554 (1983) (Russian); English transl.: Sov. Phys., Dokl. *28*, No. 9, 735–737 (1983). Zbl. 579.70014

[117] Kozlov, V.V.; Kolesnikov, N.N.: On theorems of dynamics. Prikl. Mat. Mekh. *42*, No. 1, 28–33 (1978) (Russian); English transl.: J. Appl. Math. Mech. *42*, No. 1, 26–31 (1979). Zbl. 418.70010

[118] Kozlov, V.V.; Palamodov, V.P.: On asymptotic solutions of the equations of classical mechanics. Dokl. Akad. Nauk SSSR *263*, No. 2, 285–289 (1982) (Russian); English transl.: Sov. Math., Dokl. *25*, No. 2, 335–339 (1982). Zbl. 504.70018

[119] Kravtsvov, Yu.A.; Orlov, Yu.I.: Geometric optics of inhomogeneous media. Moskva: Nauka. 304 p. (1980). (Russian)

[120] Krylov, N.M.; Bogolyubov, N.N.: Application of the methods of nonlinear mechanics to the theory of stationary oscillations. Collected works of N.N. Bogolyubov. Vol. I, 257–337. Kiev: Naukova Dumka. (1969). (Russian)

[121] Lamb, H.: Hydrodynamics. 6th ed. New York: Dover. (1945)

[122] Lazutkin, V.F.: Concerning a theorem of Moser on invariant curves. Probl. Dyn. Theory Propag. Seism. Waves *14*, 109–120. Leningrad: Nauka. (1974). (Russian)

[123] Lazutkin, V.F.: Convex billiards and eigenfunctions of the Laplace operator. Leningrad: Izdatel'stvo Leningradskogo Universiteta. 196 p. (1981). (Russian). Zbl. 532.58031

[124] Lenard, A.: Adiabatic invariance to all orders. Ann. Phys. *6*, No. 3, 261–276 (1959). Zbl. 84, 444

[125] Leontovich, A.M.: On the stability of the Lagrange periodic solution for the reduced problem of three bodies. Dokl. Akad. Nauk SSSR *143*, No. 3, 525–528 (1962) (Russian); English transl.: Sov. Math., Dokl. *3*, No. 2, 425–429 (1962). Zbl. 123, 182

[126] Levi, M.: Adiabatic invariants of the linear Hamiltonian systems with periodic coefficients. J. Differ. Equations *42*, No. 1, 47–71 (1981). Zbl. 462.58020

[127] Lidov, M.L.: On the approximate analysis of the evolution of orbits of artificial satellites. Probl. Motion Artif. Cosmic Bodies, 119–134. Moskva: Izd. Akad. Nauk SSSR. (1963). (Russian)

[128] Lidov, M.L.; Ziglin, S.L.: The analysis of restricted circular twice-averaged three body problem in the case of closed orbits. Celestial Mech. *9*, No. 2, 151–173 (1974). Zbl. 355.70007

[129] Littlejohn, R.G.: Hamiltonian theory of guiding center motion. Berkeley: Lawrence Berkeley Lab. (1980).

[130] Mackay, R.S.; Meiss, J.D.; Percival, I.C.: Transport in Hamiltonian systems. Physica D 13, 55–81 (1984). Zbl. 585.58039

[131] Markeev, A.P.: Libration points in celestial mechanics and cosmodynamics. Moskva: Nauka. (1978). (Russian)

[132] Marsden, J.; Weinstein, A.: Reduction of symplectic manifolds with symmetry. Rep. Math. Phys. *5*, No. 1, 121–130 (1974). Zbl. 327.58005

[133] Mather, J.N.: Existence of quasi-periodic orbits for twist homeomorphisms of the annulus. Topology *21*, No. 4, 457–467 (1982). Zbl. 506.58031

[134] Mather, J.N.: Concavity of the Lagrangian for quasi-periodic orbits. Comment. Math. Helv. *57*, No. 3, 356–376 (1982). Zbl. 508.58037

[135] Mather, J.N.: A criterion for the non-existence of invariant circles. Publ. Math., Inst. Hautes Étud. Sci. 63, 153–204 (1986). Zbl. 603.58028

[136] Mather, J.N.; McGehee, R.: Solutions of the collinear four body problem which become unbounded in finite time. Dyn. Syst., Theor. Appl., Battelle Seattle 1974 Renc., Lect. Notes Phys. 38, 573–597 (1975). Zbl. 331.70005

[137] Mel'nikov, V.K.: On the stability of the center for time-periodic perturbations. Tr. Mosk. Mat. O.-va *12*, 3–52 (1963). (Russian); English transl.: Trans. Mosc. Math. Soc. 1963, 1–56 (1965). Zbl. 135, 310

[138] Milnor, J.: Morse theory. Ann. Math. Stud. *51*. Princeton: Princeton University Press. VI, 153 p. (1963). Zbl. 108, 104

[139] Mitropol'skii, Yu.A.: Averaging method in nonlinear mechanics. Anal. Metody Teor. Nelinejnykh Kolebanij 1. Tr. 5 Mezhdunar. Konf. 1969, 21–39. (Russian) Kiev: Naukova Dumka. (1970). Zbl. 235.34100

[140] Mitropol'skij, Yu.A.; Lykova, O.B.: Integral manifolds in nonlinear mechanics, Moskva: Nauka. 512 p. (1973). (Russian). Zbl. 286.34067

[141] Mishchenko, A.S.; Fomenko, A.T.: Generalized Liouville method of integration of Hamiltonian systems. Funkts. Anal. Prilozh. *12*, No. 2, 46–56 (1978) (Russian); English transl.: Funct. Anal. Appl. *12*, No. 2, 113–121 (1978). Zbl. 396.58003

[142] Mishchenko, A.S.; Fomenko, A.T.: Integration of Hamiltonian systems with noncommutative symmetries. Tr. Semin. Vektorn. Tenzorn. Anal. Prilozh. Geom. Mekh. Fiz. *20*, 5–54 (1981). (Russian). Zbl. 473.58015

[143] Moltchanov, A.M.: The resonant structure of the solar system. Icarus *8*, No. 2, 203–215 (1968)

[144] Morozov, A.D.: A complete qualitative investigation of Duffing's equation. Differ. Uravn. *12*, No. 2, 241–255 (1976) (Russian); English transl.: Differ. Equations *12*, No. 2, 164–174 (1976). Zbl. 326.34049

[145] Moser, J.: The analytic invariants of an area preserving mapping near a hyperbolic fixed point. Commun. Pure Appl. Math. *9*, No. 4, 673–692 (1956). Zbl. 72, 408

[146] Moser, J.: On invariant curves of area-preserving mappings of an annulus. Nachr. Akad. Wiss. Gött., II. Math.-Phys. Kl. 1962, 1–20 (1962). Zbl. 107, 293

[147] Moser, J.: A rapidly convergent iteration method and nonlinear partial differential equations. I, II. Ann. Sc. Norm. Super. Pisa, Cl. Sci., III. Ser. *20*, No. 2, 265–315, No. 3, 499–535 (1966). Zbl. 144, 182

[148] Moser, J.: Three integrable Hamiltonian systems connected with isospectral deformations. Adv. Math. *16*, No. 2, 197–220 (1975). Zbl. 303.34019

[149] Nash, J.: The imbedding problem for Riemannian manifolds. Ann. Math., II. Ser. *63*, No. 1, 20–63 (1956). Zbl. 70, 386

[150] Nejshtadt, A.I.: Passage through a resonance in the two-frequency problem. Dokl. Akad. Nauk SSSR *221*, No. 1, 301–304 (1975) (Russian); English transl.: Sov. Phys., Dokl. *20*, No. 3, 189–191 (1975). Zbl. 325.70014

[151] Nejshtadt, A.I.: Stability of plane solutions in the doubly averaged restricted circular three-body problem. Pis'ma Astron. Zh. *1*, No. 10, 41–45 (1975) (Russian); English transl.: Sov. Astron. Lett. *1*, No. 5, 211–213 (1975)

[152] Nejshtadt, A.I.: Averaging in multifrequency systems. II. Dokl. Akad. Nauk SSSR *226*, No. 6, 1295–1298 (1976) (Russian); English transl.: Sov. Phys., Dokl. *21*, No. 2, 80–82 (1976). Zbl. 357.34041

[153] Nejshtadt, A.I.: Passage through a separatrix in a resonance problem with slowly-varying parameter. Prikl. Mat. Mekh. *39*, No. 4, 621–632 (1975) (Russian); English transl.: J. Appl. Math. Mech. *39*, No. 4, 594–605 (1975). Zbl. 356.70020

[154] Nejshtadt, A.I.: On the evolution of the rotations of a rigid body under the action of the sum of a constant and a dissipative perturbing moments, Izv. Akad. Nauk SSSR, Mekh. Tverd. Tela, No. 6, 30–36 (1980). (Russian)

[155] Nejshtadt, A.I.: On the accuracy of conservation of the adiabatic invariant. Prikl. Mat. Mekh. *45*, No. 1, 80–87 (1981) (Russian); English transl.: J. App. Math. Mech. *45*, No. 1, 58–63 (1982). Zbl. 488.70016

[156] Nejshtadt, A.I.: Estimates in the Kolmogorov theorem on conservation of conditionally periodic motions. Prikl. Mat. Mekh. *45*, No. 6, 1016–1025 (1981) (Russian); English transl.: J. Appl. Math. Mech. *45*, No. 6, 766–772 (1982). Zbl. 515. 70019

[157] Nejshtadt, A.I.: The separation of motions in systems with rapidly rotating phase. Prikl. Mat. Mekh. *48*, No. 2, 197–204 (1984) (Russian); English transl.: J. Appl. Math. Mech. *48*, No. 2, 133–139 (1984). Zbl. 571.70022

[158] Nekhoroshev, N.N.: Action-angle variables and their generalization. Tr. Mosk. Mat. O.-va *26*, 181–198 (1972) (Russian); English transl.: Trans. Mosc. Math. Soc. *26*, 180–198 (1972). Zbl. 261.70019

[159] Nekhoroshev, N.N.: Stable lower estimates for smooth mappings and for gradients of smooth functions. Mat. Sb., Nov. Ser. *90*, 432–478 (1973) (Russian); English transl.: Math. USSR, Sb. *19*, No. 3, 425–467 (1973). Zbl. 256.26011

[160] Nekhoroshev, N.N.: Exponential estimate of the stability time of near-integrable Hamiltonian systems. II. Tr. Semin. Im. I.G. Petrovskogo *5*, 5–50 (1979). (Russian) Zbl. 473.34021

[161] Northrop, T.G.: The adiabatic motion of charged particles. Interscience Tracts on Physics and Astronomy, 21. New York-London-Sydney: Interscience Publishers, a division of John Wiley & Sons. XIV, 109 p. (1963). Zbl. 119, 437

[162] Palamodov, V.P.: Stability of equilibrium in a potential field. Funkts. Anal. Prilozh. *11*, No. 4, 42–55 (1977) (Russian); English transl.: Funct. Anal. Appl. *11*, No. 4, 277–289 (1977). Zbl. 384.70006

[163] Percival, I.C.: A variational principle for invariant tori of fixed frequency. J. Phys. A *12*, L 57 – L 60 (1979). Zbl. 394.70018

[164] Percival, I.C.: Variational principles for invariant tori and cantori. Symp. on Nonlinear Dynamics and Beam-Beam Interaction. Am. Inst. Phys. Conf. Proc. No. 55, 302–310 (1980)

[165] Pidkujko, S.I.; Stepin, A.M.: Polynomial integrals of Hamiltonian systems. Dokl. Akad. Nauk SSSR *239*, No. 1, 50–53 (1978) (Russian); English transl.: Sov. Math., Dokl. *19*, No. 2, 282–286 (1978). Zbl. 419.70017

[166] Pöschel, J.: Über invariante Tori in differenzierbaren Hamiltonschen Systemen. Bonn. Math. Schr. *120*, 103 p. (1980). Zbl. 466.70021

[167] Pöschel, J.: Integrability of Hamiltonian systems on Cantor sets. Commun. Pure Appl. Math. *35*, No. 1, 653–696 (1982). Zbl. 542.58015

[168] Pronchatov, V.E.: An estimate of the error of the averaging method in a two-frequency problem. Mat. Sb., Nov. Ser. *122*, No. 2, 245–264 (1983) (Russian); English transl.: Math. USSR, Sb. *50*, No. 1, 241–258 (1985). Zbl. 539.34034

[169] Roels, J.: An extension to resonant cases of Lyapunov's theorem concerning the periodic solutions near a Hamiltonian equilibrium. J. Differ. Equations *9*, No. 2, 300–324 (1971). Zbl. 245.70022

[170] Roels, J.: Families of periodic solutions near a Hamiltonian equilibrium when the ratio of two eigenvalues is 3, J. Differ. Equations *10*, No. 3, 431–447 (1971). Zbl. 226.34038

[171] Rubin, H., Ungar, P.: Motion under a strong constraining force. Commun. Pure Appl. Math. *10*, No. 1, 65–87 (1957). Zbl. 77, 174

[172] Rüssmann, H.: Über das Verhalten analytischer Hamiltonscher Differentialgleichungen in der Nähe einer Gleichgewichtslösung. Math. Ann. *154*, 285–300 (1964). Zbl. 124, 47

[173] Rüssmann, H.: Kleine Nenner. I: Über invariante Kurven differenzierbarer Abbildungen eines Kreisringes. Nachr. Akad. Wiss. Gött., II. Math.-Phys. Kl. 1970, 67–105 (1970). Zbl. 201, 112

[174] Rüssmann, H.: Kleine Nenner. II: Bemerkungen zur Newtonschen Methode. Nachr. Akad. Wiss. Gött., II. Math.-Phys. Kl. 1972, 1–10 (1972). Zbl. 255.30003

[175] Rüssmann, H.: On optimal estimates for the solutions of linear partial differential equations of first order with constant coefficients on the torus. Dyn. Syst., Theor. Appl., Battelle Seattle 1974 Renc., Lect. Notes Phys. 38, 598–624 (1975). Zbl. 319.35017

[176] Sanders, J.: Are higher order resonances really interesting? Celestial Mech. *16*, No. 4, 421–440 (1977). Zbl. 377.70015

[177] Seifert, H.: Periodische Bewegungen mechanischer Systeme. Math. Z. *51*, No. 2, 197–216 (1948). Zbl. 30, 221

[178] Slutskin, A.A.: Motion of a one-dimensional nonlinear oscillator under adiabatic conditions. Zh. Ehksp. Teor. Fiz. *45*, No. 4, 978–988 (1963) (Russian); English transl.: Sov. Phys., JETP *18*, No. 3, 676–682 (1964)

[179] Sokol'skij, A.G.: On the stability of an autonomous Hamiltonian system with two degrees of freedom in the case of equal frequencies. Prikl. Mat. Mekh. *38*, No. 5, 791–799 (1974) (Russian); English transl.: J. Appl. Math. Mech. *38*, No. 5, 741–749 (1974). Zbl. 336.70017

[180] Sokol'skij, A.G.: On stability of an autonomous Hamiltonian system with two degrees of freedom under first-order resonance. Prikl. Mat. Mekh. *41*, No. 1, 24–33 (1977) (Russian); English transl.: J. Appl. Math. Mech. *41*, No. 1, 20–28 (1977). Zbl. 393.70025

[181] Sokol'skij, A.G.: Proof of the stability of Lagrangian solutions at a critical relation of masses. Pis'ma Astron. Zh. *4*, No. 3, 148–152 (1978) (Russian); English transl.: Sov. Astron. Lett. *4*, No. 2, 79–81 (1978)

[182] Souriau, J.-M.: Structure des systèmes dynamiques. Maîtrises de mathématiques. Paris: Dunod. XXXII, 414 p. (1970)

[183] Svanidze, A.A.: Small perturbations of an integrable dynamic system with integral invariant. Tr. Mat. Inst. Steklova *147*, 124–146 (1980) (Russian); English transl.: Proc. Steklov Inst. Math. *2*, 127–151 (1981). Zbl. 441.58008

[184] Takens, F.: Motion under the influence of a strong constraining force. Global theory of dynamical systems, Proc. int. Conf., Evanston/Ill. 1979, Lect. Notes Math. 819, 425–445 (1980). Zbl. 458.58010

[185] Tatarinov, Ya.V.: Portraits of the classical integrals of the problem of rotation of a rigid body about a fixed point. Vestn. Mosk. Univ., Ser. I 29, No. 6, 99–105 (1974). (Russian). Zbl. 291.70003

[186] Tatarinov, Ya.V.: Geometric formalism of classical dynamics: natural integrals and reduction. Vestn. Mosk. Univ., Ser. I 1978, No. 3, 109–118 (1978) (Russian); English transl.: Mosc. Univ. Mech. Bull. 33, No. 3–4, 26–33 (1978). Zbl. 379.70010

[187] Vashkov'yak, M.A.: Evolution of orbits in the restricted circular doubly-averaged three-body problem. I. Qualitative investigation. Kosm. Issled. *19*, No. 1, 5–18 (1981). (Russian)

[188] Volosov, V.M.; Morgunov, B.I.: The averaging method in the theory of nonlinear oscillatory systems. Moskva: Izdatel'stvo MGU, 507 p. (1971). (Russian). Zbl. 232.70021

[189] Yakubovich, V.A.; Starzhinskij, V.M.: Linear differential equations with periodic coefficients, and their applications. Moskva: Nauka. 720 p. (1972). (Russian); English transl.: New York-Toronto: John Wiley & Sons; Jerusalem-London: Israel Program for Scientific Translations, a Halsted Press Book. XII, 839 p. (1975). Zbl. 276.34001

[190] Young, L.C.: Lectures on the calculus of variations and optimal control theory. Philadelphia-London-Toronto: W.B. Saunders Company. XI, 331 p. (1969). Zbl. 177, 378

[191] Zaslavskij, G.M.; Chirikov, B.V.: Stochastic instability of non-linear oscillations. Usp. Fiz. Nauk *105*, No. 1, 3–40 (1971) (Russian); English transl.: Sov. Phys. Usp. *14*, No. 5, 549–568 (1972)

[192] Zehnder, E.: Generalized implicit function theorems with applications to some small divisor problems. I, II. Commun. Pure Appl. Math. *28*, No. 1, 91–140 (1975); *29*, No. 1, 49–111 (1976). Zbl. 309.58006; Zbl. 334.58009

[193] Ziglin, S.L.: Secular evolution of the orbit of a planet in a binary-star system. Pis'ma Astron. Zh. *1*, No. 9, 45–47 (1975) (Russian); English transl.: Sov. Astron. Lett. *1*, No. 5, 194–195 (1975)

[194] Ziglin, S.L.: Nonintegrability of a problem on the motion of four point vertices. Dokl. Akad. Nauk SSSR *250*, No. 6, 1296–1300 (1980) (Russian); English transl.: Sov. Math., Dokl. *21*, No. 1, 296–299 (1980). Zbl. 464.76021

[195] Ziglin, S.L.: Splitting of separatrices, branching of solutions and nonexistence of an integral in the dynamics of a solid body. Tr. Mosk. Mat. O.-va *41*, 287–302 (1980) (Russian); English transl.: Trans. Mosc. Math. Soc. 1982, No. *1*, 283–298 (1982). Zbl. 466.70009

[196] Ziglin, S.L.: Self-intersection of the complex separatrices and the nonexistence of the integrals in the Hamiltonian systems with one-and-half degrees of freedom. Prikl. Mat. Mekh. *45*, No. 3, 564–566 (1981) (Russian); English transl.: J. Appl. Math. Mech. *45*, No. 3, 411–413 (1982). Zbl. 503.70012

[197] Ziglin, S.L.: Branching of solutions and nonexistence of first integrals in Hamiltonian mechanics. I, II. Funkts. Anal. Prilozh. *16*, No. 3, 30–41 (1982) (Russian); *17*, No. 1, 8–23 (1983) (Russian); English transl.: Funct. Anal. Appl. *16*, No. 3, 181–189 (1983); *17*, No. 1, 6–17 (1983). Zbl. 524. 58013; Zbl. 518.58016

Index

Encyclopaedia of Mathematical Sciences

Editor-in-chief: R. V. Gamkrelidze

Volume 1
D. V. Anosov, V. I. Arnold (Eds.)

Dynamical Systems I

Translated from the Russian by E. R. Dawson and D. O'Shea
1988. Approx. 250 pages. Hard cover approx. DM 130,–.
ISBN 3-540-17000-6

Contents:
V. I. Arnold, Yu. S. Il'yashenko – **Ordinary Differential Equations:** Basic Concepts. – Differential Equations on Surfaces. – Singular Points of Differential Equations in a Higher Dimensional Real Phase Space. – Singular Points of Differential Equations in a Higher Dimensional Complex Phase Space. – Singular Points of Vector Fields in the Real and Complex Planes. – Cycles. – Analytic Theory of Differential Equations. – Bibliography. – Index.
D. V. Anosov, S. Kh. Aranson, I. U. Bronshtein, V. Z. Grines – **Smooth Dynamical Systems:** Basic Concepts. – Elementary Theory. – Topological Dynamics. – Flows on Two-Dimensional Manifolds. – Bibliography. – Index.

Volume 2
Ya. G. Sinai (Ed.)

Dynamical Systems II

Translated from the Russian by M. V. Yakobson
1988. ISBN 3-540-17001-4

Contents:
I. P. Kornfeld, Ya. G. Sinai, A. M. Vershik – **General Ergodic Theory of Groups and Measure Preserving Transformations:** Basic Notions of Ergodic Theory and Examples of Dynamical Systems. – Spectral Theory of Dynamical Systems. – Entropy Theory of Dynamical Systems. – Periodic Approximations and their Applications. – Trajectory Theory. – Bibliography.
L. A. Bunimovich, Ya. B. Pesin, Ya. G. Sinai, M. V. Yakobson – **Ergodic Theory of Smooth Dynamical Systems:** Stochasticity of Smooth Dynamical Systems. The Elements of KAM-Theory. – General Theory of Smooth Hyperbolic Dynamical Systems. – Dynamical Systems of Hyperbolic Type with Singularities. – Ergodic Theory of One-Dimensional Mappings. – Bibliography.
R. L. Dobrushin, N. B. Maslova, Ya. G. Sinai, Yu. M. Sukhov – **Dynamical Systems of Statistical Mechanics and Kinetic Equations:** Dynamical Systems of Statistical Mechanics. – Existence and Uniqueness Theorems for the Boltzmann Equation. Bibliography.
Index.

Springer-Verlag
Berlin Heidelberg New York
London Paris Tokyo

Springer